INDEPENDENT LEARNING PROJECT FOR ADVANCED CHEMISTRY

ILPAC

ADVANCED CHEMISTRY CALCULATIONS

second edition

REVISED BY ANN LAINCHBURY JOHN STEPHENS ALEC THOMPSON

JOHN MURRAY

Original material produced by the Independent Learning Project for Advanced Chemistry, sponsored by the Inner London Education Authority and published as Advanced Chemistry Calculations first edition.
Taken from the second edition, published as ILPAC volumes 1–12 with additional material by Ann Lainchbury, John Stephens and Alec Thompson.

First edition published in 1985
by John Murray (Publishers) Ltd
50 Albemarle Street
London W1X 4BD

Reprinted 1990, 1993, 1996 (twice).
Second edition 1998.

Design by John Townson/Creation.
Layouts by Wearset, Boldon, Tyne and Wear.
Illustrations by Barking Dog Art.

Typeset in 10/12pt Times and Helvetica.
Printed and bound in Great Britain by St Edmundsbury Press, Bury St Edmunds.

A catalogue record for this book is available from the British Library.

ISBN 0 7195 7506 0

CONTENTS

Preface **vii**

Introduction **viii**

Chapter 1 **Amount of substance – the mole** **1**
Introduction 1
Molar mass 2
Amount calculations 3
Calculating the number of particles in a given amount 4
Stoichiometry 5
Working out a percentage yield 7
Calculating empirical formulae and molecular formulae 10
Calculating the empirical formula from the masses of constituents 10
Calculating the empirical formula from percentage composition by mass 11
Calculating the empirical formula from elemental analysis data 12
Molecular formula from empirical formula 14
Amounts in solution 14
Concentration of solution 14
Standard solutions 15
Calculating concentration from volume and amount 15
Calculating concentration from mass of solute and volume 16
Calculating the amount of substance in a solution 16
End-of-chapter questions 17

Chapter 2 **Volumetric analysis (titrimetry)** **19**
Introduction 19
Calculating concentration from titration data 19
A general expression 20
Redox titrations and other redox reactions 21
Redox half-reactions 21
Oxidation number 22
Oxidation number and redox reactions 24
Balancing redox equations 25
Precipitation titrations 28
End-of-chapter questions 30

Chapter 3 **Mass spectrometry and nuclear reactions** **33**
Introduction and pre-knowledge 33
Mass spectrometry 33
Calculating the relative atomic mass of an element 33
Mass spectrometry in organic chemistry 36
Calculating the number of carbon atoms in a molecule 37
Identifying minor peaks 37
Nuclear reactions 39
End-of-chapter questions 41

Chapter 4	**Chemical energetics**	**43**
	Introduction and pre-knowledge	43
	Enthalpy change	43
	Calculating $\Delta H^{\ominus}$ from experimental data	44
	Calculating an enthalpy change of solution	47
	Hess' Law	48
	Use of energy cycles	50
	Calculating enthalpy of formation from enthalpy of combustion	52
	Uses of standard enthalpy changes of formation	54
	Bond energy term (Bond enthalpy term)	56
	Calculating bond energy terms	56
	Determining two bond energy terms simultaneously	58
	Using bond energy terms to estimate enthalpy changes	59
	Lattice energy (Lattice enthalpy)	64
	Calculating lattice energy using a Born–Haber cycle	64
	Lattice energy and stoichiometry	69
	Enthalpy of hydration	69
	Thermometric titrations	71
	The direction of change	73
	Entropy, S	74
	Calculating entropy changes	75
	Using $\Delta S^{\ominus}$ to calculate $\Delta G^{\ominus}$	76
	Temperature and spontaneous processes	77
	End-of-chapter questions	78
Chapter 5	**Gases**	**82**
	Introduction and pre-knowledge	82
	The gas laws	82
	Boyle's Law and Charles' Law	82
	The combined gas law	84
	Avogadro's theory and its applications	85
	Using Avogadro's theory to calculate reacting volumes	85
	Determining the formula of a gaseous hydrocarbon	86
	Molar volume of gases	88
	The ideal gas equation	90
	The gas constant, R	90
	Determination of molar mass of a gas	93
	Graham's law of effusion	94
	Dalton's Law of partial pressures	95
	End-of-chapter questions	97
Chapter 6	**Equilibrium I – principles**	**100**
	Introduction and pre-knowledge	100
	The equilibrium law	100
	Calculating the value of an equilibrium constant	102
	When volume can be omitted from an equilibrium law expression	104
	Various numerical problems involving equilibrium constants	105
	Problems involving quadratic equations	110
	The equilibrium constant, K_p	113
	The relationship between K_p and K_c	115
	The effect on an equilibrium system of changing pressure	116
	Average molar mass of a gaseous mixture	119
	The variation of equilibrium constant with temperature	120

	Solubility products	124
	Calculating solubility product from solubility	124
	Calculating solubility from solubility product	125
	The common ion effect	126
	Will precipitation occur?	128
	Distribution equilibrium	129
	End-of-chapter questions	131
Chapter 7	**Equilibrium II – acids and bases**	**134**
	Introduction and pre-knowledge	134
	Ionisation of water	134
	Using the expression for the ionic product of water	136
	pH, the hydrogen ion exponent	137
	Calculating pH from $[H^+(aq)]$	139
	Calculating $[H^+(aq)]$ from pH	140
	Relative strengths of acids	142
	Dissociation constants of acids	142
	Calculating K_a from $[H^+(aq)]$ (or from pH)	144
	Calculating pH from K_a	145
	pH changes during titrations	147
	pH curves by calculation	148
	Buffer solutions	150
	Buffering regions of titration curves	150
	The relationship between pH, K_a and composition of a buffer	151
	pH of a buffer solution	152
	Composition of a buffer solution	153
	Basic buffers	154
	Acid–base indicators	156
	Calculating the pH range of an indicator	157
	End-of-chapter questions	158
Chapter 8	**Equilibrium III – redox reactions**	**161**
	Introduction and pre-knowledge	161
	Standard electrode potentials	161
	The standard hydrogen electrode	162
	Calculating cell e.m.f. from tabulated $E^\ominus$ values	165
	Using $E^\ominus$ values to predict spontaneous reactions	166
	Limitations of predictions made using $E^\ominus$ values	167
	The 'anti-clockwise rule'	167
	Summary	168
	The effect of concentration on electrode potential and cell e.m.f.	169
	Relationships between $\Delta E^\ominus$, $\Delta G^\ominus$ and K_c	170
	Calculating the energy generated by a redox reaction	170
	The relationship between $\Delta E^\ominus$ and K_c	171
	End-of-chapter questions	172
Chapter 9	**Vapour pressure and Raoult's Law**	**176**
	Introduction and pre-knowledge	176
	The vapour pressure of two-component systems	176
	Raoult's Law	177
	Steam distillation	179
	End-of-chapter questions	181

Chapter 10 **Reaction kinetics** **182**
Introduction and pre-knowledge 182
The rate of a reaction 182
Rate equations, rate constants and orders of reaction 184
Rate equations involving higher orders of reaction 185
Half-life of first order reactions 186
The order of reaction with respect to individual reactants 188
Activation energy and the effect of temperature on reaction rate 193
Activation energy 193
Fraction of particles with energy greater than E_a 193
The Arrhenius equation 195
Using the Arrhenius equation 196
Autocatalysis 199
End-of-chapter questions 200

Appendix **Significant figures and scientific measurements** **206**
Zeros 206
Addition and subtraction 207
Multiplication and division 208

Answers
Chapter 1 209
Chapter 2 217
Chapter 3 222
Chapter 4 225
Chapter 5 240
Chapter 6 248
Chapter 7 261
Chapter 8 266
Chapter 9 270
Chapter 10 272
Appendix 279
Table of relative atomic mass values 280

PREFACE

This book, which covers the numerical aspects of all the major A-level Chemistry syllabuses, has been compiled from the ILPAC books (Independent Learning Project in Advanced Chemistry).

ILPAC was originally produced by an Inner London Education Authority team and all the materials were tested in schools and colleges, both inside and outside London. A revised edition of ILPAC was published by John Murray in 1983, followed by an extensively revised and updated edition in 1995.

An important feature of the complete ILPAC scheme is that it enables students to work more effectively on their own and at their own pace. This feature has been retained, as far as is possible in a single volume, by developing concepts progressively with simple exercises leading on to more complex problems.

To aid the student's understanding, detailed answers to exercises are given, including full methods of working and some descriptive aspects. At the end of each chapter, a representative collection of problems is presented to test the student's knowledge of a broad topic and, for these questions, only the numerical answers are given.

ACKNOWLEDGEMENTS

Thanks are due to the following examination boards for kind permission to reproduce questions from past A-level papers: Associated Examining Board (AEB); Joint Matriculation Board; Northern Examinations and Assessment Board (NEAB); Northern Ireland Council for the Curriculum Examinations and Assessment (NICCEA); Oxford and Cambridge Schools Examination Board (O&C); Southern Examining Group (SEG); Southern Universities Joint Board; University of Cambridge Local Examinations Syndicate (UCLES); University of London Examinations and Assessment Council (ULEAC); University of Oxford Delegacy of Local Examinations; Welsh Joint Education Committee (WJEC).

(The examination boards accept no responsibility whatsoever for the accuracy or method of working in the answers given).

The publishers have made every effort to trace the copyright holders, but it they have inadvertently overlooked any, they will be pleased to make the necessary arrangements at the earliest opportunity.

INTRODUCTION

This book is more than just a collection of numerical problems. All the relevant theoretical background for solving the problems is given, either as reminders of the basic work you will have already done before you use the book, or in the course of a logical development of the topic.

You will find that the exercises at the beginning of the main sections of each chapter are easy in order to encourage you to work on but, as you do so, they tend to become more difficult. Many A-level questions, either complete or in part, are included but, since syllabuses have changed considerably in recent years, you should check with your teacher whether they, and the accompanying theory, are suitable for you. The same applies to the other exercises but they have been chosen to reflect recent trends in examining.

To help you work effectively on your own we have included worked examples for many different types of question and all the exercises have answers in a separate section at the back of the book (pages 209–279). Most of these answers are very detailed and include a full account of the method of working. Use them wisely – if you have difficulty with a problem and you cannot relate it to a worked example, look at the first part of our answer just to help you get started and then try again.

You will need a data book for many of the exercises. Do not be surprised, however, if you find that your book gives values for some physical quantities which differ slightly from those we have used in our answers. Refinement of techniques means that data is constantly being revised and universal agreement is, therefore, not possible.

At the end of each chapter you will find a selection of questions which we feel are representative of the whole topic. You can use these as a test of your overall knowledge of the topic – we give only the numerical answers for these exercises.

The symbols below, together with the contents list on pages iii–vi, should help you to find your way around the book.

A — A-level Question

A (P) — A-level Question (part only)

A (SP) — A-level Question (Special Paper)

CHAPTER 1

AMOUNT OF SUBSTANCE – THE MOLE

INTRODUCTION

'Amount' is another physical quantity like mass, volume, length, time, etc. It gives us an alternative, and very useful, way of expressing how much there is of a substance.

We can express how much there is of a substance as a mass (by reference to a standard mass), as a volume (by reference to a standard volume), or as an amount (by reference to a standard **counting unit**).

As with all other physical quantities, an amount is written as a number times an associated unit:

mass = 10.2 kg
time = 42.1 s
temperature = 273 K
length = 0.7 m
volume = 24.2 cm^3
amount = 19 dozen

As objects get smaller, the number in a unit amount gets larger; for example, we buy a pair of socks, a dozen eggs and a ream (500 sheets) of paper.

In counting atoms we also need a convenient unit, large enough to be seen and handled. Since atoms are so small, there are a great many of them in a convenient unit.

The counting unit for atoms, molecules and ions is the mole (symbol: mol). **The mole is defined as the amount of substance that contains as many elementary particles as there are atoms in exactly 0.012 kg (12 g) of carbon-12.** You must learn this definition.

The mass of an atom of carbon-12 is 1.99252×10^{-23} g. So the number of atoms in 12 g of carbon-12 is given by

$$\frac{12 \text{ g}}{1.99252 \times 10^{-23} \text{ g}} = 6.02252 \times 10^{23}$$

Note that in this context, 12 is taken to be an integer and it does not therefore limit the significant figures in the answer to two.

The Avogadro constant (symbol: L) relates the number of particles to the amount. It is represented as $L = 6.02252 \times 10^{23}$ mol^{-1} or, to three significant figures,

$$\boldsymbol{L = 6.02 \times 10^{23} \text{ mol}^{-1}}$$

The following examples try to convey the magnitude of the Avogadro constant.

If you were to have 6.02×10^{23} tiny grains of pollen they would cover the City of London to a depth of 1 mile.

Figure 1.1

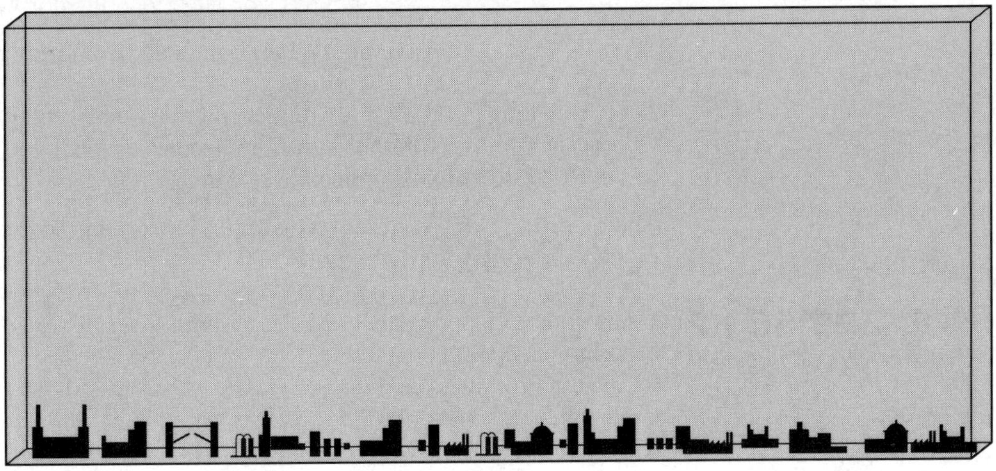

If 6.02×10^{23} marshmallows were spread over the United States of America this would yield a blanket of marshmallows more than 600 miles deep!

Computers can count about 10 million times per second. At this rate 6.02×10^{23} counts would require almost 2 billion years.

A similar example is included in the following Exercise.

EXERCISE 1.1
Answers on page 209

A 5 cm^3 spoon can hold 1.67×10^{23} molecules of water.

a How long would it take to remove them one at a time at a rate of one molecule per second?

b How many years is this?

Molar mass

The mass per unit amount of substance is called its molar mass (symbol: M) and is the mass per mole of that substance. We usually use the unit: g mol^{-1}.

The molar mass of an element is the mass per mole. It follows from the definition of the mole that the molar mass of carbon is 12.0 g mol^{-1}. Similarly, since the relative atomic mass of uranium is 238, $M = 238$ g mol^{-1}.

The term molar mass applies not only to elements in the atomic state but also to all chemical species – atoms, molecules, ions, etc.

For ethane, the molar mass is calculated from its formula, C_2H_6, which indicates that one molecule contains two atoms of carbon and six atoms of hydrogen. The relative atomic masses are: C = 12.0, H = 1.0. The relative mass of a molecule on the same scale is therefore given by:

$$C_2H_6$$

$$(2 \times 12.0) + (6 \times 1.0) = 30.0$$

Thus, the relative molecular mass of ethane is 30.0 and its molar mass, $M = 30.0$ g mol^{-1}.

There are two important points which you must bear in mind when dealing with amounts of substances:

1. You must specify exactly what entity the amount refers to. The phrase '1 mol of chlorine', for instance, has two possible meanings because it does not specify whether it refers to atoms or molecules. To avoid confusion, you must always specify the entity, either by formula or in words:

   1.0 mol of Cl **or** one mole of chlorine atoms

   1.0 mol of Cl_2 **or** one mole of chlorine molecules

2. By weighing out the same number of grams as the relative atomic mass or the relative molecular mass (whether atoms, molecules or ions) you have measured out one mole, i.e. 6.02×10^{23} atoms, molecules or ions.

The following two Exercises test your understanding of these concepts and give you practice in using your data book.

EXERCISE 1.2
Answers on page 209

Using your data book, calculate the molar masses of:

a ammonia, NH_3,

b calcium bromide, $CaBr_2$,

c phosphoric(V) acid, H_3PO_4,

d sodium sulphate-10-water, $Na_2SO_4{\cdot}10H_2O$.

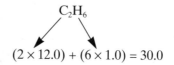

EXERCISE 1.3
Answers on page 209

What is the mass of 1.00 mol of:
a chlorine atoms,
b chlorine, Cl_2,
c phosphorus, P,
d phosphorus, P_4,
e iodide ions, I^-?

Amount calculations

In science, the amount of substance in a given sample is defined as the mass of the sample divided by the mass per mole (i.e. the molar mass):

$$\textbf{amount of substance} = \frac{\textbf{mass}}{\textbf{molar mass}}$$

The symbol for amount is n and, therefore, in symbols:

$$n = \frac{m}{M}$$

The following Worked Example shows you how to use this expression in mole calculations.

Reminder: the unit for amount is mole, for mass is gram, and for molar mass is grams per mole (abbreviation: g mol^{-1}).

WORKED EXAMPLE

A sample of carbon weighs 180 g. What amount of carbon is present?

Solution

Calculate the amount by substituting into the key expression:

$$n = \frac{m}{M}$$

where m = 180 g and M = 12.0 g mol^{-1}.

$$\therefore n = \frac{m}{M} = \frac{180\text{ g}}{12.0\text{ g mol}^{-1}} = 15.0\text{ mol}$$

At this point you should make sure that you are not using the formula in a mechanical way but understand what you are doing. So, here is an alternative method which relies on first principles. **You should never substitute into an expression without first understanding how it was derived or defined.**

Alternative Solution

1. Write down the mass of 1.00 mol of the substance in the form of a sentence:

   12.0 g is the mass of 1.00 mol of carbon.

2. Scale down to the amount of carbon in 1.00 g by dividing by 12.0 throughout:

$$\frac{12.0\text{ g}}{12.0} \text{ is the mass of } \frac{1.00\text{ mol}}{12.0} \text{ of carbon}$$

3. Scale up to the amount of carbon in 180 g by multiplying by 180 throughout:

$$180 \times \frac{12.0 \text{ g}}{12.0} \text{ is the mass of } 180 \times \frac{1.00 \text{ mol}}{12.0} \text{ of carbon}$$

i.e. 180 g is the mass of 15.0 mol of carbon
∴ the amount of carbon in 180 g is 15.0 mol.

Now attempt the following Exercises. (We use the method of substituting into the expression in our answers, largely because it takes up less space. You can try the 'sentence method' if you get stuck.)

EXERCISE 1.4
Answers on page 209

Calculate the amount in each of the following:
a 30.0 g of oxygen molecules, O_2,
b 31.0 g of phosphorus molecules, P_4,
c 50.0 g of calcium carbonate, $CaCO_3$.

You must also be prepared for questions which ask you to calculate the mass of substance in a given amount. To do this in the following Exercises, use the expression:

$$n = \frac{m}{M} \quad \text{in the form} \quad m = nM$$

EXERCISE 1.5
Answers on page 209

Calculate the mass of each of the following:
a 1.00 mol of hydrogen, H_2,
b 0.500 mol of sodium chloride, NaCl,
c 0.250 mol of carbon dioxide, CO_2.

EXERCISE 1.6
Answers on page 209

A sample of ammonia, NH_3, weighs 1.00 g.
a What amount of ammonia is contained in this sample?
b What mass of sulphur dioxide, SO_2, contains the same number of molecules as are in 1.00 g of ammonia?

We now consider another type of amount calculation.

Calculating the number of particles in a given amount

Sometimes you are required to calculate the number of particles in a given amount of substance. This is easy because you know the number of particles in 1.00 mol (6.02×10^{23}). Use the expression:

$$N = nL$$

where N = the number of particles, n = the amount, L = the Avogadro constant.

EXERCISE 1.7
Answers on page 210

Calculate the number of atoms in:
a 18.0 g of carbon, C,
b 18.0 g of copper, Cu,
c 7.20 g of sulphur, S_8.

EXERCISE 1.8
Answers on page 211

Calculate the number of molecules in:
a 1.00 g of ammonia, NH_3,
b 3.28 g of sulphur dioxide, SO_2,
c 7.20 g of sulphur, S_8.

EXERCISE 1.9
Answers on page 211

Calculate the number of ions present in:
a 0.500 mol of sodium chloride, NaCl (Na^+, Cl^-),
b 14.6 g of sodium chloride, NaCl,
c 18.5 g of calcium chloride, $CaCl_2$.

STOICHIOMETRY

Stoichiometry (pronounced stoy-key-om-i-tree) in its broadest sense includes all the quantitative relationships in chemical reactions. It has to do with how much of one substance will react with another. A chemical equation such as

$$N_2(g) + 3H_2(g) \rightarrow 2NH_3(g)$$

is a kind of chemical balance sheet; it states that one mole of nitrogen reacts with three moles of hydrogen to yield two moles of ammonia. (It does **not** tell us about the rate of the reaction or the conditions necessary to bring it about.) The numbers 1, 3 and 2 are called the stoichiometric coefficients. Such an equation is an essential starting point for many experiments and calculations; it tells us the proportions in which the substances react and in which the products are formed.

We start with a Worked Example.

WORKED EXAMPLE

What mass of iodine will react completely with 10.0 g of aluminium?

This problem is a little more complicated than those you have done previously, because it involves several steps. Each step is very simple, but you may not immediately see where to start. Before we present a detailed solution, let us look at a way of approaching multi-step problems. Even if you find this problem easy, the approach may be useful to you in more difficult problems. We suggest that you ask yourself three questions:

1. What do I know?
In this case, the answer should be:
a the equation for the reaction;
b the mass of aluminium.

In some problems you may be given the equation; in this one, you are expected to write it down from your general chemical knowledge. In nearly every problem about a reaction, the equation provides vital information.

2. What can I get from what I know?
a From the equation, I can find the ratio of reacting amounts.
b From the mass of aluminium, I can calculate the amount, provided I look up the molar mass.

3. Can I now see how to get the final answer?
In most cases the answer will be 'Yes', but you may have to ask the second question again, now that you know more than you did at the start.
a From the amount of aluminium and the ratio of reacting amounts, I can calculate the amount of iodine.
b From the amount of iodine, I can get the mass, using the molar mass.

Instead of writing answers to these questions, you can summarise your thinking in a flow diagram.

KNOWN → intermediate → ANSWER

EQUATION → relative amounts ↘
amount of I → MASS OF I
MASS OF Al → amount of Al ↗

An alternative strategy is to work backwards from the answer towards the given information, or you may use a combination of both strategies, meeting up in the middle. In any case, the steps that you use will probably be the same, although the order in which you take them may be different. Now we go through each step, using the first strategy, and doing the necessary calculations.

Solution

1. Write the balanced equation for the reaction:

$$2Al(s) + 3I_2(s) \rightarrow 2AlI_3(s)$$

This equation tells us that 2 mol of Al reacts with 3 mol of I_2; so we write the ratio:

$$\frac{\text{amount of Al}}{\text{amount of } I_2} = \frac{2}{3}$$

2. Calculate the amount of aluminium using the expression:

$$n = \frac{m}{M}$$

$$\therefore\ n = \frac{10.0\text{ g}}{27.0\text{ g mol}^{-1}} = 0.370\text{ mol}$$

3. Calculate the amount of iodine which reacts with this amount of aluminium by substituting into the expression based on the equation:

$$\frac{\text{amount of Al}}{\text{amount of } I_2} = \frac{2}{3}$$

$$\therefore\ \text{amount of } I_2 = \frac{3}{2} \times \text{amount of Al}$$

$$= \frac{3}{2} \times 0.370\text{ mol} = 0.555\text{ mol}$$

4. Calculate the mass of iodine from the amount using the expression:

$$n = \frac{m}{M} \quad \text{in the form} \quad m = nM$$

$$\therefore m = nM = 0.555\text{ mol} \times 254\text{ g mol}^{-1} = 141\text{ g}$$

Now try some similar problems for yourself.

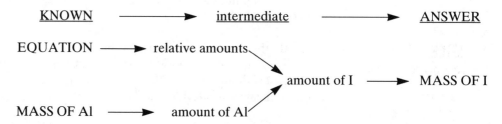

EXERCISE 1.10
Answers on page 210

a What mass of magnesium would react completely with 16.0 g of sulphur?

$$Mg(s) + S(s) \rightarrow MgS(s)$$

b What mass of oxygen would be produced by completely decomposing 4.25 g of sodium nitrate, $NaNO_3$?

$$2NaNO_3(s) \rightarrow 2NaNO_2(s) + O_2(g)$$

EXERCISE 1.11
Answer on page 211

What mass of phosphorus(V) oxide, P_2O_5, would be formed by complete oxidation of 4.00 g of phosphorus?

$$4P(s) + 5O_2(g) \rightarrow 2P_2O_5(s)$$

EXERCISE 1.12
Answer on page 211

When 0.27 g of aluminium is added to excess copper(II) sulphate solution, 0.96 g of copper is precipitated. Deduce the equation for the reaction which takes place.

In the next Exercise you have to decide first which one of two reactants is present in excess.

EXERCISE 1.13
Answer on page 211

A mixture containing 2.80 g of iron and 2.00 g of sulphur is heated together. What mass of iron(II) sulphide, FeS, is produced?

$$Fe(s) + S(s) \rightarrow FeS(s)$$

Another important application of the mole concept is in working out a percentage yield.

Working out a percentage yield

Organic preparations rarely give 100% yield. It is not uncommon for yields to be around 60% or less. This is due to the occurrence of other reactions giving different products, to reversible reactions leading to equilibrium, and to the losses that are involved in the separations. This means you rarely obtain the amount of product you would calculate from the equation. In the Worked Example which follows we show you how to work out the yield of a product from given quantities of reactants and products.

The percentage yield is given by the following expression:

$$\textbf{\% yield} = \frac{\textbf{actual mass of product}}{\textbf{maximum mass of product}} \times \textbf{100}$$

WORKED EXAMPLE

A student hydrolysed 10 g of 1-bromobutane and obtained 3.9 g of butan-1-ol.

$$\underset{\text{1-bromobutane}}{C_4H_9Br} + NaOH \rightarrow \underset{\text{butan-1-ol}}{C_4H_9OH} + NaBr$$

Calculate the % yield of butan-1-ol.

Solution

We know the actual mass of product obtained was 3.9 g and we now need to work out the maximum mass of product which would have been obtained if all the 1-bromobutane had been converted into butan-1-ol.

1. Calculate the amount of reactant (i.e. 1-bromobutane) actually used by means of the expression

$$n = \frac{m}{M}; \quad n = \frac{10\text{ g}}{137\text{ g mol}^{-1}} = \textbf{0.073 mol}$$

2. The equation tells you that 1 mol of 1-bromobutane produces 1 mol of butan-1-ol. We should therefore expect to produce 0.073 mol of butan-1-ol from 0.073 mol of 1-bromobutane.
3. Calculate the maximum mass of product, i.e. butan-1-ol, from the amount using:

$$m = nM = 0.073\ \text{mol} \times 74\ \text{g mol}^{-1} = \mathbf{5.4\ g}$$

4. Substitute into the expression:

$$\%\ \text{yield} = \frac{\text{actual mass of product}}{\text{maximum mass of product}} \times 100$$

$$= \frac{3.9\ \text{g}}{5.4\ \text{g}} \times 100 = \mathbf{72\%}$$

(Yields are usually given to the nearest whole number.)

You should now be able to do the next Exercise. Since you usually measure out volumes of liquids rather than weigh them, the first part of the exercise requires you to convert a volume into mass.

EXERCISE 1.14

Answer on page 211

In an experiment, 5.1 g of iodoethane was prepared by refluxing 5.0 cm^3 of ethanol with red phosphorus and iodine.

$$\underset{\text{ethanol}}{C_2H_5OH} \xrightarrow[\text{heat}]{\text{P, } I_2} \underset{\text{iodoethane}}{C_2H_5I}$$

What was the percentage yield of the product? (Density of ethanol = 0.79 g cm^{-3}.)

In a synthesis with several steps, if they all have poor yields, the product may effectively disappear before the synthesis is complete.

Consider the following three-step synthesis:

$$\underset{1\ \text{mol}}{A} \xrightarrow{40\%} \underset{0.4\ \text{mol}}{B} \xrightarrow{40\%} \underset{0.16\ \text{mol}}{C} \xrightarrow{40\%} \underset{0.064\ \text{mol}}{D}$$

If each step had a 40% yield,

$$\text{the overall yield} = \frac{40}{100} \times \frac{40}{100} \times 40\% = 6.4\%,$$

a very poor yield. This is one reason why preparations are chosen with as few steps as possible. Not only do multi-stage procedures produce very low yields but they also require separation at each step.

In the next Exercise you are expected to estimate suitable quantities of reagents to prepare a product at a given yield. This is really the reverse of our previous Worked Example on pages 7–8 (above). We now show you how this is done.

WORKED EXAMPLE

A student is asked to prepare 6.0 g of butan-1-ol assuming a yield of 60% and is required to estimate suitable quantities of reagents.

$$C_4H_9Br + NaOH \rightarrow C_4H_9OH + NaBr$$

(Density of 1-bromobutane = 1.28 g cm^{-3}.)

Solution

1. Calculate the amount of product to be obtained by substituting into the expression:

$$n = \frac{m}{M} = \frac{6.0\ \text{g}}{74\ \text{g mol}^{-1}} = \mathbf{0.081\ mol}$$

2. From the equation, 1 mol of butan-1-ol is produced from 1 mol of 1-bromobutane, so amount of reactant required (i.e. 1-bromobutane) assuming a 100% yield = 0.081 mol.
3. Calculate the amount of reactant needed for the given percentage yield. For a 60% yield we should need more reagents to compensate for the loss. So amount of 1-bromobutane required

$$= \frac{100}{60} \times 0.081\ \text{mol} = \mathbf{0.135\ mol}$$

4. Convert amount to mass using $m = nM = 0.135\ \text{mol} \times 137\ \text{g mol}^{-1} = \mathbf{18.5\ g}$.
5. For liquids convert mass to volume using the expression:

$$\text{density} = \frac{\text{mass}}{\text{volume}}$$

So:

$$\text{volume of 1-bromobutane} = \frac{18.5\ \text{g}}{1.28\ \text{g cm}^{-3}} = \mathbf{14.5\ cm^3\ 1\text{-}bromobutane}$$

6. Calculate the amounts of other reagents, i.e. NaOH. From the equation we should also expect to need 0.135 mol NaOH.
7. Convert amounts to mass using $m = nM = 0.135\ \text{mol} \times 40\ \text{g mol}^{-1} = 5.4$ g **sodium hydroxide**. This represents the minimum amount of sodium hydroxide we should use but in practice we use more than this, say 50% more, to ensure complete conversion of 1-bromobutane. You should now be able to do the next Exercise.

EXERCISE 1.15
Answers on page 212

2-Chloro-2-methylpropane can be obtained in a yield of 62% by the following reaction:

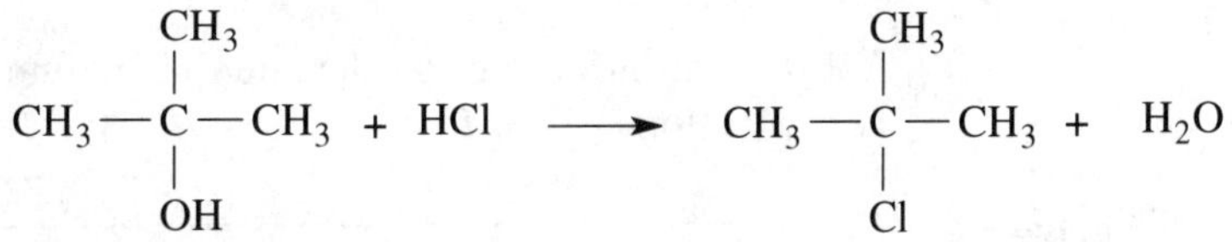

2-methylpropan-2-ol 2-chloro-2-methylpropane

Calculate suitable quantities of all reagents to prepare 5.5 g of pure 2-chloro-2-methylpropane using this method.

Density of hydrochloric acid (32% acid in water) = 1.16 g cm^{-3}
Density of 2-methylpropan-2-ol = 0.79 g cm^{-3}

Another important application of the mole concept is in the calculation of the empirical formulae of substances.

CALCULATING EMPIRICAL FORMULAE AND MOLECULAR FORMULAE

The empirical formula of a compound is the simplest form of the ratio of the atoms of different elements in it. The molecular formula tells the actual number of each kind of atom in a molecule of the substance. For example, the molecular formula of phosphorus(V) oxide is P_4O_{10}, whereas its empirical formula is P_2O_5.

Calculating the empirical formula from the masses of constituents

To determine the empirical formula of a compound, we must first calculate the amount of each substance present in a sample and then calculate the simplest whole-number ratio of the amounts.

It is convenient to set out the results in tabular form and we suggest that you use this method. However, in the following Worked Example, we will go through the procedure step by step, establishing the table as we go.

WORKED EXAMPLE An 18.3 g sample of a hydrated compound contained 4.0 g of calcium, 7.1 g of chlorine and 7.2 g of water only. Calculate its empirical formula.

Solution

1. List the mass of each component and its molar mass. Although water is a molecule, in the calculation we treat it in the same way as we do atoms.

Table 1.1

	Ca	Cl	H_2O
Mass/g	4.0	7.1	7.2
Molar mass/g mol^{-1}	40.0	35.5	18.0

2. From this information calculate the amount of each substance present using the expression $n = m/M$.

Table 1.2

	Ca	Cl	H_2O
Amount/mol	$\frac{4.0}{40.0} = 0.10$	$\frac{7.1}{35.5} = 0.20$	$\frac{7.2}{18.0} = 0.40$

(For simplicity, we have omitted the units from the calculation – they cancel anyway.) This result means that in the given sample there is 0.10 mol of calcium, 0.20 mol of chlorine and 0.40 mol of water.

3. Calculate the **relative** amount of each substance by dividing each amount by the smallest amount.

Table 1.3

	Ca	Cl	H_2O
Amount/smallest amount = relative amount	$\frac{0.10}{0.10} = 1.0$	$\frac{0.20}{0.10} = 2.0$	$\frac{0.40}{0.10} = 4.0$

The relative amounts are in the simple ratio 1:2:4.
From this result you can see that the empirical formula is $CaCl_2{\cdot}4H_2O$.
To see if you understand this procedure, try Exercise 1.16.

EXERCISE 1.16
Answer on page 212

A sample of a hydrated compound was analysed and found to contain 2.10 g of cobalt, 1.14 g of sulphur, 2.28 g of oxygen and 4.50 g of water. Calculate its empirical formula.

A modification of this type of problem is to determine the ratio of the amount of water to the amount of anhydrous compound. You have practice in this type of problem in the next Exercise.

EXERCISE 1.17
Answer on page 212

10.00 g of hydrated barium chloride is heated until all the water is driven off. The mass of anhydrous compound is 8.53 g. Determine the value of x in $BaCl_2{\cdot}xH_2O$.

You should be prepared for variations to this type of problem. The following part of an A-level question illustrates such a variation.

EXERCISE 1.18
Answer on page 212

When 585 mg of the salt $UO(C_2O_4){\cdot}6H_2O$ was left in a vacuum desiccator for 48 hours, the mass changed to 535 mg. What formula would you predict for the resulting substance?

Now we look at another way of calculating the empirical formula.

Calculating the empirical formula from percentage composition by mass

The result of the analysis of a compound may also be given in terms of the percentage composition by mass. Study the following Worked Example which deals with this type of problem.

WORKED EXAMPLE

An organic compound was analysed and was found to have the following percentage composition by mass: 48.8% carbon, 13.5% hydrogen and 37.7% nitrogen. Calculate the empirical formula of the compound.

Solution

If we consider a sample with mass 100.0 g, we can write immediately the mass of each substance: 48.8 g of carbon, 13.5 g of hydrogen and 37.7 g of nitrogen. Then we set up a table as before. The instructions between each step are omitted this time, but you should check our calculations.

Table 1.4

	C	H	N
Mass/g	48.8	13.5	37.7
Molar mass/g mol^{-1}	12.0	1.00	14.0
Amount/mol	4.07	13.5	2.69
$\frac{\text{Amount}}{\text{Smallest amount}}$	$\frac{4.07}{2.69} = 1.51$	$\frac{13.5}{2.69} = 5.02$	$\frac{2.69}{2.69} = 1.00$
Simplest ratio of relative amounts	3	10	2

Empirical formula = $C_3H_{10}N_2$

In the preceding Exercises you 'rounded off' values close to whole numbers in order to get a simple ratio. This is justified because small differences from whole numbers are probably due to experimental errors. Here, however, we cannot justify rounding off 1.51 to 1 or 2, but we can obtain a simple ratio by multiplying the relative amounts by two.

Now attempt the following Exercises where you must decide whether to round off or multiply by a factor.

EXERCISE 1.19
Answer on page 213

A compound of carbon, hydrogen and oxygen contains 40.0% carbon, 6.6% hydrogen and 53.4% oxygen. Calculate its empirical formula.

EXERCISE 1.20
Answer on page 213

Determine the formula of a mineral with the following mass composition: Na = 12.1%, Al = 14.2%, Si = 22.1%, O = 42.1%, H_2O = 9.48%.

EXERCISE 1.21
Answer on page 213

A 10.00 g sample of a compound contains 3.91 g of carbon, 0.87 g of hydrogen and the remainder is oxygen. Calculate the empirical formula of the compound.

Calculating the empirical formula from elemental analysis data

Having shown you how to work out an empirical formula from mass percentage data, we now show you how it is determined from the results of elemental analysis. You need only do this if it is specified on your syllabus. First, we do a Worked Example for a compound containing carbon and hydrogen only.

WORKED EXAMPLE

A 1.00 g sample of hydrocarbon was burnt in excess oxygen. The carbon dioxide and water produced were absorbed in previously weighed tubes containing magnesium chlorate(VII) and soda-lime respectively. The tubes were found to have gained 3.38 g of carbon dioxide and 0.692 g of water. Determine the empirical formula of the compound.

Solution

1. Calculate the amount of carbon in the sample from the mass of carbon dioxide absorbed.

$$\text{amount of } CO_2 = \frac{\text{mass}}{\text{molar mass}} = \frac{3.38\text{ g}}{44.0\text{ g mol}^{-1}} = 0.0768\text{ mol}$$

   In 1 mol of CO_2 there is 1 mol of C,
   ∴ amount of C = amount of CO_2 = 0.0768 mol
2. Calculate the amount of hydrogen in the sample from the mass of water absorbed.

$$\text{amount of } H_2O = \frac{\text{mass}}{\text{molar mass}} = \frac{0.692\text{ g}}{18.0\text{ g mol}^{-1}} = 0.0384\text{ mol}$$

   In 1 mol of H_2O there are 2 mol of H atoms
   ∴ amount of H = 2 × amount of H_2O = 0.0768 mol
3. Calculate the empirical formula from the ratio C : H. In this case, the amounts of C and H are equal. Therefore, the empirical formula is C_1H_1, i.e. **CH**.

When oxygen is present, as well as carbon and hydrogen, the amount must be determined by difference. We now show this method.

WORKED EXAMPLE

A 1.00 g sample of compound A was burnt in excess oxygen producing 2.52 g of CO_2 and 0.443 g of H_2O. Calculate the empirical formula of the compound.

Solution As before, determine the amounts of carbon and hydrogen atoms.

1. Amount of $CO_2 = \dfrac{\text{mass}}{\text{molar mass}} = \dfrac{2.52\ \text{g}}{44.0\ \text{g mol}^{-1}} = 0.0573\ \text{mol}$

   $\therefore$ amount of C = 0.0573 mol

2. Amount of $H_2O = \dfrac{\text{mass}}{\text{molar mass}} = \dfrac{0.443\ \text{g}}{18.0\ \text{g mol}^{-1}} = 0.0246\ \text{mol}$

   $\therefore$ amount of H = 2 × 0.0246 mol = 0.0492 mol

3. Now calculate the mass of carbon and mass of hydrogen.
   Mass of C = amount × molar mass = 0.0573 mol × 12.0 g mol^{-1} = 0.688 g
   Mass of H = amount × molar mass = 0.0492 mol × 1.00 g mol^{-1} = 0.0492 g
4. The mass of oxygen = mass of sample − (mass of C + mass of H)
   = 1.00 g − (0.688 g + 0.0492 g) = 0.263 g
5. Now construct a table:

Table 1.5

	C	H	O
Mass/g			0.263
Molar mass/g mol^{-1}			16.0
Amount/mol	0.0573	0.0492	0.0164
$\dfrac{\text{Amount}}{\text{Smallest amount}}$	$\dfrac{0.0573}{0.0164} = 3.49$	$\dfrac{0.0492}{0.0164} = 3.00$	$\dfrac{0.0164}{0.0164} = 1.00$
Simplest ratio of relative amounts	7	6	2

Empirical formula = $\mathbf{C_7H_6O_2}$.

Use similar methods in the next two Exercises.

EXERCISE 1.22
Answer on page 213

Complete combustion of a hydrocarbon, Z, gives 0.66 g of carbon dioxide and 0.36 g of water. What is the empirical formula of Z?

EXERCISE 1.23
Answer on page 214

The combustion of 0.146 g of compound B gave 0.374 g of carbon dioxide and 0.154 g of water. Assuming that B contains carbon, hydrogen and oxygen only, determine its empirical formula.

One of the ways of determining the halogen content of a compound is to make from it a precipitate of silver halide, which is filtered off, washed, dried and weighed. The next Exercise includes information that will enable you to determine the empirical formula of a halogeno-compound.

EXERCISE 1.24
Answer on page 214

0.2243 g of compound C gave 0.3771 g of carbon dioxide, 0.0643 g of water and 0.2685 g of silver bromide. Determine the empirical formula of the compound.

Having established the empirical formula of a compound, the next step is to determine its molecular formula.

Molecular formula from empirical formula

In order to determine the molecular formula of a compound not only do we need to know its empirical formula, we also need to know its molar mass.

To give you an idea of how we use empirical formula and molar mass to determine molecular formula we give you a simple Worked Example.

WORKED EXAMPLE

The empirical formula of compound X is found from quantitative analysis to be C_2H_5. Mass spectrometry gave a relative molecular mass of 58. What is the molecular formula of X?

Solution

1. Determine the relative molecular mass corresponding to the empirical formula, C_2H_5.

$$M_r(C_2H_5) = (2 \times 12) + (5 \times 1) = 29$$

2. Substitute into the expression:

molecular formula = (empirical formula)$_n$

$$\text{where } n = \frac{M_r(\text{compound X})}{M_r(C_2H_5)} = \frac{58}{29} = 2$$

molecular formula = $(C_2H_5)_2$ = $\mathbf{C_4H_{10}}$

You should now be able to do the next two Exercises.

EXERCISE 1.25
Answers on page 215

A

An organic compound A has a relative molecular mass of 178 and contains 74.2% C, 7.9% H and 17.9% O. Determine:

a the empirical formula of A,
b the molecular formula of A.

EXERCISE 1.26
Answers on page 215

Substance Q has the elemental composition C = 60.0%, H = 13.3%, O = 26.7% (by mass) and a relative molecular mass of 60.1.

a Calculate the empirical formula of Q.
b What is its molecular formula?

AMOUNTS IN SOLUTION

So far we have shown how to calculate amount of substance from the mass of a substance and its molar mass. However, most chemical reactions take place in solution. If we want to know the amount of substance in solution, then we must know the concentration of the solution and its volume.

Concentration of solution

We express the concentration of a solution as the amount of solute dissolved in a given volume of solution; i.e.

$$\textbf{concentration} = \frac{\textbf{amount}}{\textbf{volume}} \quad \textbf{or, in symbols,} \quad c = \frac{n}{V}$$

Normally, we measure amount in mol, and volume in dm^3, so the usual unit of concentration is **mol dm^{-3}**.

Standard solutions

A standard solution is one of known concentration. We can 'know' the concentration either by preparing the solution according to a given recipe or by analysing it.

Let us suppose we dissolve 0.15 mol (27.0 g) of glucose, $C_6H_{12}O_6$, in enough water to make 1.00 dm^3 of solution. Then its concentration is given as $c = 0.15\ mol\ dm^{-3}$. The letter M is sometimes used as an abbreviation for mol dm^{-3}, but you should use it **only** in conjunction with a formula. For example, 0.0100 M NaOH means a solution of sodium hydroxide, NaOH, having a concentration of 0.0100 mol dm^{-3}.

In this chapter, we go through the main types of problems you are likely to meet which involve solutions. For each one, read the Worked Example, then try the Exercises which follow it.

Calculating concentration from volume and amount

WORKED EXAMPLE Calculate the concentration of a solution which is made by dissolving 0.500 mol of sodium hydroxide, NaOH, in 200 cm^3 of solution.

Solution Calculate the concentration by substituting into the key expression:

$$c = \frac{n}{V}$$

where $n = 0.500$ mol and $V = \left(\frac{200}{1000}\right) dm^3$.

$$c = \frac{n}{V} = \frac{0.500\ mol}{0.200\ dm^3} = 2.50\ mol\ dm^{-3}$$

Now attempt the following Exercise.

EXERCISE 1.27
Answers on page 215

Assume that 0.100 mol of $CuSO_4{\cdot}5H_2O$ is placed in each of the volumetric flasks shown (Fig. 1.2) and is properly diluted to the volumes shown. Calculate the concentration of each solution.

Figure 1.2

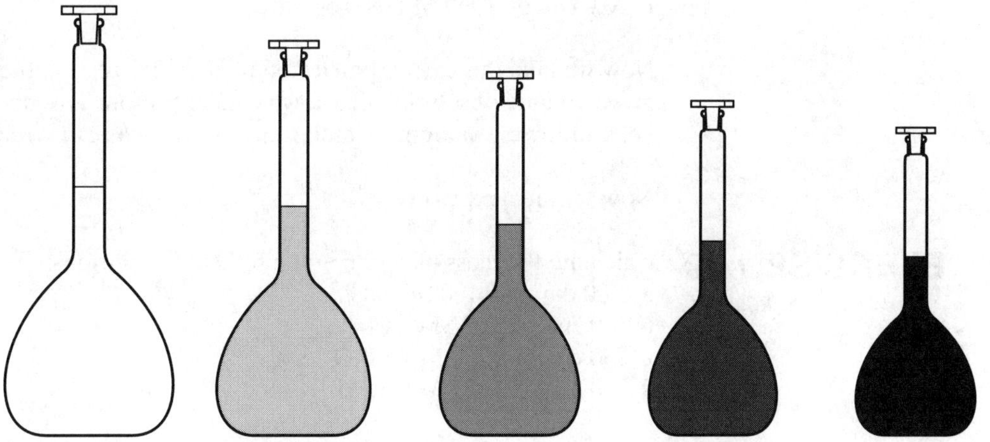

(a) $V = 2000\ cm^3$ (b) $V = 1000\ cm^3$ (c) $V = 500\ cm^3$ (d) $V = 250\ cm^3$ (e) $V = 100\ cm^3$

Calculating concentration from mass of solute and volume

Now we take this calculation a step further – be prepared for problems which give the mass of a substance, not the amount. These need an extra step at the start, i.e. dividing mass by molar mass to get amount using:

$$n = \frac{m}{M}$$

Now attempt the following Exercise.

EXERCISE 1.28
Answers on page 216

The table below indicates the masses of various compounds that were used to prepare the solutions of the stated volumes. Calculate the concentrations of these solutions.

Table 1.6

Compound	Mass/g	Volume/cm^3
a $AgNO_3$	8.50	1000
b KIO_3	10.7	250
c $Pb(NO_3)_2$	11.2	50.0
d $K_2Cr_2O_7$	14.3	250
e $CuSO_4{\cdot}5H_2O$	11.9	500

Calculating the amount of substance in a solution

Often you want to know the amount of substance contained in a given volume of solution of known concentration.

To do this you substitute values for c and V into the expression:

$$c = \frac{n}{V} \quad \text{in the form} \quad n = cV$$

You use this in the next Exercise.

EXERCISE 1.29
Answers on page 216

Calculate the amount of solute in each of the following solutions:

a 4.00 dm^3 of 5.00 M NaOH,
b 1.00 dm^3 of 2.50 M HCl,
c 20.0 cm^3 of 0.439 M HNO_3.

Now we take this calculation a step further, calculating the mass of solute contained in a given volume of solution of known concentration. This needs an extra step at the finish, i.e. multiplying amount by molar mass to get mass, $m = nM$.

Now try the next Exercise.

EXERCISE 1.30
Answers on page 216

Calculate the mass of solute in the following solutions:

a 1.00 dm^3 of 0.100 M NaCl,
b 500 cm^3 of 1.00 M $CaCl_2$,
c 250 cm^3 of 0.200 M $KMnO_4$,
d 200 cm^3 of 0.117 M NaOH.

END-OF-CHAPTER QUESTIONS

Answers on page 216

1.1 What mass of material is there in each of the following?
a 2.00 mol of SO_3
b 0.0300 mol of Cl
c 9.00 mol of SO_4^{2-}
d 0.150 mol of $MgSO_4{\cdot}7H_2O$

1.2 What amount of each substance is contained in the following?
a 31.0 g of P_4
b 1.00×10^{22} atoms of Cu
c 70.0 g of Fe^{2+}
d 9.00×10^{24} molecules of C_2H_5OH

1.3 The mass of one molecule of a compound is 2.19×10^{-22} g. What is the molar mass of the compound?

1.4 What mass of aluminium, Al, is required to produce 1000 g of iron, Fe, according to the equation:

$$3Fe_3O_4(s) + 8Al(s) \rightarrow 4Al_2O_3(s) + 9Fe(s) \ ?$$

1.5 A solution is made containing 2.38 g of magnesium chloride, $MgCl_2$, in 500 cm^3 of solution.
a What is the concentration of magnesium chloride, $MgCl_2$, in this solution?
b What is the concentration of chloride ions in this solution?

1.6 A solution is made by dissolving 8.50 g of sodium nitrate, $NaNO_3$, and 16.40 g of calcium nitrate, $Ca(NO_3)_2$, in enough water to give 2000 cm^3 of solution.
a What is the concentration of sodium ions in the solution?
b What is the concentration of nitrate ions in the solution?

1.7 A hydrated aluminium sulphate, $Al_2(SO_4)_3{\cdot}xH_2O$, contains 8.10% of aluminium by mass. Find the value of *x*.

1.8 Tin(IV) iodide (stannic iodide) is prepared by direct combination of the elements, as follows.

Add 2.0 g of granulated tin to a solution of 6.35 g of iodine in 25 cm^3 of a suitable solvent in a 100 cm^3 flask fitted with a reflux condenser. Reflux gently until reaction is complete. Filter through a pre-heated funnel, wash the residue with 10 cm^3 of hot solvent and add the washings to the filtrate. Cool in ice until orange crystals of tin(IV) iodide separate. Filter off the crystals and dry. The melting point of tin(IV) iodide is 144 °C.

a Write an equation for the reaction.
b Suggest reasons why a solvent is used.
c Which reactant is in excess?
d Calculate the maximum theoretical yield of product.

1.9 Compound A was analysed and found to contain 52.2% carbon, 13.0% hydrogen and 34.8%oxygen. Determine its empirical formula.

1.10 A liquid X contains by mass 31.5% of carbon, 5.3% of hydrogen and 63.2% oxygen. Calculate the empirical formula of X.

1.11 Compound A is a liquid hydrocarbon of molecular weight (relative molecular mass) 78 and contains 92.3% carbon. Calculate the molecular formula of A.

1.12 A compound, P, contains carbon 59.4%, hydrogen 10.9%, nitrogen 13.9% and oxygen 15.8%, by mass. Calculate the empirical formula of P.

1.13 The composition of X by mass is C 66.7%, H 11.1%, O 22.2%.
The molar mass of X is 72. Calculate the molecular formula of X.

1.14 When preparing esters in the laboratory, moderate yields may be obtained by reaction of an alcohol with a carboxylic acid. The reaction is usually slow and incomplete.

$$\underset{\text{ethanoic acid}}{CH_3CO_2H} + \underset{\text{butan-1-ol}}{C_4H_9OH} \rightarrow \underset{\text{butyl ethanoate}}{CH_3CO_2C_4H_9} + \underset{\text{water}}{H_2O}$$

Calculate the quantities of starting materials required to produce 10 g of butyl ethanoate, assuming a 50% yield.

1.15 The following scheme indicates a sequence of single-stage reactions to prepare aspirin from methylbenzene:

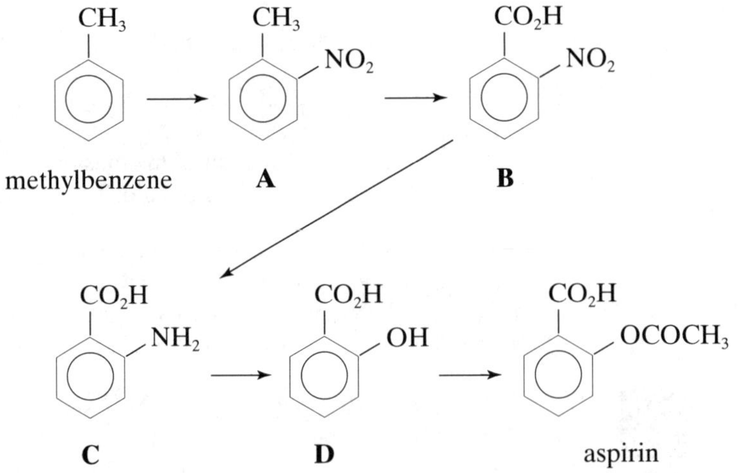

In a preparation 4.60 g of methylbenzene gave 0.90 g of aspirin.

a Calculate the percentage yield obtained.

b What feature of the proposed synthesis is mainly responsible for the low yield?

CHAPTER 2

VOLUMETRIC ANALYSIS (TITRIMETRY)

INTRODUCTION

A titration is a laboratory procedure where a measured volume of one solution is added to a known volume of another reagent until the reaction is complete. This operation is an example of volumetric (titrimetric) analysis. The stoichiometric point (equivalence point) is usually shown by the colour change of an indicator, and is then known as the end-point.

Volumetric analysis is a powerful technique which is used in a variety of ways by chemists in many different fields.

We consider three types of titrations in this chapter:

- acid–base titrations
- redox titrations
- precipitation titrations.

CALCULATING CONCENTRATION FROM TITRATION DATA

First we give a Worked Example to illustrate titrimetric calculations. For the Worked Example, we use an acid–base titration, but the method is applicable to all titrations.

WORKED EXAMPLE

We placed 20.0 cm^3 of a solution of barium hydroxide, $Ba(OH)_2$, of unknown concentration in a conical flask and titrated with a solution of hydrochloric acid, HCl, which had a concentration of 0.0600 mol dm^{-3}. The volume of acid required was 25.0 cm^3. Calculate the concentration of the barium hydroxide solution.

$$Ba(OH)_2(aq) + 2HCl(aq) \rightarrow BaCl_2(aq) + 2H_2O(l)$$

Solution

This is a multi-step calculation. You may find it helpful to look again at the advice we gave on such calculations on page 5. A summarised 'flow-chart' for the solution to this problem is:

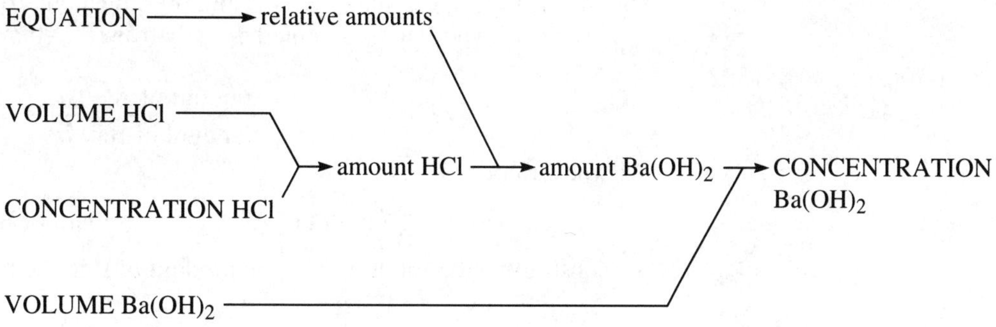

Over the page, we look at the calculations, step by step.

1. Calculate the amount of HCl delivered (the solution of known concentration) by substitution into the expression:

$$c = \frac{n}{V} \quad \text{in the form} \quad n = cV$$

where $c = 0.0600 \text{ mol dm}^{-3}$ and $V = (25.0/1000)\text{dm}^3 = 0.0250 \text{ dm}^3$
$\therefore\ n = cV = 0.0600 \text{ mol dm}^{-3} \times 0.0250 \text{ dm}^3 = 1.50 \times 10^{-3} \text{ mol}$

2. Calculate the amount of $Ba(OH)_2$ (the solution of unknown concentration) which reacts with this amount of HCl by substituting into the expression derived from the equation:

$$\frac{\text{amount of Ba(OH)}_2}{\text{amount of HCl}} = \frac{1}{2}$$

$$\therefore \text{amount of Ba(OH)}_2 = \frac{1}{2} \text{ amount of HCl*}$$

$$= \frac{1}{2}(1.50 \times 10^{-3} \text{ mol})$$

$$= 7.50 \times 10^{-4} \text{ mol}$$

(*Always check this step carefully – it is easy to put the '1/2' in the wrong place.)

3. Calculate the concentration of $Ba(OH)_2$ by substitution into the expression $c = n/V$, where $n = 7.50 \times 10^{-4}$ mol and $V = 0.0200 \text{ dm}^3$

$$\therefore c = \frac{n}{V} = \frac{7.50 \times 10^{-4} \text{ mol}}{0.0200 \text{ dm}^3} = 0.0375 \text{ mol dm}^{-3}$$

It is possible to derive a general expression to solve this problem and many like it. Remember, however, that you should not use an expression unless you understand its derivation.

A general expression

Consider the general equation:

$$a\text{ A} + b\text{ B} \rightarrow \text{Products}$$

We can derive an expression relating concentration of A (c_A), concentration of B (c_B), volume of A (V_A), volume of B (V_B) and the stoichiometric coefficients a and b.

We start with a relation obtained directly from the chemical equation. It is:

$$\frac{\text{amount of A}}{\text{amount of B}} = \frac{a}{b}$$

We know that:

$$\text{amount of A} = c_A V_A \quad \text{amount of B} = c_B V_B$$

So, substituting for amount of A and amount of B in the first expression gives:

$$\frac{c_A V_A}{c_B V_B} = \frac{a}{b}$$

To help you use this expression correctly, remember that both A and *a* are on the top and both B and *b* are on the bottom.

This expression is very useful for solving titrimetric problems and we use it in our answers to such problems wherever appropriate. However, it may not **always** provide the **best** way to tackle a particular problem.

Now we give you some practice in doing volumetric calculations.

EXERCISE 2.1
Answers on page 217

a Solve the Worked Example on page 19 using the expression we derived.
b Why is it not necessary to convert volumes in cm^3 to dm^3 when using the expression?

EXERCISE 2.2
Answers on page 217

What volume of sodium hydroxide solution, 0.500 M NaOH, is needed to neutralise
a 50.0 cm^3 of nitric acid, 0.100 M HNO_3:

$$NaOH(aq) + HNO_3(aq) \rightarrow NaNO_3(aq) + H_2O(l)$$

b 22.5 cm^3 of sulphuric acid, 0.262 M H_2SO_4:

$$2NaOH(aq) + H_2SO_4(aq) \rightarrow Na_2SO_4(aq) + 2H_2O(l)$$

Although you most often start your calculations knowing the equation for the reaction, you may sometimes have to derive the equation, as in the next Exercise.

EXERCISE 2.3
Answers on page 217

As a result of a titration, it was found that 25.0 cm^3 of silver nitrate solution, 0.50 M $AgNO_3$, reacted with 31.3 cm^3 of barium chloride solution, 0.20 M $BaCl_2$. Use these results to determine the stoichiometric coefficients, *a* and *b*, in the equation:

$$a\ AgNO_3(aq) + b\ BaCl_2(aq) \rightarrow \text{Products}$$

REDOX TITRATIONS AND OTHER REDOX REACTIONS

Reactions that involve both *RED*uction and *OX*idation are called *REDOX* reactions. Before we deal with calculations concerning redox titrations, we show you how to separate the equation for a redox reaction into two half-equations, one for oxidation and one for reduction.

Redox half-reactions

Every redox reaction involves the transfer of electrons. For example, consider the following reaction:

$$Zn(s) + Cu^{2+}(aq) \rightarrow Zn^{2+}(aq) + Cu(s)$$

The result is that two electrons from each zinc atom are transferred to a copper ion. We can split the overall reaction into an oxidation reaction and a reduction reaction:

$$Zn(s) \rightarrow Zn^{2+}(aq) + 2e^- \qquad \textbf{oxidation}$$
$$Cu^{2+}(aq) + 2e^- \rightarrow Cu(s) \qquad \textbf{reduction}$$

These are the two half-reactions for this reaction: one which shows electron loss and the other electron gain. Any redox reaction can be written as two half-reactions.

If you experience difficulty in remembering the difference between oxidation and reduction, the following mnemonic may help you (Fig. 2.1).

Figure 2.1

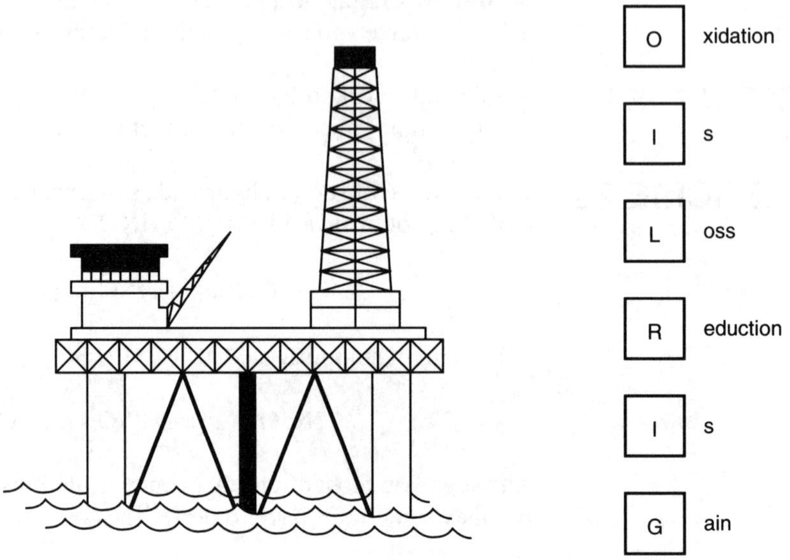

Sometimes, it is difficult to 'pin down' electrons and to say where they are or to which atoms they 'belong'. But, if we are not sure where the electrons are, how can we say anything useful about electron transfer? The concept of **oxidation number** has been developed, using ideas of electronegativity, to overcome this difficulty and enable us to identify redox reactions more readily.

Oxidation number

By assigning every bonding electron in a compound to a particular atom according to certain rules, we give each atom a number called an oxidation number. A transfer of electrons then involves a change in oxidation number.

The oxidation number of an atom in a simple ion is equal to the charge on the ion. Use this definition in the next Exercise.

EXERCISE 2.4
Answers on page 217

What are the oxidation numbers of the following elements?

a Na in Na^+
b O in O^{2-}
c Al in Al^{3+}
d H in H^+
e H in H^-

You have learned that the s-block metals, with the exception of beryllium, always form ionic compounds, and the charge on the positive ion is always the same as the group number. This fact enables you to do the next Exercise.

EXERCISE 2.5
Answers on page 217

What are the oxidation numbers of the elements in the following compounds?

a KCl
b Na_2O
c BaF_2
d Na_3P
e Mg_3N_2
f CsH

In the last Exercise you made use of the fact that the sum of the ion charges in an ionic compound is zero. Since ion charge is, by definition, the same as oxidation number, it follows that the sum of the oxidation numbers for each atom in a compound is zero. This rule also applies to covalent compounds, as we now show in a Worked Example.

WORKED EXAMPLE

What are the oxidation numbers of the elements in these compounds?

a HCl

b H_2S

Solution to a

1. Consider the relative electronegativities of hydrogen and chlorine.
   Electronegativity of H = 2.1
   Electronegativity of Cl = 3.0
   (You should **know** that chlorine is more electronegative than hydrogen, without having to look up values.)
2. Imagine that the more electronegative atom gains both the bonding electrons so that ions are formed.

$$\overset{\delta+}{H}\text{—}\overset{\delta-}{Cl} \text{ imagined as } H^+ \quad Cl^-$$

3. Write down oxidation numbers from the ion charges.

$$Ox(H) = +1; \quad Ox(Cl) = -1$$

Solution to b

1. Consider the relative electronegativities of hydrogen and sulphur.
   Electronegativity of H = 2.1
   Electronegativity of S = 2.5
2. Imagine the partial separation of charge to be completed so that ions are formed.

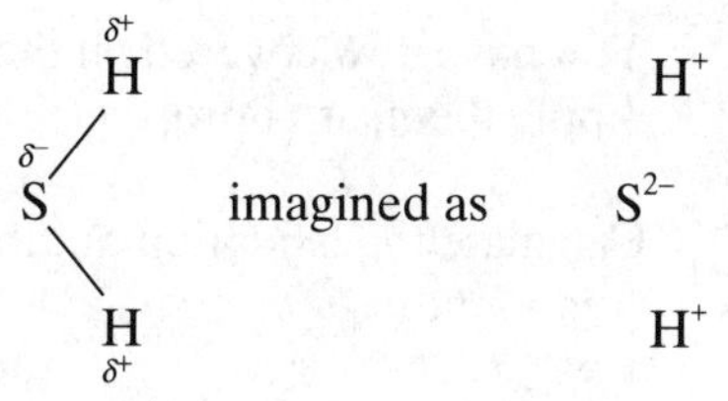

3. Write down oxidation numbers from the ion charges.

$$Ox(H) = +1; \quad Ox(S) = -2$$

Note that the **sum** of the oxidation numbers for **each atom** is zero, because there are two hydrogen atoms:

$$2 \times (+1) + (-2) = 0$$

Now do the following Exercise in the same way.

EXERCISE 2.6
Answers on page 217

What are the oxidation numbers of the elements in these compounds?

a NH_3

b Cl_2O

Repeated application of the method used above has shown that certain elements in compounds always (or nearly always) have the same oxidation number.

Group I metals – always + 1
Group II metals – always + 2

Fluorine – always –1
Hydrogen – always +1, except in ionic hydrides where it is –1
Oxygen – always –2, except in peroxides (which contain an O–O linkage), and when combined with fluorine

In the next Exercise, you consider some cases where oxygen does not have an oxidation number of –2.

EXERCISE 2.7
Answers on page 218

What is the oxidation number of oxygen in the following molecules?

a H_2O_2 (H—O—O—H)
b F_2O

The knowledge that some elements have fixed oxidation numbers enables us to take a short cut in finding the oxidation numbers of other elements, because the total of oxidation numbers for all the atoms in a formula equals the charge on the formula. We show this in a Worked Example.

WORKED EXAMPLE

What is the oxidation number of As in H_3AsO_3?

Solution

Write $Ox(\text{As}) + Ox(\text{all other atoms}) = 0$

$$\therefore\ Ox(\text{As}) = -Ox(\text{all other atoms})$$
$$= -[3\,Ox(\text{H}) + 3\,Ox(\text{O})]$$
$$= -[3(+1) + 3(-2)] = -3 + 6 = \mathbf{+3}$$

Use the same method in the following Exercise.

EXERCISE 2.8
Answers on page 218

What are the oxidation numbers of the following atoms?

a As in H_3AsO_4
b Cr in $HCr_2O_7^-$

You have now covered all the simple rules which help you to assign oxidation numbers. Apply them, as appropriate, in the next Exercise.

EXERCISE 2.9
Answers on page 218

Calculate the oxidation numbers of the underlined elements in the following compounds:

a $K_2\underline{Cr_2}O_7$,
b $Pb\underline{S}O_4$,
c $H\underline{N}O_3$,
d $Na_2\underline{S_2}O_3$,
e $Ca\underline{C}O_3$,
f $Na_8\underline{Ta_6}O_{19}$,
g $\underline{N}H_4\underline{N}O_3$,
h $Na_2\underline{S_4}O_6$,
i $\underline{Fe_3}O_4$.

■ Oxidation number and redox reactions

We bring together two statements which should by now be familiar to you:

Oxidation is a loss of electrons

Oxidation number is the number of electrons 'lost' by an atom in going from its uncombined state to a particular combined state.

The combination of these two statements gives an alternative definition of oxidation:

Oxidation involves an increase in oxidation number

Similarly,

Reduction involves a decrease in oxidation number

You can apply these alternative definitions in the next Exercise.

EXERCISE 2.10
Answers on page 218

By assigning oxidation numbers to the elements in both reactants and products, decide whether or not the following are redox reactions.

a $Fe(s) + CuSO_4(aq) \rightarrow FeSO_4(aq) + Cu(s)$
b $2KBr(aq) + Cl_2(aq) \rightarrow 2KCl(aq) + Br_2(aq)$
c $CuSO_4(aq) + Pb(NO_3)_2(aq) \rightarrow Cu(NO_3)_2(aq) + PbSO_4(s)$
d $C_2H_4(g) + H_2(g) \rightarrow C_2H_6(g)$
e $MnO_4^-(aq) + 5Fe^{2+}(aq) + 8H^+(aq) \rightarrow Mn^{2+}(aq) + 5Fe^{3+}(aq) + 4H_2O(l)$

EXERCISE 2.11
Answers on page 218

Study the equations below and answer the questions which follow.

i) $H_2S(aq) + Br_2(aq) \rightarrow 2HBr(aq) + S(s)$
ii) $Al(s) + 1\frac{1}{2}I_2(s) \rightarrow AlI_3(s)$
iii) $Cl_2(g) + H_2O(l) \rightarrow HCl(aq) + HClO(aq)$
iv) $2Na_2S_2O_3(aq) + I_2(aq) \rightarrow Na_2S_4O_6(aq) + 2NaI(aq)$

a Write down the oxidation number for each atom appearing in the above equations.
b For each redox reaction, state which species is oxidised and which is reduced.

You learned from the Worked Example on pages 19–20 that, in order to solve a titrimetric calculation, you need to know the balanced equation for the reaction. Equations for redox reactions may be difficult to balance by trial and error, so we now show you how a knowledge of oxidation numbers and electron transfer can be used to balance equations.

Balancing redox equations

The following Worked Example shows you how to balance a fairly cumbersome redox equation by two different methods.

WORKED EXAMPLE

Balance the following redox equation:

$$ClO_3^-(aq) + I^-(aq) + H^+(aq) \rightarrow I_2(aq) + Cl^-(aq) + H_2O(l)$$

Solution

1. Identify which species is oxidised and which is reduced. This can be done by looking at oxidation numbers.

$$\overset{+5}{Cl}O_3(aq) + \overset{-1}{I^-}(aq) + H^+(aq) \rightarrow \overset{0}{I_2}(aq) + \overset{-1}{Cl^-}(aq) + H_2O(l)$$

∴ the species oxidised is I^-, $[Ox(I) -1 \rightarrow 0]$
and the species reduced is ClO_3^- $[Ox(Cl) +5 \rightarrow -1]$

2. Construct two balanced half-equations, one representing oxidation and one representing reduction.

   a Balance the elements in each half-equation: if necessary, include H^+, OH^- and H_2O, as appropriate.

   Oxidation $2I^-(aq) \rightarrow I_2(aq)$

   Reduction $ClO_3^-(aq) \rightarrow Cl^-(aq)$

   The three oxygen atoms become incorporated in water molecules, i.e.

   $$ClO_3^-(aq) \rightarrow Cl^-(aq) + 3H_2O(l)$$

   Since the reaction occurs in an acid medium, include H^+ ions to balance the hydrogen in the water molecules, i.e.

   $$ClO_3^-(aq) + 6H^+(aq) \rightarrow Cl^-(aq) + 3H_2O(l)$$

   b Balance the charge in each half-equation. This can be done either by looking at the change in oxidation number or by balancing the charges on the ions.

Oxidation number method

$$2I^-(aq) \rightarrow I_2(aq)$$

Ox(I) changes from –1 to 0, i.e. each I atom loses one electron. Since there are two iodide ions, two electrons must be released in forming each iodine molecule.

Balancing charge method

$$2I^-(aq) \rightarrow I_2(aq)$$

Count the charges on each side of the half-equations:

2 negative charges → zero charge

Therefore, two electrons must be released from two iodide ions in order to form an iodine molecule.

$$2I^-(aq) \rightarrow I_2(aq) + 2e^-$$

Oxidation number method

Similarly, for

$$ClO_3^-(aq) + 6H^+(aq) \rightarrow Cl^-(aq) + 3H_2O(l)$$

Ox(Cl) changes from +5 to –1, i.e. 6 units. Therefore, for each ClO_3^- ion reduced, 6 electrons are required.

Balancing charge method

Similarly, for

$$ClO_3^-(aq) + 6H^+(aq) \rightarrow Cl^-(aq) + 3H_2O(l)$$

5 positive charges (6 – 1) → 1 negative charge

Therefore, 6 electrons are needed for each ClO_3^- ion reduced.

$$ClO_3^-(aq) + 6H^+(aq) + 6e^- \rightarrow Cl^-(aq) + 3H_2O(l)$$

3. Scale the half-equations up or down, so that the number of electrons required by one half-equation is the same as the number provided by the other.

   $$ClO_3^-(aq) + 6H^+(aq) + 6e^- \rightarrow Cl^-(aq) + 3H_2O(l)$$

   $$2I^-(aq) \rightarrow I_2(aq) + 2e^- \text{ (multiply this equation by 3)}$$

   $$6I^-(aq) \rightarrow 3I_2(aq) + 6e^-$$

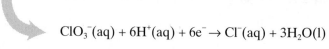

4. Combine the two half-equations by adding the left sides together and the right sides together.

$$ClO_3^-(aq) + 6H^+(aq) + 6I^-(aq) \rightarrow 3I_2(aq) + Cl^-(aq) + 3H_2O(l)$$

Decide which of the two methods you prefer and attempt the next Exercise. We use the balancing charge method in the answers, but you should be aware of both methods.

EXERCISE 2.12
Answers on page 219

Balance the following equations.

a $IO_3^-(aq) + I^-(aq) + H^+(aq) \rightarrow I_2(aq) + H_2O(l)$
b $S_2O_3^{2-}(aq) + I_2(aq) \rightarrow I^-(aq) + S_4O_6^{2-}(aq)$
c $MnO_4^-(aq) + H^+(aq) + Fe^{2+}(aq) \rightarrow Mn^{2+}(aq) + Fe^{3+}(aq) + H_2O(l)$
d $MnO_4^-(aq) + H^+(aq) + C_2O_4^{2-}(aq) \rightarrow Mn^{2+}(aq) + CO_2(g) + H_2O(l)$
e $MnO_4^-(aq) + H^+(aq) + H_2O_2(aq) \rightarrow Mn^{2+}(aq) + O_2(g) + H_2O(l)$

Now attempt the following titrimetric calculations that are based on redox reactions. Remember that the procedure for solving these problems is the same as for acid–base volumetric calculations.

EXERCISE 2.13
Answer on page 219

9.80 g of pure ammonium iron(II) sulphate, $FeSO_4{\cdot}(NH_4)_2SO_4{\cdot}6H_2O$, was made up to 250 cm^3 with dilute sulphuric acid. 25.0 cm^3 of the solution reacted completely with 24.6 cm^3 of potassium manganate(VII) solution. Calculate the concentration of potassium manganate(VII) solution.

EXERCISE 2.14
Answers on page 219

25.0 cm^3 of a solution containing ethanedioic acid, $H_2C_2O_4$, and sodium ethanedioate, $Na_2C_2O_4$, required 14.8 cm^3 of 0.100 M sodium hydroxide solution for neutralisation. A further 25.0 cm^3 of the $H_2C_2O_4/Na_2C_2O_4$ solution required 29.5 cm^3 of 0.0200 M potassium manganate(VII) solution for oxidation in acidic conditions at about 60 °C. Calculate the mass of each anhydrous constituent per dm^3 of the solution.

You may sometimes have to derive the equation, as in the following two Exercises.

EXERCISE 2.15
Answers on page 219

A

a What mass of potassium iodate(V) (KIO_3) would be required to make 250 cm^3 of a solution containing one-sixtieth of a mole per dm^3?
b When 25.0 cm^3 of the solution of potassium iodate(V) of the concentration in **a** was added to excess of acidified potassium iodide solution, the iodine liberated reacted with 20.0 cm^3 of a solution of sodium thiosulphate. Calculate the concentration of the thiosulphate solution in moles per dm^3.

$$IO_3^- + 5I^- + 6H^+ \rightarrow 3I_2 + 3H_2O$$

$$I_2 + 2S_2O_3^{2-} \rightarrow 2I^- + S_4O_6^{2-}$$

c State how you would use the iodate(V)/iodide reaction and a standard solution of sodium thiosulphate to find the concentration of a solution of hydrochloric acid.
d 50 cm^3 of a solution containing 0.10 mol dm^{-3} of bromine (Br_2) was added to 10 cm^3 of a 0.10 mol dm^{-3} sodium thiosulphate solution. Excess potassium iodide was then added to the solution; the bromine which was left over from the first reaction liberated enough iodine to react with exactly 20 cm^3 of the same thiosulphate solution.
i) What is the apparent oxidation number of S in the thiosulphate ion ($S_2O_3^{2-}$)?
ii) From the figures given, calculate the number of moles of bromine which reacted directly with one mole of thiosulphate.
iii) To what oxidation number of the S did the bromine oxidise the thiosulphate?
iv) Derive the equation for the reaction between bromine molecules and thiosulphate ions.

EXERCISE 2.16
Answers on page 220

Under certain concentration and temperature conditions, 0.8 g of iron is found to react with 0.9 g of pure nitric acid, evolving an oxide of nitrogen (N_2O, or NO, or N_2O_4).

a Calculate the molar ratio of iron and nitric acid that react.

b Complete:

Gain of .... electrons

$$\text{........ Fe(s)} + \text{........ } HNO_3(aq) \longrightarrow \text{........ } Fe^{3+}(aq) + \begin{bmatrix} N_2O \\ \text{or NO} \\ \text{or } N_2O_4 \end{bmatrix}$$

Loss of .... electrons

c From the number of electrons gained by the nitric acid, deduce the change in oxidation number of nitrogen and decide which of the three oxides of nitrogen is formed.

d Complete:

$$\text{........ Fe(s)} - \text{........ } e^- \rightarrow \text{........ } Fe^{3+}(aq)$$

$$\text{........ } H^+ + 1HNO_3(aq) + \text{........ } e^- \rightarrow \begin{bmatrix} \tfrac{1}{2}N_2O \\ \text{or 1NO} \\ \text{or } \tfrac{1}{2}N_2O_4 \end{bmatrix} + \text{........ } H_2O$$

and so, by addition, balance the equation for the reaction between iron and nitric acid under these conditions.

PRECIPITATION TITRATIONS

The most common precipitation titrations involve the reaction between silver ions and chloride ions in solution, according to the equation:

$$Ag^+(aq) + Cl^-(aq) \rightarrow AgCl(s)$$

The necessary data for the calculations are the results of titrations in which the indicator is usually potassium chromate(VI). The end-point is the first trace of a red precipitate of silver chromate, Ag_2CrO_4, which appears as soon as precipitation of colourless silver chloride is complete. Alternatively, an adsorption indicator such as fluorescein may be used.

EXERCISE 2.17
Answers on page 220

0.767 g of a chloride of phosphorus was dissolved in water and made up to 250 cm^3 of solution. After neutralisation, 25.0 cm^3 of this solution required 18.4 cm^3 of 0.100 M $AgNO_3$ for complete precipitation of the chloride.

a Calculate the amount of chloride ions in 25.0 cm^3 of solution.

b Calculate the amount and hence the mass of chloride ions in the whole sample.

c Calculate the mass and hence the amount of phosphorus in the whole sample.

d Calculate the amount of chlorine combined with one mole of phosphorus.

e Hence state the simplest formula of the chloride.

EXERCISE 2.18
Answers on page 220

1.420 g of a chloride of sulphur was dissolved in water and the solution made up to a volume of 250 cm^3. 25.0 cm^3 of this solution, after neutralisation, needed 21.0 cm^3 of 0.100 M $AgNO_3$ for complete precipitation of the chloride.

a Calculate the number of moles of chlorine combined with one mole of sulphur.

b If the relative molar mass of the chloride is 135.2 what is its molecular formula?

The previous two Exercises gave you some help by indicating some of the steps in the calculation. However, for some A-level questions you are expected to devise your own calculation method. What may be required of you is indicated in the following Worked Example.

WORKED EXAMPLE 1.58 g of hydrated barium chloride $BaCl_2.xH_2O$ was dissolved in water and made up to 250 cm^3. 10.0 cm^3 of this solution required 10.2 cm^3 of 0.0506 M silver nitrate ($AgNO_3$) solution for complete reaction. Calculate the value of x in the formula $BaCl_2.xH_2O$.

Solution There is more than one way of tackling this problem. Here is one suggestion.

First calculate the masses of $BaCl_2$ and H_2O in each sample. The mass of $BaCl_2$ can be calculated indirectly from titration data and the mass of H_2O by difference.

Next convert masses to relative amounts to determine x. We explain this step by step.

1. From the volume and concentration of silver nitrate, calculate the amount of chloride ions present in a 10.0 cm^3 sample (as Cl^-).

$$Ag^+(aq) + Cl^-(aq) \rightarrow AgCl(s)$$

$$\therefore \text{ amount of Cl} = \text{amount of } AgNO_3$$
$$= cV = 0.0506 \text{ mol dm}^{-3} \times 0.0102 \text{ dm}^3$$
$$= 5.16 \times 10^{-4} \text{ mol}$$

2. Calculate the mass of anhydrous barium chloride, $BaCl_2$, present in a sample.

$$BaCl_2(s) + aq \rightarrow Ba^{2+}(aq) + 2Cl^-(aq)$$

$$\therefore \text{ amount of } BaCl_2 = \tfrac{1}{2} \times \text{amount of } Cl^-$$
$$= \tfrac{1}{2} \times 5.16 \times 10^{-4} \text{ mol} = 2.58 \times 10^{-4} \text{ mol}$$
$$\text{mass of } BaCl_2 = nM = 2.58 \times 10^{-4} \text{ mol} \times 208 \text{ g mol}^{-1}$$
$$= 0.0537 \text{ g}$$

3. Calculate the mass of water present by subtracting the mass of $BaCl_2$ from the mass of $BaCl_2{\cdot}xH_2O$.

$$\text{mass of } H_2O = \text{mass of } BaCl_2 \cdot xH_2O - \text{mass of } BaCl_2$$
$$= 0.0632 \text{ g} - 0.0537 \text{ g} = 0.0095 \text{ g}$$

$$\text{amount of } H_2O = \frac{m}{M} = \frac{0.0095 \text{ g}}{18.0 \text{ g mol}^{-1}} = 5.3 \times 10^{-4} \text{ mol}$$

4. Determine the ratio of amount of $BaCl_2$ to amount of H_2O and thus the value of x.

	$BaCl_2$	**H_2O**
Amount/mol	2.58×10^{-4}	5.3×10^{-4}
Amount/ smallest amount = relative amount	$\frac{2.58 \times 10^{-4}}{2.58 \times 10^{-4}} = 1.00$	$\frac{5.3 \times 10^{-4}}{2.58 \times 10^{-4}} = 2.1$

The relative amounts are very close to the integers 1 and 2.
$\therefore$ $x = 2$ and the formula is $BaCl_2{\cdot}2H_2O$.

A flow-chart for this multi-step calculation is as follows:

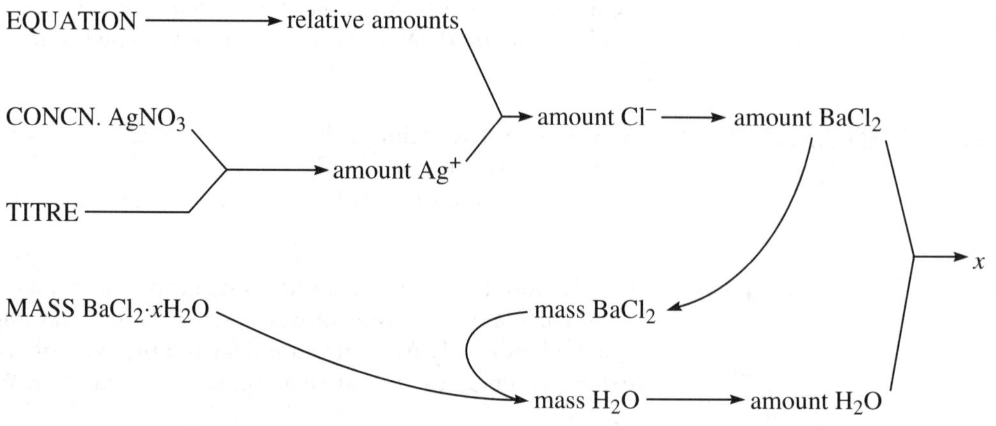

The Worked Example above should help you to do the next Exercise, which is part of an A-level question. We suggest that if you find it difficult, you look at our introduction to the Worked Example on page 5 and sketch out a flow-chart before you begin.

EXERCISE 2.19
Answers on page 221

When 0.203 g of hydrated magnesium chloride, $MgCl_m{\cdot}nH_2O$, was dissolved in water and titrated with 0.100 M silver nitrate ($AgNO_3$) solution, 20.0 cm^3 of the latter was required. A sample of the hydrated chloride lost 53.2% of its mass when heated in a stream of hydrogen chloride, leaving a residue of anhydrous magnesium chloride. From these figures, calculate the values of *m* and *n*.

END-OF-CHAPTER QUESTIONS

Answers on page 222

2.1 100 cm^3 of concentrated hydrochloric acid was diluted accurately to 1000 cm^3 with distilled water. The diluted acid was titrated against 25.0 cm^3 of 0.500 M sodium carbonate, which required 27.2 cm^3 of acid for neutralisation, using methyl orange indicator. What was the concentration of the original acid?

2.2 50.0 cm^3 of a 0.400 M hydrochloric acid solution was added to 25.0 cm^3 of a solution of sodium hydroxide and the resulting solution required 21.6 cm^3 of 0.200 M sodium hyroxide solution for neutralisation. Calculate the concentration of the original sodium hydroxide solution.

2.3 Iodine dissolves in hot concentrated solutions of sodium hydroxide according to the equation

$$3I_2(s) + 6NaOH(aq) \rightarrow NaIO_3(aq) + 5NaI(aq) + 3H_2O(l)$$

In one experiment 3.81 g of iodine was dissolved in 4.00 M sodium hydroxide solution.

a How many moles of iodine were used?

b What volume of 4.00 M sodium hydroxide would be just sufficient to react with the iodine?

2.4 White phosphorus reacts with dilute aqueous solutions of copper(II) sulphate to deposit metallic copper and produce a strongly acidic solution.

In an experiment to investigate this reaction, 0.31 g of white phosphorus reacted in excess aqueous copper(II) sulphate giving 1.60 g of metallic copper.

a i) Calculate the number of moles of phosphorus atoms used.
ii) Calculate the number of moles of copper produced.
iii) Hence calculate the number of moles of copper deposited by one mole of phosphorus atoms.

b i) State the change in oxidation number of the copper in this reaction.
ii) Calculate the new oxidation number of the phosphorus after the reaction.
iii) In the reaction the phosphorus forms an acid, HPO_n. What is the value of n?

c Now write a balanced equation showing the action of white phosphorus on copper(II) sulphate in the presence of water.

2.5 **a** Deduce the oxidation state of nitrogen in:
i) hydrazine, N_2H_4, ii) nitric acid, HNO_3.

b Ammonia dissolves in water to form an alkaline solution.
i) Write a balanced equation to illustrate this.
ii) Write a balanced equation for the neutralisation of aqueous ammonia by nitric acid.
iii) Calculate the volume of 0.20 mol dm^{-3} nitric acid required to react completely with 20 cm^3 of 1.0 mol dm^{-3} aqueous ammonia.

2.6 For reasons of environmental safety the concentration of ammonia in the air downwind of an ammonia production plant was measured by the following procedure.

A 20 000 litre (measured at s.t.p.) sample of the air was slowly bubbled through an excess of dilute hydrochloric acid. The resulting solution was made alkaline and heated, the ammonia liberated being dissolved in exactly 50 cm^3 of 0.1 M hydrochloric acid, which is a large excess. 40.00 cm^3 of 0.1 M sodium hydroxide solution were required to neutralise the excess of acid.

Calculate the concentration of ammonia in the air in units of moles of ammonia per litre of air.

2.7 Sulphur dioxide is a major pollutant in the atmosphere. The quantity of sulphur dioxide in a sample of air can be determined by reaction with hydrogen peroxide and titrating the sulphuric acid formed with standard alkali.

a Write the equation for the reaction between sulphur dioxide and hydrogen peroxide.

b In a typical experiment 200 m^3 of polluted air was bubbled slowly through 25.0 cm^3 of a solution of hydrogen peroxide. The resulting solution was neutralised by 15.3 cm^3 of sodium hydroxide solution of concentration 0.100 mol dm^{-3}.

Calculate the sulphur dioxide concentration, in g m^{-3}, in the sample of air.

2.8 Sodium disulphate(IV) dissolves in water to form HSO_3^- ions as follows:

$$Na_2S_2O_5(s) + H_2O(l) \rightarrow 2Na^+(aq) + 2HSO_3^-(aq)$$

Potassium dichromate(VI) can be used to oxidise HSO_3^- ions to SO_4^{2-} ions. The ionic equation for this reaction is:

$$Cr_2O_7^{2-} + 3HSO_3^- + 5H^+ \rightarrow 2Cr^{3+} + 3SO_4^{2-} + 4H_2O$$

A tablet of sodium disulphate(IV) ($M_r = 190.0$), when dissolved in water, reacted with 17.55 cm^3 of 0.1000 M potassium dichromate(VI) solution. Use this information to determine the mass of sodium disulphate(IV) in the tablet.

2.9 When a solution containing a high concentration of sulphate ions is electrolysed under carefully controlled conditions, the ion $S_2O_8^{2-}$ is produced at the anode. Give a balanced half-equation for this reaction.

In this reaction either the sulphur atoms or the oxygen atoms might have been oxidised. Calculate the average oxidation numbers of sulphur and oxygen in the ion $S_2O_8^{2-}$ for both of these possibilities.

The oxidising properties of the ion, $S_2O_8^{2-}$, were investigated. 25.0 cm^3 of a solution containing 0.0500 mol dm^{-3} of $S_2O_8^{2-}$ was added to 10.0 cm^3 of a solution of potassium iodide, KI, of concentration 1.00 mol dm^{-3}. The iodine released required 20.0 cm^3 of a solution of sodium thiosulphate, $Na_2S_2O_3$, of concentration 0.125 mol dm^{-3} for complete reaction. Calculate whether $S_2O_8^{2-}$ or iodide ions were present in excess and hence deduce the equation for the reaction of $S_2O_8^{2-}$ with iodide ions.

2.10 This question concerns a hydrated double salt *A*, $Cu_w(NH_4)_x(SO_4)_y \cdot zH_2O$, whose formula may be determined from the following experimental data.

a 2.0 g of salt *A* was boiled with excess sodium hydroxide and the ammonia expelled collected by absorption in 40 cm^3 of 0.50 M hydrochloric acid in a cooled flask. Subsequently, this solution required 20 cm^3 of 0.50 M sodium hydroxide for neutralisation.
i) Calculate the number of moles of ammonium ions in 2 g of salt *A*.
ii) Calculate the mass of ammonium ions present in 2 g of salt *A*.

b A second sample of 2 g of salt *A* was dissolved in water and treated with an excess of barium chloride solution. The mass of the precipitate formed, after drying, was found to be 2.33 g.
i) Calculate the number of moles of sulphate ions in 2 g of salt *A*.
ii) Calculate the mass of sulphate ions in 2 g of salt *A*.

c A third sample of 10 g of salt *A* was dissolved to give 250 cm^3 of solution. 25 cm^3 of this solution, treated with an excess of potassium iodide, gave iodine equivalent to 25 cm^3 of 0.1 M sodium thiosulphate solution.

$$2Cu^{2+} + 4I^- \rightarrow 2CuI + I_2$$

$$I_2 + 2S_2O_3^{2-} \rightarrow 2I^- + S_4O_6^{2-}$$

i) Calculate the number of moles of copper(II) ions in 2 g of salt *A*.
ii) Calculate the mass of copper(II) ions in 2 g of salt *A*.

d i) Calculate the mass of water of crystallisation in 2 g of salt *A*.
ii) Calculate the number of moles of water of crystallisation in 2 g of salt *A*.

e Determine the formula of the hydrated salt *A*.

2.11 A solution made from pure iron(II) sulphate became partially oxidised on standing. 25.0 cm^3 of this solution gave a precipitate of mass 0.829 g when treated with excess barium chloride solution. A further 25.0 cm^3 was acidified and titrated with 0.0200 M potassium manganate(VII) solution, of which 26.2 cm^3 was required to give a permanent pink colour. Calculate the relative percentages of Fe^{2+} and Fe^{3+} ions in the partially oxidised solution.

2.12 Iron pyrites (iron disulphide, FeS_2) is oxidised by roasting in air to form iron(III) oxide, used to make iron, and sulphur dioxide which is used to make sulphuric acid. If 80% of the sulphur in iron pyrites is ultimately converted to pure sulphuric acid, calculate the mass of H_2SO_4 produced from 1.00 tonne of pyrites.

CHAPTER 3

MASS SPECTROMETRY AND NUCLEAR REACTIONS

INTRODUCTION AND PRE-KNOWLEDGE

We assume that you have read about the mass spectrometer, and know how the instrument produces beams of ions which can be detected separately according to their mass/charge ratios. Measurement of the relative intensities of the beams enables us to determine the relative atomic masses of elements and, more importantly, the relative molecular masses of compounds, particularly organic ones.

MASS SPECTROMETRY

When a beam of ions strikes the detector in a mass spectrometer, it produces an electrical impulse, which is amplified and fed into a recorder. The mass of the isotope and its relative abundance are then shown by a peak on the chart. A set of such peaks is a mass spectrum.

Figure 3.1 shows a mass spectrum for rubidium. The horizontal axis shows the mass/charge ratio of the ions entering the detector. If it is assumed that all the ions carry a single positive charge, the horizontal axis can also be labelled 'mass number', 'isotopic mass' or 'relative atomic mass'. The vertical axis shows the abundance of the ions. It can be labelled 'detector current', 'relative abundance' or 'ion intensity'.

Figure 3.1

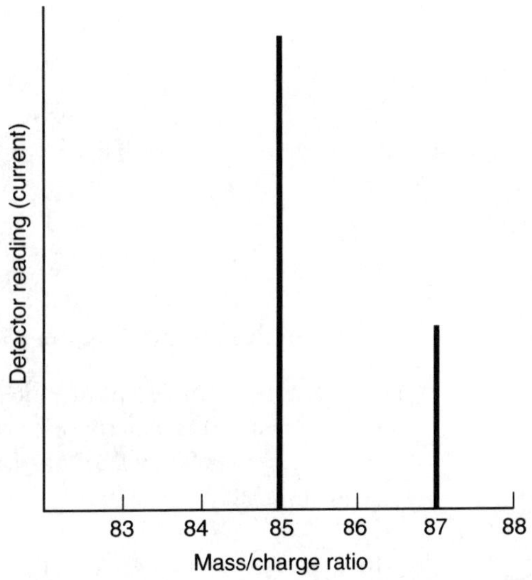

We can use the information from mass spectra to calculate the relative atomic masses of elements.

Calculating the relative atomic mass of an element

We start with a Worked Example, using the mass spectrum shown in Fig. 3.1.

WORKED EXAMPLE Use Fig. 3.1 to calculate the relative atomic mass of rubidium.

Solution

1. Measure the height of each peak. The height is proportional to the amount of each isotope present.

Height of rubidium-85 peak = 5.82 cm
Height of rubidium-87 peak = 2.25 cm

Therefore, the relative amounts of ^{85}Rb and ^{87}Rb are 5.82 and 2.25, respectively.

2. Express each relative amount as a percentage of the total amount. This gives the percentage abundance.

$$\% \text{ abundance} = \frac{\text{amount of isotope}}{\text{total amount of all isotopes}} \times 100$$

$$\% \text{ abundance of } ^{85}\text{Rb} = \frac{5.82}{5.82 + 2.25} \times 100 = 72.1\%$$

$$\% \text{ abundance of } ^{87}\text{Rb} = \frac{2.25}{5.82 + 2.25} \times 100 = 27.9\%$$

The percentage abundance figures mean that for every 100 atoms, about 72 are the ^{85}Rb isotope and about 28 are the ^{87}Rb isotope.

3. Find the total mass of an average sample of 100 atoms.

$$\text{Total mass} = \left((85 \times 72.1) + (87 \times 27.9)\right) \text{ amu}$$

$$= (6129 + 2427) \text{ amu} = 8556 \text{ amu}$$

$$= 8.56 \times 10^3 \text{ amu (3 sig. figs)}$$

4. Find the average mass. This gives the relative atomic mass of rubidium.

$$\text{Average mass} = \frac{\text{total mass}}{\text{number of atoms}} = \frac{8.56 \times 10^3 \text{ amu}}{100} = 85.6 \text{ amu}$$

$\therefore$ relative atomic mass = 85.6

The letters amu represent atomic mass unit; 1 amu = 1.660×10^{-27} kg. This value was chosen because it is exactly one twelfth the mass of an atom of the carbon-12 isotope, which constitutes 98.89% of natural carbon.

Now try a similar calculation for yourself by doing the next Exercise.

EXERCISE 3.1

Answers on page 222

Use the mass spectrum shown in Fig. 3.2 to calculate:

a the percentage of each isotope present in a sample of naturally occurring lithium,
b the relative atomic mass of lithium.

Figure 3.2

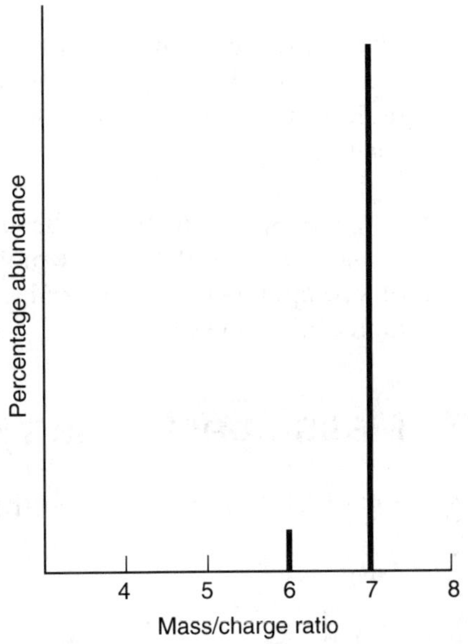

The next Exercise is similar, but you are given data rather than a diagram of a mass spectrum.

EXERCISE 3.2
Answers on page 222

The mass spectrum of neon consists of three lines corresponding to mass/charge ratios of 20, 21 and 22 with relative intensities of 0.910, 0.0026 and 0.088, respectively. Explain the significance of these data and, hence, calculate the relative atomic mass of neon.

You could also be asked to do the reverse of this type of problem – sketch a spectrum from percentage abundance data. Have a try at this by doing the next Exercise.

EXERCISE 3.3
Answers on page 223

a Look up the percentage abundances of the stable isotopes of chromium.
b Sketch the mass spectrum that would be obtained from naturally occurring chromium. (Let 10.0 cm represent 100% on the vertical scale.)
c Calculate the relative atomic mass of chromium, correct to three significant figures.
d Label each peak on the mass spectrum using isotopic symbols.

You may be asked to calculate relative atomic masses directly from percentage abundance data, as you have just done in part **c** of Exercise 3.3. For more practice at this type of problem, see the End-of-chapter questions.

The next Exercise concerns a molecular element.

EXERCISE 3.4
Answers on page 223

Figure 3.3

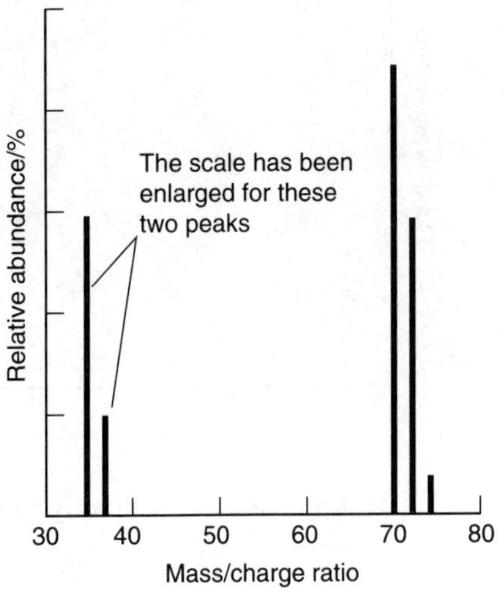

The element chlorine has isotopes of mass number 35 and 37 in the approximate proportion 3:1. Interpret the mass spectrum of gaseous chlorine shown in Fig. 3.3, indicating the formula (including mass number) and charge of the ion responsible for each peak.

The mass spectrometer can be used to determine not only the relative atomic mass of an element but also the relative molecular mass of a compound. In addition, the structure of a compound can be determined by identifying the fragments formed when the molecule is ionised.

Mass spectrometry in organic chemistry

EXERCISE 3.5
Answers on page 223

Figure 3.4 shows the mass spectrum of ethanol. Study it and try the questions that follow.

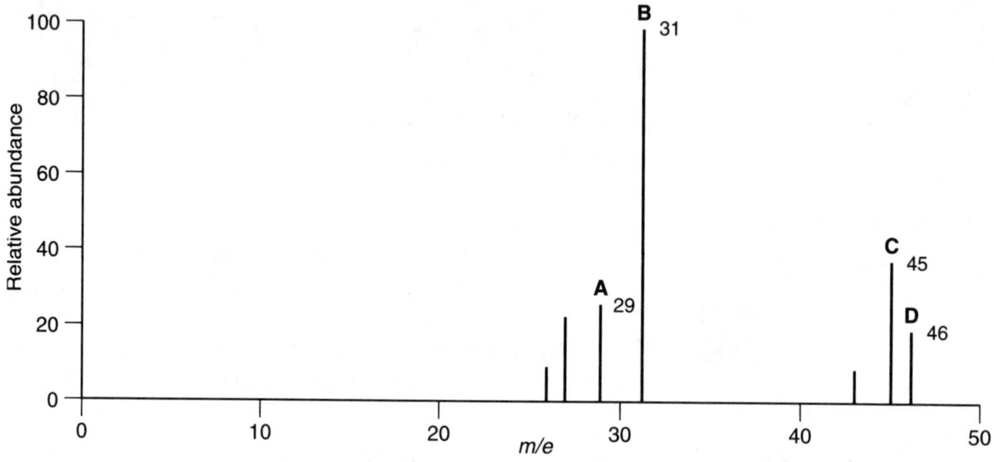

Figure 3.4
Mass spectrum of ethanol.

a How can a single substance such as ethanol give several peaks on the mass spectrum?
b Write the molecular formulae for the species giving traces at **A**, **B**, **C** and **D**.

In the last Exercise, you saw how the highest mass/charge ratio in the spectrum corresponds to the single-positively charged ion of the complete molecule, known as the molecular ion, M^+. However, in some spectra there is a very small peak at $(M + 1)$ due to the presence of molecules containing the ^{13}C isotope, as shown in Fig. 3.5.

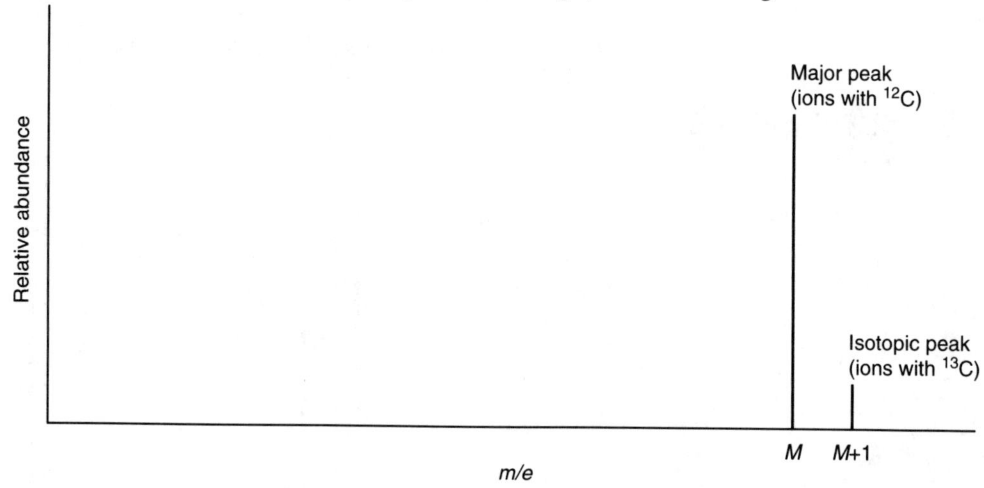

Figure 3.5
$(M + 1)$ peak.

The isotopic peak is relatively small because the ^{13}C isotope has a natural abundance of

1.1%. The more carbon atoms there are in a molecule the greater the chance that *one* of them will be a ^{13}C atom. The chances that two (or more) are ^{13}C also increases of course but is nearly always negligibly small.

Generally, for an organic compound containing N carbon atoms, the height of the $(M + 1)$ peak is 1.1 N% of the height of the M peak. We can make use of this in calculating the number of carbon atoms in a molecule.

Calculating the number of carbon atoms in a molecule

We start with a Worked Example using the mass spectrum shown in Fig. 3.5 above.

WORKED EXAMPLE

Calculate the number of carbon atoms in a molecule that gives a mass spectrum whose M and $(M + 1)$ peaks are shown in Fig. 3.5 above.

Solution

1. Measure the heights of the M and $(M + 1)$ peaks.
   height of M peak = 4.2 cm
   height of $(M + 1)$ peak = 0.5 cm

2. Substitute into the expression:

$$\frac{\text{height of } (M+1) \text{ peak}}{\text{height of } M} = \frac{1.1N}{100}$$

(where N = number of carbon atoms)

This arises from the fact that the ^{13}C isotope has a natural abundance of only 1.1%. For an organic compound containing N carbon atoms, the height of the $(M + 1)$ peak is 1.1N% of the height of the M peak.

$$\therefore N = \frac{\text{height of } (M+1) \text{ peak}}{\text{height of } M \text{ peak}} \times \frac{100}{1.1}$$

$$= \frac{0.5 \text{ cm}}{4.2 \text{ cm}} \times \frac{100}{1.1} = 10.8 = 11 \text{ (to the nearest whole number)}$$

Now try a similar Exercise for yourself. The first part of the next Exercise requires you to calculate the empirical formula from percentage composition by mass as shown in the Worked Example on pages 11–12.

EXERCISE 3.6
Answer on page 223

A hydrocarbon X contains 92.4% carbon by mass. Its mass spectrum gives a peak at M and a smaller peak at $(M + 1)$. There are no significant peaks at higher mass/charge ratio than this. The peak at M is 11.5 times as intense as the peak at $(M + 1)$. What is the molecular formula of X?

Identifying minor peaks

In the next two Exercises, you look at the spectra of two hydrocarbons and explain why each peak is accompanied by several smaller peaks.

EXERCISE 3.7

Answers on page 224

Figure 3.6
Mass spectra of two hydrocarbons.

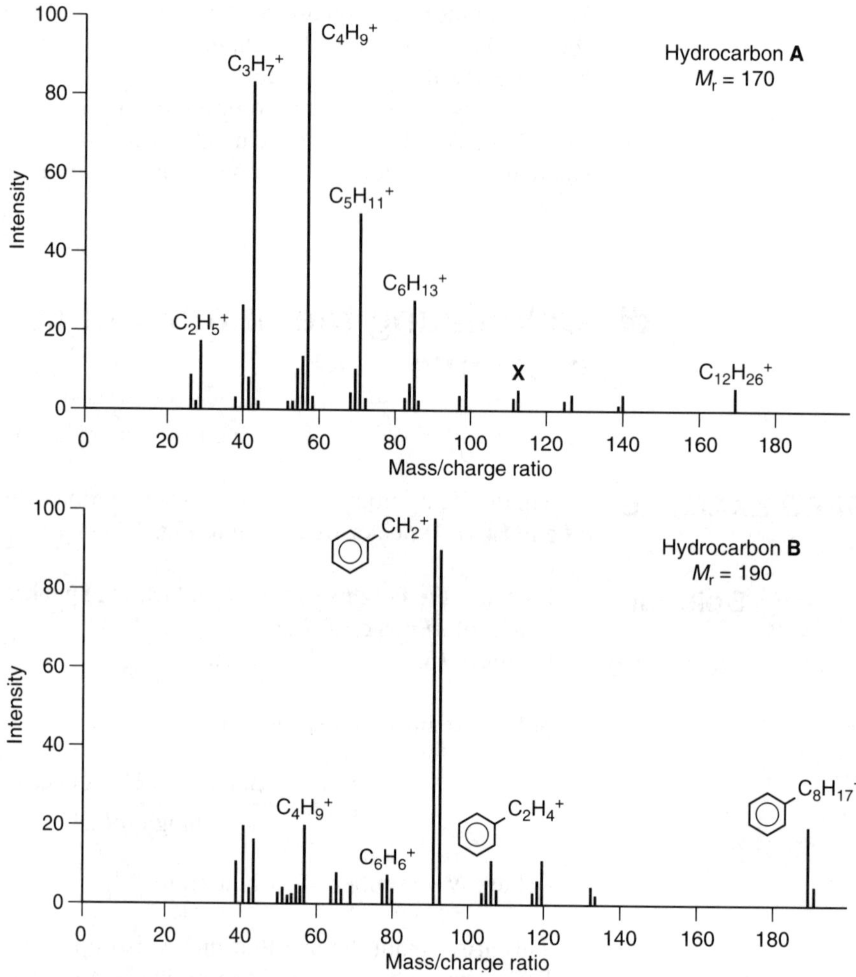

a What is the connection between the species of highest mass/charge ratio and the substance being analysed?

b Explain why each peak is accompanied by several smaller peaks.

c What is the probable formula for the species giving the trace marked **X** in the spectrogram for **A**?

EXERCISE 3.8

Answer on page 224

Identify the minor peaks associated with the major peak $C_5H_{11}^+$, shown in Fig. 3.6.

The mass spectra of compounds containing chlorine or bromine show peaks at $(M + 2)$ as you will see in the next two Exercises.

EXERCISE 3.9
Answers on page 224

Examine the mass spectrum of chlorobenzene shown in Fig. 3.7 and answer the questions that follow.

Figure 3.7

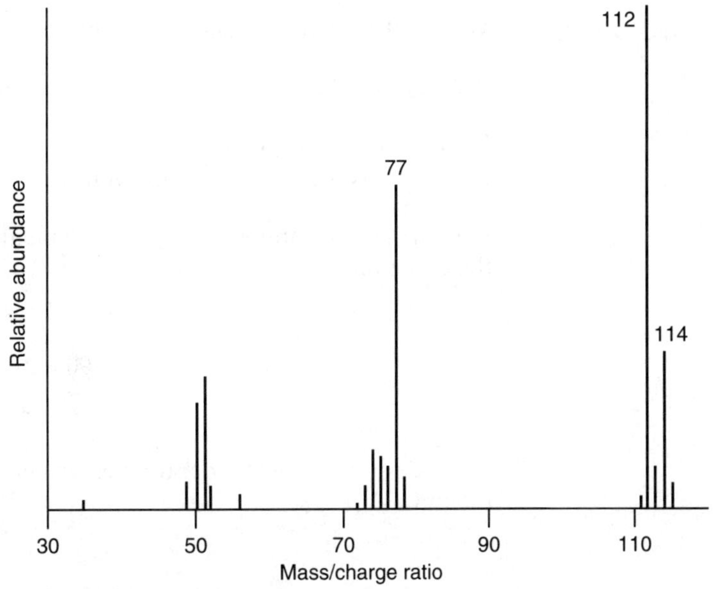

a Identify and explain the presence of two molecular ion peaks.
b Suggest an explanation for the presence of the ions of mass/charge ratios 113 and 115.
c Identify, with a reason, the fragment responsible for the peak of mass/charge ratio 77.

For compounds containing a bromine atom the M and $(M + 2)$ peaks are almost the same height. This is due to the fact that bromine is a mixture of approximately equal amounts of ^{79}Br and ^{81}Br.

A compound with two chlorine or two bromine atoms also shows a peak at $(M + 4)$ as you will see in the next Exercise.

EXERCISE 3.10
Answers on page 224

Bromine consists of a mixture of isotopes ^{79}Br and ^{81}Br. Assuming no bonds are broken in the mass spectrometer and that hydrogen has only one significant isotope, how many peaks will be given by the compound CH_2Br_2 and what will each peak be due to?

The second half of this chapter concerns nuclear reactions and the simple calculations required in order to write nuclear equations.

NUCLEAR REACTIONS

Whereas chemical reactions leave the nucleus untouched and involve only the electrons surrounding it, nuclear reactions rearrange the particles within the nucleus. Also, different elements may be formed, in a process known as transmutation, which does not, of course, occur in ordinary chemical reactions.

The basic nuclear reactions are α and β particle emission, often called α and β decay. An α particle can be defined as a helium nucleus ($^{4}_{2}He^{2+}$). A β particle can be defined as an electron ($^{0}_{-1}e$). We go through α decay in detail in the following Worked Example.

WORKED EXAMPLE When thorium-228 decays, each atom emits one α particle. Write a balanced nuclear equation to show the process.

Solution

1. Write out the equation, putting in letters for the unknowns:

$$^{228}_{90}\text{Th} \rightarrow {}^{4}_{2}\text{He} + {}^{A}_{Z}\text{X}$$

In these equations, the atomic numbers and mass numbers of each side must balance; this allows you to calculate the unknowns, A and Z and X.

2. Find the atomic number of the new element, X, using the atomic numbers shown in the equation.

$$_{90}\text{Th} \rightarrow {}_{2}\text{He} + {}_{Z}\text{X}$$

$$90 = 2 + Z$$

$$Z = 90 - 2 = 88$$

3. Use a Periodic Table to identify the new element: element with atomic number 88 is radium.

$$\therefore \text{X} = \text{Ra}$$

4. Find the mass number of the particular isotope of radium, using the mass numbers in the equation:

$$^{228}\text{Th} \rightarrow {}^{4}\text{He} + {}^{A}\text{X}$$

$$228 = 4 + A$$

$$\therefore A = 224$$

5. Write out the complete equation:

$$^{228}_{90}\text{Th} \rightarrow {}^{4}_{2}\text{He} + {}^{224}_{88}\text{Ra}$$

Note that ionic charges are not shown in nuclear equations. You may wonder, for example, why we do not write:

$$^{4}_{2}\text{He}^{2+}$$

In fact, the positive ions tend to pick up stray electrons and so the charges are usually omitted as a simplification. This also focuses attention on changes in the nucleus.

Now try some calculations yourself, by doing the next Exercise.

EXERCISE 3.11
Answers on page 224

Write balanced nuclear equations to show the α decay of:

a $^{212}_{84}\text{Po}$,

b $^{220}_{86}\text{Rn}$.

You can adapt the method given in the last Worked Example to write an equation for β decay. The following equation illustrates how a nucleus consisting of protons and neutrons can emit electrons:

$$^{1}_{0}\text{n} \rightarrow {}^{1}_{1}\text{H} + {}^{0}_{-1}\text{e}$$

You make use of this idea in the next Exercise.

EXERCISE 3.12
Answers on page 224

Write balanced nuclear equations to show the β decay of:

a $^{212}_{84}Po$,

b $^{24}_{11}Na$,

c $^{108}_{47}Ag$.

For further practice in writing nuclear equations, try the next Exercise.

EXERCISE 3.13
Answers on page 224

Write the nuclear equations which represent:

a the loss of an α particle by radium-226,
b the loss of a β particle by potassium-43,
c the loss of an α particle by the product of **b**.

END-OF-CHAPTER QUESTIONS

Answers on page 224

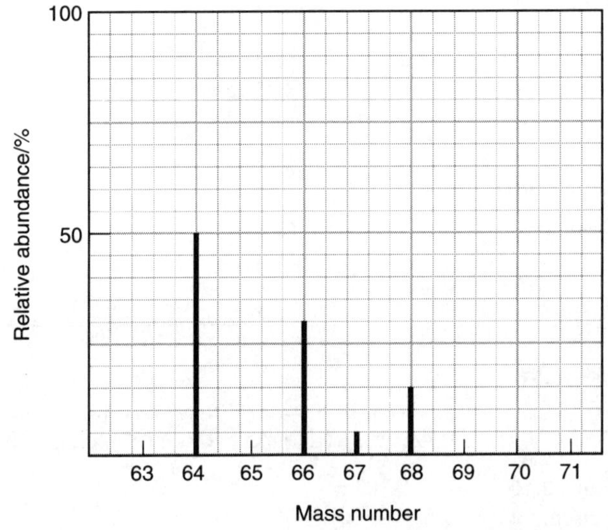

Figure 3.8

3.1 The graph in Fig. 3.8 shows the relative abundances of four isotopes of a certain element. Use it to calculate the relative atomic mass of the element (to the nearest whole number).

3.2 **a** Calculate the relative atomic mass of copper assuming it to contain 70% of ^{63}Cu and 30% of ^{65}Cu.
b Complete the following equations:

$$^{220}_{86}Rn \rightarrow {}^{216}_{84}Po +$$

$$^{214}_{82}Pb \rightarrow {}^{214}_{83}Bi +$$

3.3 Chlorine consists of two isotopes having mass numbers 35 and 37 of relative abundance 75% and 25%, respectively. The questions below relate to the mass spectrum of a sample of gaseous chlorine.
a How many peaks would you expect in the *m*/*e* range 34–38? What would be the *m*/*e* value of the most intense peak in this range? (Consider singly charged ions only.)

b How many peaks corresponding to ions of type Cl_2^+ would you expect to observe? (Consider singly charged ions only.) Give the *m*/*e* value of each and identify the most intense peak.

3.4 The mass spectrum of an element enables the relative abundance of each isotope of the element to be determined. Data relating to the mass spectrum of bromine, atomic number 35, appear below.

Mass number of isotope	79	81
% Relative abundance	50.5	49.5

a Calculate the relative atomic mass of bromine to three significant figures.

b i) Make a copy of Fig. 3.9 and on it sketch the expected spectrum of bromine vapour in the *m*/*e* region shown.

ii) Explain the origin of each peak in your spectrum.

Figure 3.9

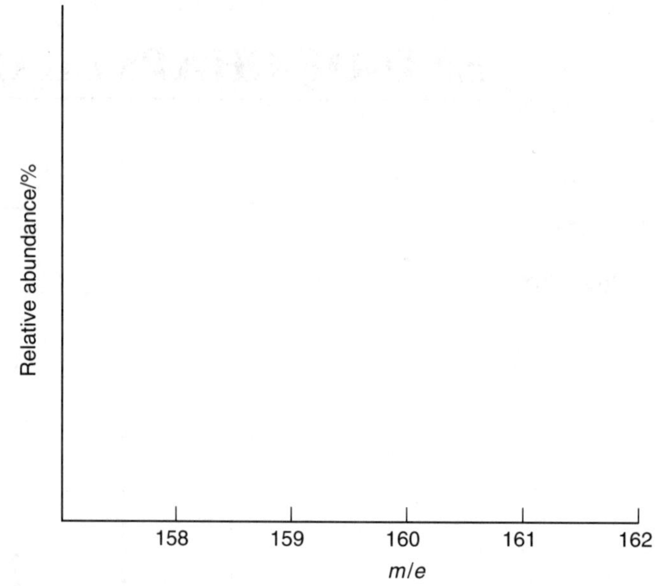

3.5 Compound B contains C, H and O only. It has a composition by mass of carbon 48.6%, hydrogen 8.1%, and oxygen 43.3%. The mass spectrum is shown below.

Figure 3.10
Mass spectrum for compound B.

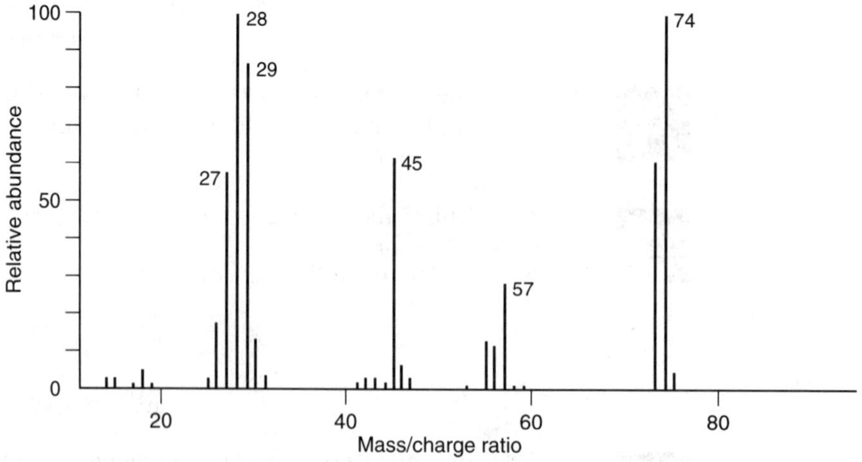

Use this information to determine:

a the empirical formula of B,

b the molecular formula of B,

c the possible identity of the main fragments responsible for the peaks of given mass/charge ratio.

CHAPTER 4

CHEMICAL ENERGETICS

INTRODUCTION AND PRE-KNOWLEDGE

In any chemical reaction (and in many physical changes too) heat energy tends to be transferred, either from the system to the surroundings, or in the opposite direction. This transfer increases or decreases the quantity of energy stored in the chemical concerned. If we specify that all the reactants and products are in their standard states and at standard conditions, the quantity of energy transferred under the condition of constant pressure defines what is known as the *standard enthalpy change of reaction*, $\Delta H^{\ominus}$. A positive sign indicates transfer *from* the surroundings.

ENTHALPY CHANGE

Information about the transfer of heat energy in a reaction can be summarised in two ways, by a *thermo*chemical equation or by an *energy (or enthalpy) level diagram*. For instance, when two moles of hydrogen react with one mole of oxygen to produce two moles of liquid water, with both initial and final states at 298 K and 1 atm, 572 kJ of heat energy are released to the surroundings. The thermochemical equation for this reaction is:

$$2H_2(g) + O_2(g) \rightarrow 2H_2O(l); \quad \Delta H^{\ominus} = -572 \text{ kJ mol}^{-1}$$

Notice the use of the symbol $^{\ominus}$ to indicate standard conditions. If half the amounts are used, then the energy is halved also:

$$H_2(g) + \tfrac{1}{2}O_2(g) \rightarrow H_2O(l); \quad \Delta H^{\ominus} = -286 \text{ kJ mol}^{-1}$$

Notice that, in both examples, the unit for standard enthalpy change is kJ mol^{-1}. You can consider that the term 'per mole' refers to all the amounts specified in the equation. You may find it useful to think of the term 'per mole' as referring to 'a mole of equation' if, as in the two examples, the stoichiometric coefficients are not all equal to one.

It is important to include the state symbols (e.g. g, l, s, aq) because the energy change depends on the state of the substance. If the water produced were not allowed to condense, then we would write a thermochemical equation with a different value of the enthalpy change:

$$2H_2(g) + O_2(g) \rightarrow 2H_2O(g); \quad \Delta H^{\ominus} = -484 \text{ kJ mol}^{-1}$$

An enthalpy-level diagram corresponding to these equations is shown below.

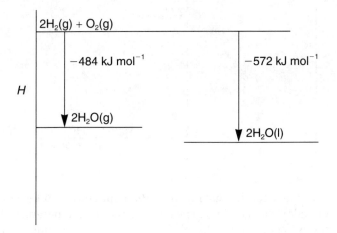

The standard enthalpy changes for particular types of reaction are often given special names, e.g. the standard enthalpy change of formation, $\Delta H^\ominus_f$, and the standard enthalpy change of combustion, $\Delta H^\ominus_c$. You should be able to write definitions for these based on the information above.

Try the next Exercise to check that you have grasped the basic ideas.

EXERCISE 4.1
Answers on page 225

a Write a thermochemical equation showing that when 1.00 mol of carbon burns completely in oxygen, 394 kJ of heat is liberated.
b Calculate the enthalpy change on complete combustion of:
i) 10.0 mol of carbon,
ii) 0.25 mol of carbon,
iii) 18.0 g of carbon.
c What mass of carbon would have to be burnt to produce:
i) 197 kJ,
ii) 1000 kJ?

■ Calculating $\Delta H^\ominus$ from experimental data

In the introduction, we spoke of heat transfers to the surroundings. It is not easy to measure these transfers directly. In practice, it is much easier to measure the enthalpy change of a reaction using a calorimeter in which the system is insulated from the surroundings. For an exothermic reaction, the energy that would otherwise be given to the surroundings results in an increase in the temperature of the system.

If the maximum temperature change of the system is recorded and if the heat capacity of the system is known, it is easy to calculate the quantity of heat which would have to be taken from the system in order to restore it to its initial temperature. This quantity of heat is the enthalpy change. The sequence is summarised in Fig. 4.1.

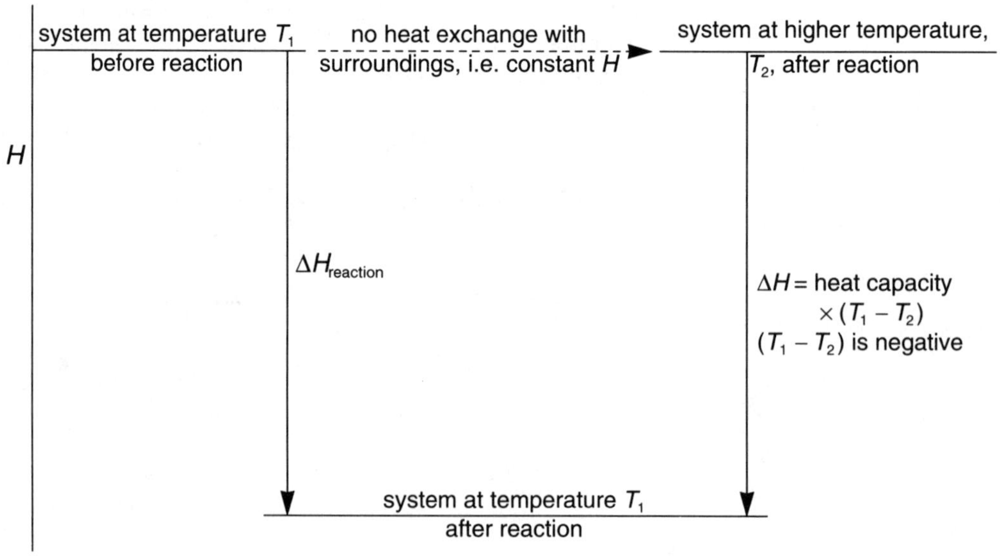

Figure 4.1
Measuring ΔH for an exothermic reaction.

For an endothermic reaction, the temperature of the system would drop, but again the enthalpy change can be calculated from the temperature change if the heat capacity of the system is known. In this case the sequence is summarised in Fig. 4.2.

Figure 4.2
Measuring ΔH for an endothermic reaction.

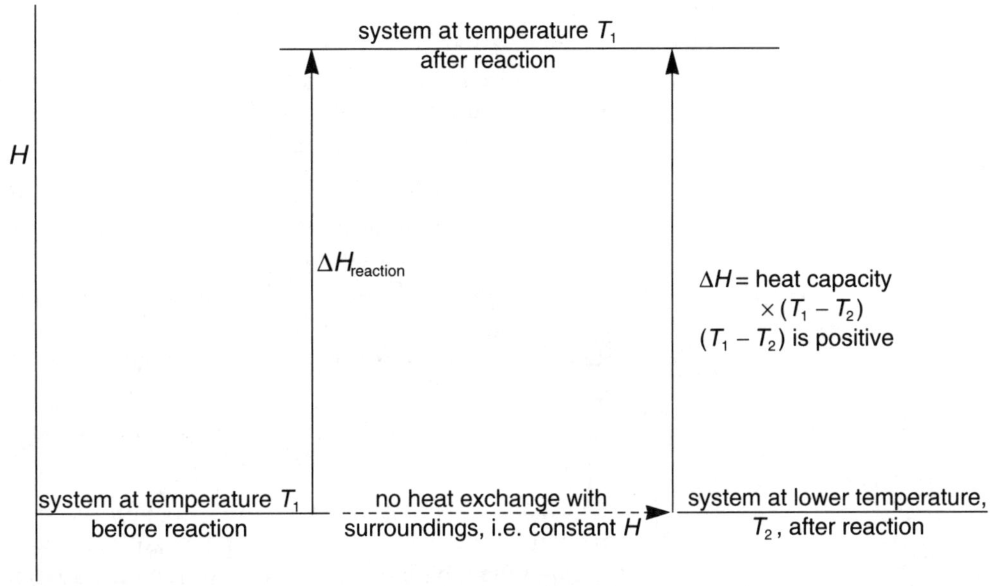

There are several types of insulated calorimeter you can use for simple experiments, as illustrated in Fig. 4.3. These calorimeters insulate the system from the surroundings so that all energy changes occur within the system. The energy exchanged with the surroundings is usually small enough to be ignored.

Figure 4.3
Various insulated calorimeters.

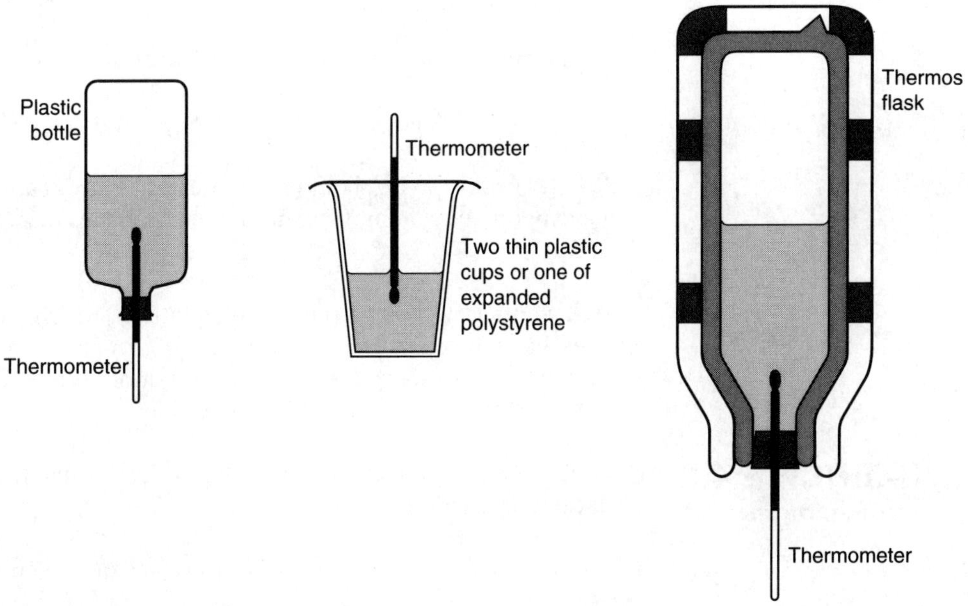

We now give a Worked Example showing how the enthalpy change of reaction can be calculated from experimental data.

WORKED EXAMPLE

An excess of zinc powder was added to 50.0 cm^3 of 0.100 M $AgNO_3$ in a polystyrene cup. Initially, the temperature was 21.10 °C and it rose to 25.40 °C. Calculate the enthalpy change for the reaction:

$$Zn(s) + 2Ag^+(aq) \rightarrow Zn^{2+}(aq) + 2Ag(s)$$

Assume that the density of the solution is 1.00 g cm^{-3} and its specific heat capacity is 4.18 kJ kg^{-1} K^{-1}. Ignore the heat capacity of the metals.

Solution

1. Since the polystyrene cup is an insulator and its heat capacity is almost zero, you can assume that no energy is exchanged between system and surroundings. All the chemical energy released in the reaction is transformed into heat energy which raises the temperature of the solution. This energy must be removed in restoring the solution to the original temperature.

$$\begin{bmatrix}\text{enthalpy change, }\Delta H\text{, due} \\ \text{to reaction (at constant } T)\end{bmatrix} = \begin{bmatrix}\text{enthalpy change in restoring} \\ \text{solution to original temperature}\end{bmatrix}$$

$$= [\text{mass} \times \text{specific heat capacity} \times \text{temperature change}] = mc_p{}^* \,\Delta T$$

(Remember that ΔT is negative in restoring the original temperature.)

$$\therefore \Delta H = mc_p\Delta T = \frac{50.0}{1000}\text{kg} \times 4.18 \text{ kJ kg}^{-1}\text{ K}^{-1} \times (-4.30 \text{ K}) = -0.899 \text{ kJ}$$

2. The value –0.899 kJ is the enthalpy change for the amounts used in the experiment. To obtain a value for the enthalpy change of reaction, compare the amounts used in the experiment with the amounts shown in the equation:

$$\begin{array}{l}Zn(s) + 2Ag^+(aq) \rightarrow Zn^{2+}(aq) + Ag(s) \\ 1 \text{ mol} \quad\ \ 2 \text{ mol}\end{array}$$

The amount of silver ions used = $0.0500 \text{ dm}^3 \times 0.100 \text{ mol dm}^{-3} = 5.00 \times 10^{-3}$ mol

$$\therefore \text{ the enthalpy change using 2 mol of } Ag^+ = -0.899 \text{ kJ} \times \frac{2.00 \text{ mol}}{5.00 \times 10^{-3} \text{ mol}} = -360 \text{ kJ}$$

3. Now write the complete thermochemical equation:

$$Zn(s) + 2Ag^+(aq) \rightarrow Zn^{2+}(aq) + Ag(s); \quad \Delta H = -360 \text{ kJ mol}^{-1}.$$

Strictly speaking, you should not write ΔH° in this case because the conditions of the experiment were not standard, but the values of ΔH and ΔH° would be very close.

Note that enthalpy changes related to equations, which include all standard enthalpy changes, have the unit kJ mol^{-1}, where mol^{-1} means 'per mole of whatever is written in the equation'. For example, in step 3 above, mol^{-1} refers to a mole of units, **each** consisting of one zinc atom and two silver ions.

EXERCISE 4.2
Answers on page 225

The following experiment was performed to determine the enthalpy change for the displacement reaction:

$$Zn(s) + Cu^{2+}(aq) \rightarrow Cu(s) + Zn^{2+}(aq)$$

25.0 cm^3 of 1.00 M copper(II) sulphate solution was placed in a polystyrene cup and the temperature of the solution was recorded to the nearest 0.2 °C every 30 seconds for 2½ minutes. At precisely 3 minutes, excess zinc powder was added, with continual stirring, and the temperature was recorded for an additional 6 minutes. The results of the experiment were as shown in Table 4.1.

a Plot the temperature (y-axis) against time (x-axis).
b Extrapolate the curve to 3.0 minutes to establish the maximum temperature rise as shown in Fig. 4.4.

*c_p is the symbol commonly used for specific heat capacity. The 'p' refers to constant pressure conditions (as does ΔH).

Table 4.1

Time/min	0.0	0.5	1.0	1.5	2.0	2.5	3.0	3.5	4.0	4.5
Temperature/°C	27.0	27.0	27.2	27.2	27.2	27.2	–	66.0	71.4	71.8
Time/min	5.0	5.5	6.0	6.5	7.0	7.5	8.0	8.5	9.0	9.5
Temperature/°C	70.2	68.0	66.2	64.4	62.8	61.0	59.5	58.0	56.6	55.1

Figure 4.4

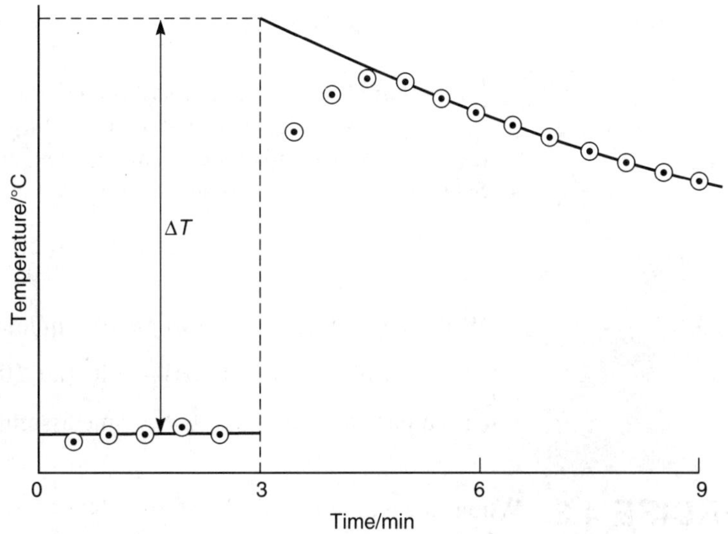

c Calculate the enthalpy change for the quantities used, making the same assumptions as in the preceding Worked Example.

d Calculate the enthalpy change for one mole of Zn and $CuSO_4$(aq), and write the thermochemical equation for the reaction.

Calculating an enthalpy change of solution

This is the enthalpy change which occurs when 1 mole of solute is dissolved in a specified amount of solvent under standard conditions. The abbreviation is $\Delta H^{\ominus}_{soln}$. A Worked Example illustrates the calculations.

WORKED EXAMPLE When 0.85 g of anhydrous lithium chloride, LiCl, was added to 36.0 g of water at 25.0 °C in a polystyrene cup, the final temperature of the solution was 29.7 °C. Calculate the enthalpy change of solution for one mole of lithium chloride.

Solution

1. In problems involving heat of solution, the amount of solvent is important so calculate the ratio: amount of LiCl/amount of H_2O. Then you can write the correct equation describing the process for one mole of the salt.

$$\text{amount of LiCl} = \frac{m}{M} = \frac{0.85\ \text{g}}{42.4\ \text{g mol}^{-1}} = 0.020\ \text{mol}$$

$$\text{amount of H}_2\text{O} = \frac{m}{M} = \frac{36.0\ \text{g}}{18.0\ \text{g mol}^{-1}} = 2.00\ \text{mol}$$

$$\therefore \frac{\text{amount of LiCl}}{\text{amount of H}_2\text{O}} = \frac{0.020}{2.00} = \frac{1}{100}$$

2. Write the equation:

$$LiCl(s) + 100H_2O(l) \rightarrow LiCl(aq, 100H_2O)$$

3. Calculate the enthalpy change for the amounts used in the experiment:

$$\begin{bmatrix}\text{enthalpy change, } \Delta H, \\ \text{on dissolving LiCl}\end{bmatrix} = \begin{bmatrix}\text{enthalpy change in restoring} \\ \text{solution to original temperature}\end{bmatrix}$$

$$\therefore\ \Delta H = mc_p\Delta T = \frac{36.0}{1000}\,\text{kg} \times 4.18\ \text{kJ kg}^{-1}\ \text{K}^{-1} \times (-4.7\ \text{K}) = -0.71\ \text{kJ}$$

[Note that although the mass of the solution is greater than 36.0 g (36.85 g) its specific heat capacity is slightly **less** than 4.18 kJ kg^{-1} K^{-1}. The error arising from considering only the water in this type of calculation is therefore small.]

4. Scale up to the amounts shown in the equation, as in the last Worked Example:

$$\Delta H = -0.71\ \text{kJ} \times \frac{1\ \text{mol}}{0.020\ \text{mol}} = -35\ \text{kJ}$$

5. Write the complete thermochemical equation:

$$LiCl(s) + 100H_2O(l) \rightarrow LiCl(aq, 100H_2O);\quad \Delta H_{soln} = -35\ \text{kJ mol}^{-1}$$

Use a similar method, making the same assumptions, in the next Exercise.

EXERCISE 4.3
Answers on page 226

When 2.67 g of ammonium chloride, NH_4Cl, was added to 90.0 g of water at 24.5 °C in a polystyrene cup, the final temperature of the solution was 22.7 °C. Calculate the enthalpy change of solution and write a thermochemical equation.

■ Hess' Law

Hess' Law (sometimes called Hess' Law of constant heat summation) is a corollary to the law of conservation of energy. We use Hess' Law to calculate enthalpy changes which cannot be measured directly. It states that the enthalpy change in converting reactants, A and B, to products, X and Y, is the same, regardless of the route by which the chemical change occurs, provided the initial and final conditions are the same.

It is not possible to measure the enthalpy change for the reaction:

$$C(s) + \tfrac{1}{2}O_2(g) \rightarrow CO(g)$$

directly in a calorimeter because it represents an incomplete combustion which cannot be controlled. However, it is possible to measure the enthalpy change for the complete combustion of both graphite and carbon monoxide, and we can **calculate** the heat of formation of carbon monoxide from these results. We illustrate Hess' Law by showing this calculation, using an energy-level diagram, in a Worked Example.

WORKED EXAMPLE

Calculate the standard enthalpy change for the reaction

$$C(s) + \tfrac{1}{2}O_2(g) \rightarrow CO(g)$$

given the following information:

$$C(s) + O_2(g) \rightarrow CO_2(g);\quad \Delta H^\ominus = -394\ \text{kJ mol}^{-1}$$

$$CO(g) + \tfrac{1}{2}O_2(g) \rightarrow CO_2(g);\quad \Delta H^\ominus = -283\ \text{kJ mol}^{-1}$$

Solution

1. Draw an energy-level diagram to summarise the information given. Start by drawing a line to represent the enthalpy of 1 mol of graphite and 1 mol of oxygen at standard conditions.

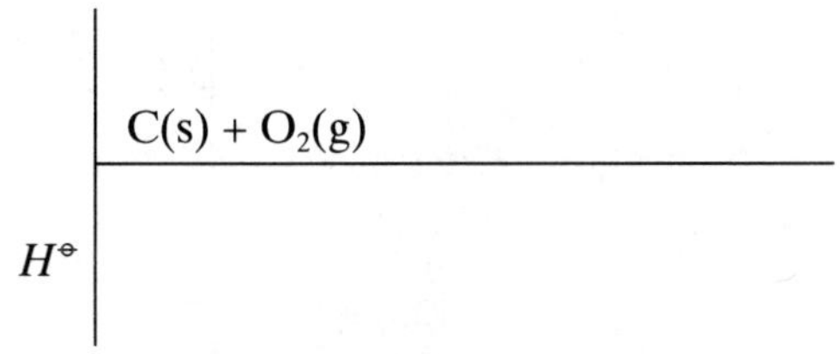

2. The combustion of 1 mol of graphite produces CO_2 and releases 394 kJ. Show this on the energy-level diagram (not necessarily to scale):

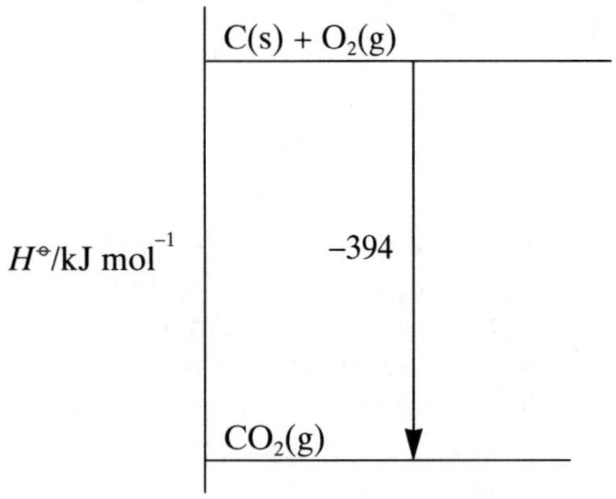

3. Now consider the combustion of CO, which gives the same product, 1 mol of CO_2. The energy released is 283 kJ mol^{-1}. Show this on the diagram:

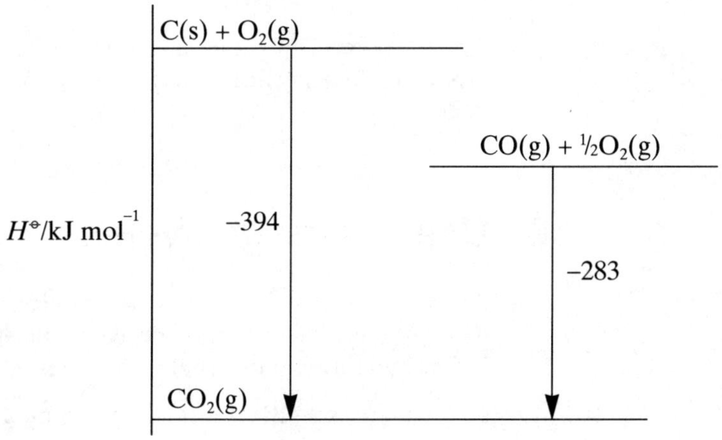

4. The gap between the two upper lines represents the heat of formation of carbon monoxide. Show this on the diagram:

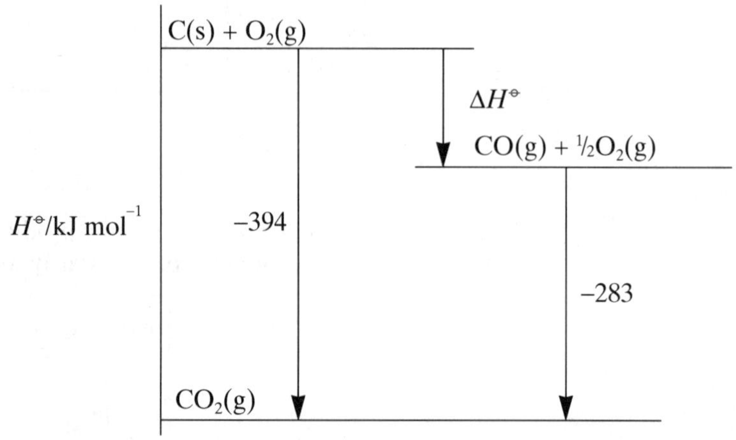

Don't worry about the 'extra' $\frac{1}{2}O_2$ in the two upper levels. Of course, the actual energy levels for $C(s) + \frac{1}{2}O_2(g)$ and for $CO(g)$ would be a little lower, but the **difference** between them would be exactly the same.

5. So you have two routes to the formation of CO_2. One is represented by the heat of formation of $CO_2(g)$ and the other by the sum of two processes: the heat of formation of $CO(g)$ plus the heat of combustion of $CO(g)$. Hess' law states that the energy changes for the two routes are equal and you can see that this must be so from the energy-level diagram. Thus:

$$-394 \text{ kJ mol}^{-1} = \Delta H_f^{\ominus}[CO(g)] + (-283 \text{ kJ mol}^{-1})$$

$$\therefore \Delta H_f^{\ominus}[CO(g)] = (-394 + 283) \text{ kJ mol}^{-1} = \mathbf{-111 \text{ kJ mol}^{-1}}$$

An energy-level diagram clearly establishes the validity of Hess' law but, for solving problems, you may find it simpler to use an alternative method based on energy cycles.

■ Use of energy cycles

An energy cycle simply shows two different routes between initial and final states, without reference to energy levels. We illustrate this by using the same example as before and showing two routes for the production of CO_2.

1. First represent the direct combination of elements:

$$\boxed{C(s) + O_2(g)} \longrightarrow \boxed{CO_2(g)}$$

2. Write the two-step process:

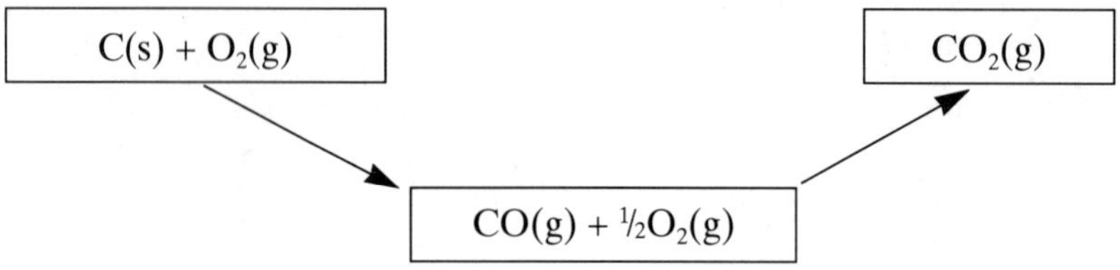

3. Hess' law tells us that the enthalpy change via one route ('route 1') must equal the enthalpy change via the two-step route ('route 2').

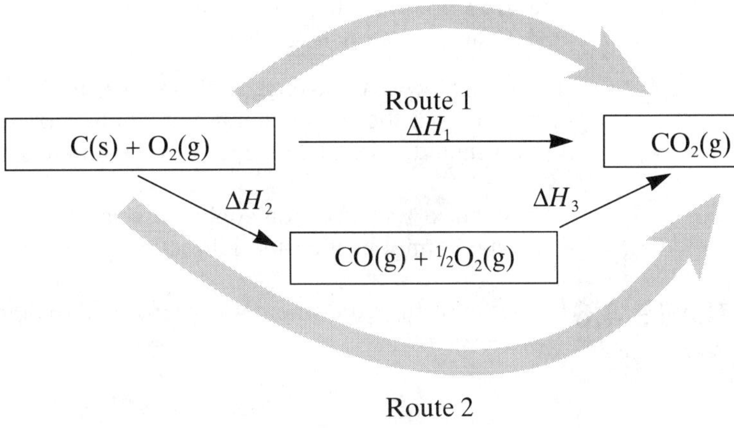

$$\therefore \Delta H_1 = \Delta H_2 + \Delta H_3$$

4. Calculate the unknown by substituting the known values:

$$\Delta H_2 = \Delta H_1 - \Delta H_3$$

$$\therefore \Delta H_f^{\ominus}[CO(g)] = -394 \text{ kJ mol}^{-1} - (-283 \text{ kJ mol}^{-1}) = \mathbf{-111 \text{ kJ mol}^{-1}}$$

Both energy cycles and energy-level diagrams are very useful in solving problems. In the next exercise you use both methods: be careful to distinguish between the two. In an energy-level diagram, the arrows should be drawn vertically, and preferably to scale, to represent both direction and extent of enthalpy change. In an energy cycle, the arrows simply indicate a change from one state to another, and to avoid confusion we suggest you do not use vertical arrows in energy cycles.

EXERCISE 4.4

Answers on page 226

Calculate the standard enthalpy change for the reaction

$$2NO_2(g) \rightarrow N_2O_4(g)$$

given the thermochemical equations:

$$N_2(g) + 2O_2(g) \rightarrow 2NO_2(g); \quad \Delta H^{\ominus} = +33.2 \text{ kJ mol}^{-1}$$

$$N_2(g) + 2O_2(g) \rightarrow N_2O_4(g); \quad \Delta H^{\ominus} = +9.2 \text{ kJ mol}^{-1}$$

a by drawing an energy-level diagram,
b by drawing an energy cycle.

In the next Exercise, you use and justify yet another way of applying Hess' law.

EXERCISE 4.5

Answers on page 226

Two of the possible methods of preparing a solution of ammonium chloride containing 1.00 mol of NH_4Cl in 200 mol of H_2O are summarised in the thermochemical equations below:

Method 1

$NH_3(g) + HCl(g) \rightarrow NH_4Cl(s); \quad \Delta H^{\ominus} = -175.3 \text{ kJ mol}^{-1}$

$NH_4Cl(s) + 200H_2O(l) \rightarrow NH_4Cl(aq, 200H_2O); \quad \Delta H^{\ominus} = +16.3 \text{ kJ mol}^{-1}$

Method 2

$NH_3(g) + 100H_2O(l) \rightarrow NH_3(aq, 100H_2O); \quad \Delta H^{\ominus} = -35.6 \text{ kJ mol}^{-1}$

$HCl(g) + 100H_2O(l) \rightarrow HCl(aq, 100H_2O); \quad \Delta H^{\ominus} = -73.2 \text{ kJ mol}^{-1}$

$NH_3(aq, 100H_2O) + HCl(aq, 100H_2O) \rightarrow NH_4Cl(aq, 200H_2O); \quad \Delta H^{\ominus} = -50.2 \text{ kJ mol}^{-1}$

a Add the equations in Method 1, and simplify the result.
b Add the equations in Method 2, and simplify the result.
c How does this illustrate Hess' law?
d Draw an energy-level diagram and an energy cycle.

We usually use the energy cycle method in the solutions we provide to Exercises because it is the most general, but you may prefer to add equations as in Exercise 4.5. You should always be able to justify your method by means of an energy-level diagram.

In the next Exercise, you apply Hess' law to another reaction for which the enthalpy change cannot be measured directly – the hydration of magnesium sulphate, $MgSO_4$.

EXERCISE 4.6
Answers on page 227

The following experiment was performed to determine the enthalpy change for the process:

$$MgSO_4(s) + 7H_2O(l) \rightarrow MgSO_4{\cdot}7H_2O(s)$$

The experiment was divided into two parts. The first part measured the enthalpy change of solution for the process:

$$MgSO_4(s) + 100H_2O(l) \rightarrow MgSO_4(aq, 100H_2O)$$

The second part measured the enthalpy change of solution for:

$$MgSO_4{\cdot}7H_2O(s) + 93H_2O(l) \rightarrow MgSO_4(aq, 100H_2O)$$

Part 1 Solution of $MgSO_4(s)$
45.00 g of water was placed in a polystyrene cup and the temperature of the water was found to be 24.1 °C. 3.01 g of $MgSO_4$ was dissolved in the water and, after careful stirring, the maximum temperature was found to be 35.4 °C.

Part 2 Solution of $MgSO_4{\cdot}7H_2O(s)$
41.85 g of water was placed in a polystyrene cup and the temperature of the water was found to be 24.8 °C. 6.16 g of $MgSO_4{\cdot}7H_2O$ was dissolved in the water and the minimum temperature was found to be 23.4 °C.

a Calculate the enthalpy change of solution for one mole of $MgSO_4$.
Assume $c_p = 4.18\ kJ\ kg^{-1}\ K^{-1}$.
b Similarly, calculate the enthalpy change of solution for one mole of $MgSO_4{\cdot}7H_2O$.
c By means of an enthalpy cycle, calculate the enthalpy change for the reaction:

$$MgSO_4(s) + 7H_2O(l) \rightarrow MgSO_4{\cdot}7H_2O$$

d Plot the results on an enthalpy-level diagram.

Now we look at another application of Hess' law.

Calculating enthalpy of formation from enthalpy of combustion

As you will see later, it is particularly useful to list enthalpy changes of formation for compounds. However, most compounds cannot be formed directly from the elements, so it is impossible to measure these enthalpy changes in a calorimeter.

One of the most useful applications of Hess' law is in calculating heat of formation from heats of combustion, which can be measured directly.

WORKED EXAMPLE   Calculate the standard enthalpy of formation of ethane, C_2H_6, given:

$$C(s) + O_2(g) \rightarrow CO_2(g); \quad \Delta H^\ominus = -394 \text{ kJ mol}^{-1}.$$

$$H_2(g) + \tfrac{1}{2}O_2(g) \rightarrow H_2O(l); \quad \Delta H^\ominus = -286 \text{ kJ mol}^{-1}.$$

$$C_2H_6(g) + 3\tfrac{1}{2}O_2(g) \rightarrow 2CO_2(g) + 3H_2O \ (l); \quad \Delta H^\ominus = -1560 \text{ kJ mol}^{-1}.$$

Solution   1. Starting with the reaction for which you want to calculate $\Delta H^\ominus_f$ begin to construct an energy cycle:

2. Put in the combustion reactions:

$$\boxed{2C(s) + 3H_2(g)} \xrightarrow{\quad \Delta H_1 \quad} \boxed{C_2H_6(g)}$$

Note that the inclusion of $3\tfrac{1}{2}O_2(g)$, which is necessary for the combustion, makes no

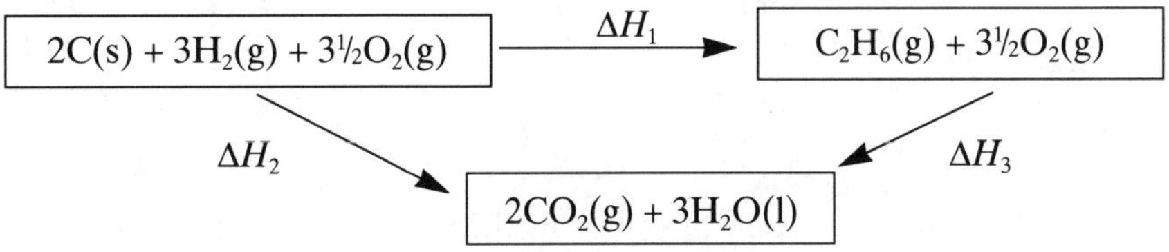

difference to ΔH_1 (see step 4 of the Worked Example on pages 48–50).

3. Use Hess' law to equate the enthalpy changes.

$$\Delta H_2 = \Delta H_1 + \Delta H_3$$

or

$$\Delta H_1 = \Delta H_2 - \Delta H_3$$

4. Substitute numerical values for ΔH_2 and ΔH_3.

$$\Delta H_2 = 2\Delta H^\ominus_c [C(s)] + 3\Delta H^\ominus_c [H_2(g)]$$

$$= 2(-394 \text{ kJ mol}^{-1}) + 3(-286 \text{ kJ mol}^{-1})$$

$$= (-788 - 858 \text{ kJ mol}^{-1}) = -1646 \text{ kJ mol}^{-1}$$

$$\Delta H_3 = \Delta H^\ominus_c [C_2H_6(g)] = -1560 \text{ kJ mol}^{-1}$$

$$\therefore \Delta H_1 = \Delta H_2 - \Delta H_3$$

$$= -1646 \text{ kJ mol}^{-1} - (-1560 \text{ kJ mol}^{-1}) = \textbf{-86 kJ mol}^{\textbf{-1}}$$

Now try the next two Exercises to test your understanding of this application of Hess' law.

EXERCISE 4.7
Answer on page 228

Calculate the standard enthalpy change of formation of carbon disulphide, CS_2, given that.

$$\Delta H^\ominus_f [CO_2(g)] = -393.5 \text{ kJ mol}^{-1},$$

$$\Delta H^\ominus_f [SO_2(g)] = -296.9 \text{ kJ mol}^{-1},$$

$$\Delta H^\ominus_c [CS_2(l)] = -1075.2 \text{ kJ mol}^{-1}.$$

EXERCISE 4.8
Answers on page 228

Calculate the standard enthalpy change of formation of the following compounds:
a ethane, C_2H_6,
b ethanol, C_2H_5OH,
c methylamine, CH_3NH_2,
  Hint: For part **c** assume that when CH_3NH_2 burns, nitrogen is released as $N_2(g)$.

Having shown you how enthalpy changes of formation can be calculated, we now consider what use can be made of the values.

Uses of standard enthalpy changes of formation

In your data book you will find lists of enthalpy changes of formation for both organic and inorganic compounds. Many of the values have been calculated from experimental results in ways similar to those we have described; now you learn how to use them.

Standard enthalpy changes of formation can be used to calculate the enthalpy changes in a reaction. The fact that we can predict the enthalpy change for any reaction is vitally important, for instance, to chemical engineers when planning a chemical plant. They need to know how much heat will be generated or absorbed during the course of a particular reaction and make adequate provision for extremes of temperature in their designs.

Before you go further make sure you are familiar with the expression:

$$\boldsymbol{\Delta H^{\ominus}_{r} = \Sigma\Delta H^{\ominus}_{f}\,[\text{products}] - \Sigma\Delta H^{\ominus}_{f}\,[\text{reactants}]}$$

($\Delta H^{\ominus}_{r}$ is the standard enthalpy change for any reaction and the symbol Σ means 'the sum of'.)

This follows from the application of Hess' law to the enthalpy cycle shown below for a generalised reaction:

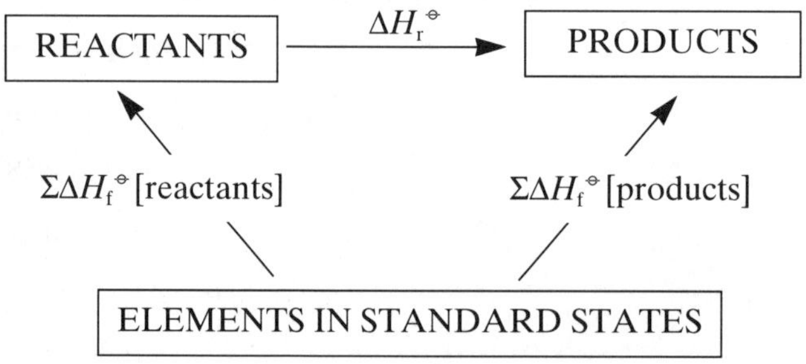

You can apply the same procedure to any reaction. However, once you have understood the derivation of the expression you can use it to solve problems without drawing an energy cycle, as we show in a Worked Example.

WORKED EXAMPLE Calculate the standard enthalpy change for the reaction:

$$2H_2S(g) + SO_2(g) \rightarrow 3S(s) + 2H_2O(l)$$

using only standard enthalpy of formation data.

Compound	**$\Delta H^{\ominus}_{f}$/kJ mol^{-1}**
$H_2S(g)$	–20.6
$SO_2(g)$	–296.9
$H_2O(l)$	–285.9

Solution 1. Write an equation, leaving room to put the value of $\Delta H^{\ominus}_{f}$ under each formula.

$$2H_2S(g) + SO_2(g) \rightarrow 3S(s) + 2H_2O(l)$$

2. Calculate the standard enthalpy change of formation for the amount specified in the equation for each substance. In this equation, 2 mol of $H_2O(l)$ are produced.

$$2\,\Delta H^{\ominus}_{f}[H_2O(l)] = -571.8\ \text{kJ mol}^{-1}$$

Similarly, for $H_2S(g)$, $2\,\Delta H^{\ominus}_{f}[H_2S(g)] = -41.2\ \text{kJ mol}^{-1}$ and for $SO_2(g)$, $\Delta H^{\ominus}_{f}[SO_2(g)] = -296.9\ \text{kJ mol}^{-1}$. Because $\Delta H^{\ominus}_{f}$ for all elements in their standard states is zero, no value appears for sulphur.

3. Put these values under the compounds to which they refer:

	$2H_2S(g)$ +	$SO_2(g) \rightarrow$	$3S(s)$ +	$2H_2O(l)$
$\Delta H^{\ominus}_{f}$/kJ mol^{-1}	−41.2	−296.9	0	−571.8

4. Add the values for the products:

$$\Sigma\Delta H^{\ominus}_{f}[\text{products}] = (0 - 571.8)\ \text{kJ mol}^{-1} = -571.8\ \text{kJ mol}^{-1}$$

Add the values for the reactants:

$$\Sigma\Delta H^{\ominus}_{f}[\text{reactants}] = (-41.2 - 296.9)\ \text{kJ mol}^{-1} = -338.1\ \text{kJ mol}^{-1}$$

5. Subtract:

$$\Delta H_r = \Sigma\Delta H^{\ominus}_{f}[\text{products}] - \Sigma\Delta H^{\ominus}_{f}[\text{reactants}]$$
$$= [-571.8 - (-338.1)]\ \text{kJ mol}^{-1} = \mathbf{-233.7\ kJ\ mol^{-1}}$$

Now try some examples for yourself. Be very careful about positive and negative signs.

EXERCISE 4.9
Answers on page 229

Calculate the standard enthalpy changes for the following reactions. Obtain values of $\Delta H^{\ominus}_{f}$ from your data book.

a $CH_3OH(l) + 1\tfrac{1}{2}O_2(g) \rightarrow CO_2(g) + 2H_2O(l)$
b $2CO(g) + O_2(g) \rightarrow 2CO_2(g)$
c $ZnCO_3(s) \rightarrow ZnO(s) + CO_2(g)$
d $2Al(s) + Fe_2O_3(s) \rightarrow 2Fe(s) + Al_2O_3(s)$

EXERCISE 4.10
Answer on page 229

The standard enthalpies of formation ($\Delta H^{\ominus}_{f}$) at 298 K for a number of compounds are, in kJ mol^{-1}:

$$CH_4(g) - 75;\ H_2O(g) - 242;\ CO(g) - 110$$

Calculate the enthalpy change for the reaction:

$$CH_4(g) + H_2O(g) \rightarrow CO(g) + 3H_2(g)$$

EXERCISE 4.11
Answer on page 229

Use your data book to find the enthalpy change for the following reaction.

$$SiO_2(s) \rightarrow Si(g) + 2O(g)$$

EXERCISE 4.12
Answers on page 229

Alumina cannot be reduced by carbon directly in a process similar to the production of iron in a blast furnace.

$$2Al_2O_3 + 3C \rightarrow 4Al + 3CO_2$$

Look up values of $\Delta H^{\ominus}_{f}$ and use them to explain why this is so.

In the next section we show how the enthalpy change for any reaction is related to the strengths of the individual bonds broken and formed in the reaction.

BOND ENERGY TERM (BOND ENTHALPY TERM)

The energy required to break a particular bond depends not only on the nature of the two bonded atoms but also on the environment of those atoms. For instance, not all the C—H bonds in ethanol, C_2H_5OH, are equally strong because the oxygen atom has more influence on those near it than on those further away.

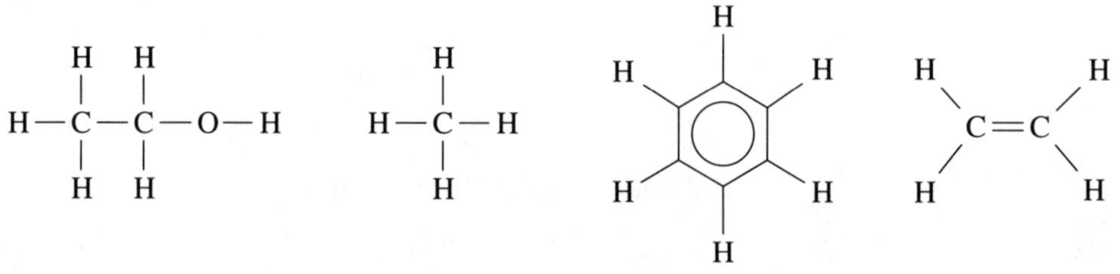

Similarly, C—H bonds in methane, CH_4, are not quite the same as those in benzene, C_6H_6, or ethene, C_2H_4. Fortunately, however, the differences are small enough for the concept of **average** bond energy (bond energy term) to be very useful.

Note that bond dissociation energy (enthalpy) refers to a particular bond in a particular molecule, as in the next Exercise.

EXERCISE 4.13
Answers on page 229

Table 4.2 gives enthalpy changes for four successive dissociation reactions:

Table 4.2

Reaction	$\Delta H^\ominus$/kJ mol^{-1}
$CH_4(g) \rightarrow CH_3(g) + H(g)$	+435
$CH_3(g) \rightarrow CH_2(g) + H(g)$	+444
$CH_2(g) \rightarrow CH(g) + H(g)$	+440
$CH(g) \rightarrow C(g) + H(g)$	+343

a Calculate the enthalpy change for the reaction:

$$CH_4(g) \rightarrow C(g) + 4H(g)$$

b Calculate the average bond energy of a C—H bond in methane.
c What name do we give each of the enthalpy changes in Table 4.2?
d Suggest a reason why each is different.

We cannot always obtain bond energy terms by the method you have just used because bond dissociation energies are often not known. In the next section we look at another method of calculation.

■ Calculating bond energy terms

Once again we can use an energy cycle and Hess' law to determine an enthalpy change which cannot be measured directly. We illustrate this method by a Worked Example below, in which we make use of tabulated values of enthalpy of formation of compounds and enthalpy of atomisation of elements.

The standard enthalpy change of atomisation of an element is the enthalpy change in the production of one mole of gaseous atoms from the element in its standard state. For example, it refers to processes such as:

$$Na(s) \rightarrow Na(g) \quad \text{and} \quad \tfrac{1}{2}Cl_2(g) \rightarrow Cl(g)$$

We deal with the atomisation of compounds later.

WORKED EXAMPLE Calculate the C—H bond energy term in methane.

Solution

1. Write the equation for the complete dissociation of methane.

$$CH_4(g) \rightarrow C(g) + 4H(g)$$

2. Construct an energy cycle by including the elements in their standard states.

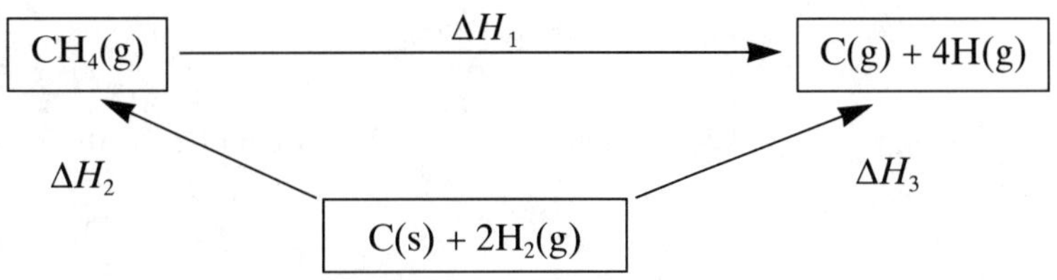

Here ΔH_2 is the standard enthalpy change of formation of methane, $\Delta H^{\ominus}_{f}[CH_4(g)]$; and ΔH_3 is the standard enthalpy change of atomisation of carbon plus four times the standard enthalpy change of atomisation of hydrogen; i.e.

$$\Delta H_3 = \Delta H^{\ominus}_{at}[C(g)] + 4\,\Delta H^{\ominus}_{at}[H(g)].$$

3. Look up these values in your data book:

$$\Delta H^{\ominus}_{f}[CH_4(g)] = -74.8 \text{ kJ mol}^{-1}$$

$$\Delta H^{\ominus}_{at}[C(g)] = +715.0 \text{ kJ mol}^{-1}$$

$$4\,\Delta H^{\ominus}_{at}[H(g)] = 4(+218.0 \text{ kJ mol}^{-1}) = 872.0 \text{ kJ mol}^{-1}$$

Note that the value listed here is 'per mole of H atoms formed', not 'per mole of H_2 molecules atomised'.

4. Apply Hess' law: $\Delta H_2 + \Delta H_1 = \Delta H_3$.

5. Substitute the appropriate values and solve for ΔH_1:

$$\Delta H_1 = \Delta H_3 - \Delta H_2$$

$$= (715.0 + 872.0 - (-74.8)) \text{ kJ mol}^{-1} = 1661.8 \text{ kJ mol}^{-1}$$

6. This is the energy required to break four bonds. We want to know the average value for one bond.

$$\therefore\ \overline{E}(C—H) = \frac{1661.8}{4} \text{ kJ mol}^{-1} = \mathbf{415\ kJ\ mol^{-1}}$$

Now try two similar problems.

EXERCISE 4.14
Answers on page 229

Determine the enthalpy change for the process:

$$H_2S(g) \rightarrow 2H(g) + S(g)$$

and so calculate the bond energy term, $\overline{E}$(H—S).

EXERCISE 4.15
Answer on page 229

Calculate $\overline{E}$(N—H), given the following data:
$\frac{1}{2}N_2(g) + \frac{3}{2}H_2(g) \rightarrow NH_3(g)$; $\Delta H^\ominus = -46$ kJ mol^{-1}
$\frac{1}{2}H_2(g) \rightarrow H(g)$; $\Delta H^\ominus = +218$ kJ mol^{-1}
$\frac{1}{2}N_2(g) \rightarrow N(g)$; $\Delta H^\ominus = +473$ kJ mol^{-1}

Notice that in all these examples and Exercises, reactants and products are in the **gaseous** state. If the substance we are decomposing exists as a liquid in its standard state, it must first be vaporised, and this requires energy. For example:

$$CCl_4(l) \rightarrow CCl_4(g);\quad \Delta H^\ominus_{vap}[CCl_4(l)] = 30.5 \text{ kJ mol}^{-1}$$

Apply this idea in the next Exercise.

EXERCISE 4.16
Answers on page 230

Calculate the enthalpy change for the process:

$$CCl_4(g) \rightarrow C(g) + 4Cl(g)$$

and calculate the bond energy term, $\overline{E}$(C—Cl).

So far you have averaged energies for **similar** bonds within **identical** molecules but bond energy terms can be applied also to similar bonds in **different** molecules.

Determining two bond energy terms simultaneously

Work through the following Revealing Exercise where we show you how to do this calculation. You need the following information:

butane: $C_4H_{10}(g) \rightarrow 4C(g) + 10H(g)$; $\Delta H^\ominus = +5165$ kJ mol^{-1}

pentane: $C_5H_{12}(g) \rightarrow 5C(g) + 12H(g)$; $\Delta H^\ominus = +6337$ kJ mol^{-1}

REVEALING EXERCISE

Q1 How many bonds are broken in the atomisation of butane and pentane?

A1 In butane, 3 C—C bonds and 10 C—H bonds are broken per molecule; in pentane, 4 C—C bonds and 12 C—H bonds are broken per molecule.

Q2 Express the enthalpy changes for the atomisation of pentane and butane as the sums of the bond energy terms.

A2 Butane: $3\overline{E}$(C—C) + $10\overline{E}$(C—H) = +5165 kJ mol^{-1}
Pentane: $4\overline{E}$(C—C) + $12\overline{E}$(C—H) = +6337 kJ mol^{-1}

Notice that for a compound, $\Delta H^\ominus_{at}$ refers to one mole of molecules, whereas for an element it refers to one mole of **atoms**.

Q3 Let $\overline{E}$ (C—C) = x kJ mol^{-1}
$\overline{E}$ (C—H) = y kJ mol^{-1}
and solve these equations simultaneously.

A3 $3x + 10y = 5165$ (1)
$4x + 12y = 6337$ (2)
Multiply equation (1) by 4 and equation (2) by 3 and subtract:
$12x + 40y = 20660$
$12x + 36y = 19011$
$4y = 1649 \therefore y = 412.25$
Substitute this value for y into either equation (1) or (2).
$3x + 10(412.25) = 5165$
$3x = 5165 - 4122.5 = 1042.5 \therefore x = 347.5$
$\therefore$ $\overline{E}$(C—C) = **347.5 kJ mol^{-1}**, $\overline{E}$(C—H) = **412.2 kJ mol^{-1}**

The next Exercise can be done in the same way.

EXERCISE 4.17
Answers on page 230

The enthalpies of complete dissociation (atomisation) of gaseous hexane and heptane are 7512 kJ mol^{-1} and 8684 kJ mol^{-1} respectively. Calculate bond enthalpy terms for C—C and C—H bonds.

Now that we have shown you how bond energy terms can be obtained we illustrate their use in estimating enthalpy changes in reactions for which there are no experimental data.

Using bond energy terms to estimate enthalpy changes

In order to obtain a bond energy term which can be tabulated and used generally, an average value is taken for a particular type of bond in a large number of compounds. The bond energy terms quoted in your data book should, therefore, not be considered as accurately stating the bond energy in any particular molecular environment. Calculations using bond energy terms can give, therefore, only approximate answers.

You have already used one method for calculating enthalpy changes that cannot be determined directly, by substituting known enthalpy data into an energy cycle and applying Hess' law. We now show you another method, this time using bond energy terms, in a Worked Example. We remind you that:

Bond breaking requires energy (endothermic)

Bond making releases energy (exothermic)

WORKED EXAMPLE

Use bond energy terms to calculate the enthalpy change for the reaction:

$$CH_4(g) + Cl_2(g) \rightarrow CH_3Cl(g) + HCl(g)$$

Solution 1. Write the equation using structural formulae:

$$
\begin{array}{ccccccc}
 & \text{H} & & & & \text{H} & \\
 & | & & & & | & \\
\text{H}-\text{C}-\text{H} & + & \text{Cl}-\text{Cl} & \longrightarrow & \text{H}-\text{C}-\text{Cl} & + & \text{H}-\text{Cl} \\
 & | & & & & | & \\
 & \text{H} & & & & \text{H} &
\end{array}
$$

2. List the bonds broken and bonds made under the equations as shown:

$$
\begin{array}{ccccccc}
 & \text{H} & & & & \text{H} & \\
 & | & & & & | & \\
\text{H}-\text{C}-\text{H} & + & \text{Cl}-\text{Cl} & \longrightarrow & \text{H}-\text{C}-\text{Cl} & + & \text{H}-\text{Cl} \\
 & | & & & & | & \\
 & \text{H} & & & & \text{H} &
\end{array}
$$

bonds broken

C — H

Cl — Cl

bonds made

C — Cl

H — Cl

3. Look up the bond energy terms for the bonds broken and made. Add them up as shown, including negative signs for the exothermic bond-making.

$$
\begin{array}{ccccccc}
 & \text{H} & & & & \text{H} & \\
 & | & & & & | & \\
\text{H}-\text{C}-\text{H} & + & \text{Cl}-\text{Cl} & \longrightarrow & \text{H}-\text{C}-\text{Cl} & + & \text{H}-\text{Cl} \\
 & | & & & & | & \\
 & \text{H} & & & & \text{H} &
\end{array}
$$

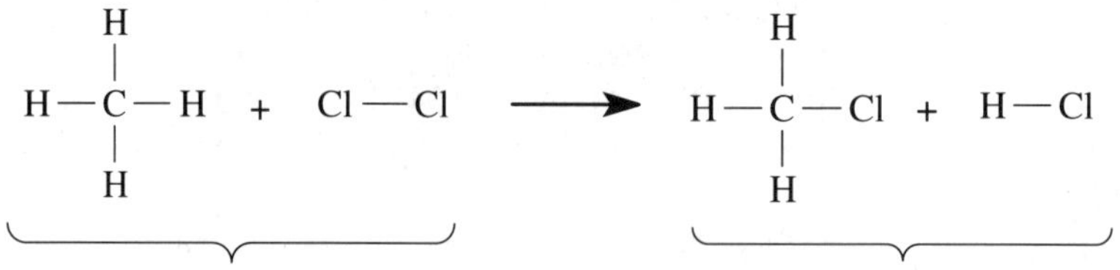

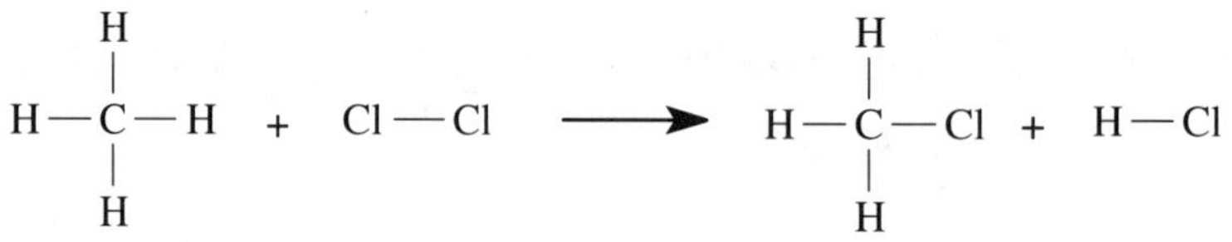

bonds broken   bonds made

$\overline{E}$ (C—H) = 435 kJ mol^{-1} $\overline{E}$ (C—Cl) = −339 kJ mol^{-1}

$\overline{E}$ (Cl—Cl) = 242 kJ mol^{-1} $\overline{E}$ (H—Cl) = −431 kJ mol^{-1}

Total   = 677 kJ mol^{-1} Total   = −770 kJ mol^{-1}

4. Add the values for bond breaking and making (including the correct signs) to obtain the required enthalpy change:

$$\Delta H^{\ominus} = 677 \text{ kJ mol}^{-1} - 770 \text{ kJ mol}^{-1} = -93 \text{ kJ mol}^{-1}$$

5. Write the complete thermochemical equation:

$$CH_4(g) + Cl_2(g) \rightarrow CH_3Cl(g) + HCl(g); \quad \Delta H^{\ominus} = \mathbf{-93 \text{ kJ mol}^{-1}}$$

The following energy-level diagram summarises the energy changes during the reaction we have just considered.

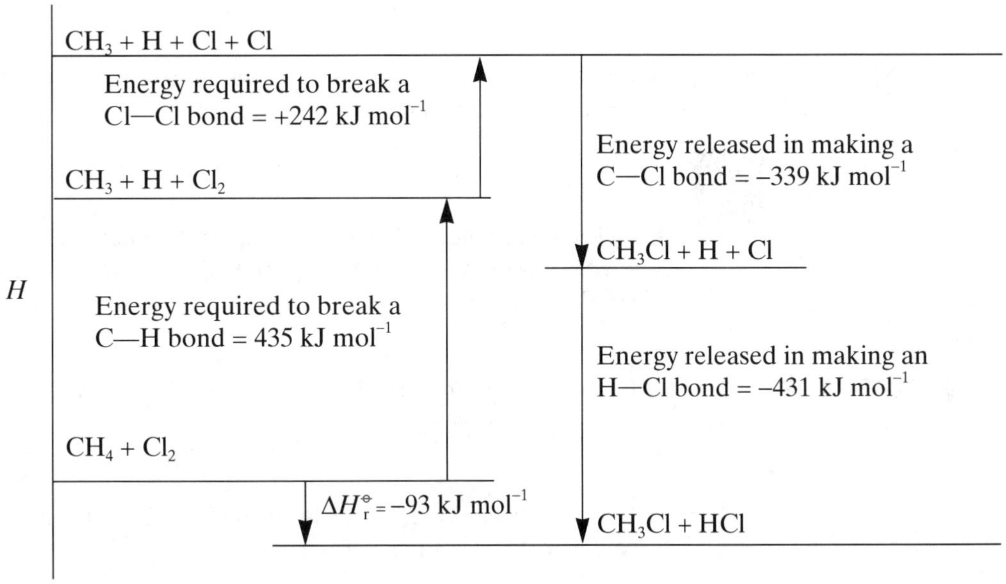

Now attempt the following Exercises, two of which are A-level questions.

EXERCISE 4.18

Answers on page 230

Use bond energy terms listed in your data book to calculate the enthalpy change for the reactions:

a $H_2(g) + Cl_2(g) \rightarrow 2HCl(g)$

b $N_2(g) + 3H_2(g) \rightarrow 2NH_3(g)$

EXERCISE 4.19

Answers on page 230

A

Some bond energy terms are listed below:

Table 4.3

Bond	H—H	C—H	C—Br	C—C	C=C	Br—Br
Bond energy /kJ mol^{-1}	435	415	284	356	598	193

a What do you understand by **bond energy term**?

b Using the given data, calculate the enthalpies of formation, from gaseous atoms, of

i) gaseous propene:

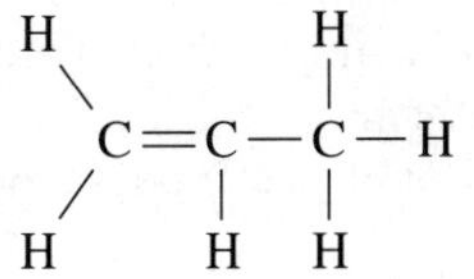

ii) gaseous 1,2-dibromopropane:

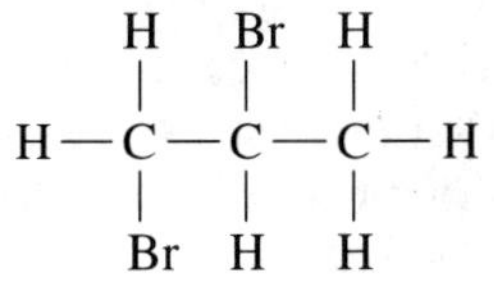

c Calculate the enthalpy change, $\Delta H^{\ominus}$, for the reaction:

$$CH_2{=}CH{-}CH_3(g) + Br_2(g) \rightarrow CH_2BrCHBrCH_3(g)$$

EXERCISE 4.20

Answers on page 231

A

a Use the values of bond energy terms contained in your data book to calculate the standard enthalpy change for the reaction:

$$C_2H_6(g) + Cl_2(g) \rightarrow C_2H_5Cl(g) + HCl(g)$$

b Calculate another value for this standard enthalpy change from heats of formation.
c Write a short account of the reasons why the two values you have calculated differ from each other.

EXERCISE 4.21

Answers on page 231

a Construct a suitable Hess' law cycle to calculate $\Delta H^{\ominus}$ for the reaction:

$6C(g) + 6H(g) \rightarrow$  (g)

b Look up mean bond energy terms for C—H, C—C and C=C bonds. Use them to calculate $\Delta H^{\ominus}$ for the reaction:

$6C(g) + 6H(g) \rightarrow$  (g)

c Which of the structures for benzene would be the more stable? What is the stabilisation energy?
d Assuming that the C—H bond energy is 412 kJ mol^{-1}, calculate the mean bond energy for the six bonds between carbon atoms in benzene. How does this compare with the mean bond energies for C—C and C=C?

EXERCISE 4.22

Answers on page 231

a Calculate the C—C bond dissociation energy in ethane, C_2H_6, given the following information:

$$2C(g) + 6H(g) \rightarrow C_2H_6(g); \quad \Delta H^{\ominus} = -2820 \text{ kJ mol}^{-1}$$

$$\bar{E}(C{-}H) = 412 \text{ kJ mol}^{-1}$$

b Calculate the C=C bond dissociation energy in ethene, C_2H_4, given the following information:

$$2C(g) + 4H(g) \rightarrow C_2H_4(g); \quad \Delta H^{\ominus} = -2260 \text{ kJ mol}^{-1}$$

c Assuming that the σ-bonds in ethane and ethene are identical (they are certainly very similar), calculate the approximate bond dissociation energy for the π-bond in ethene.
d Would you expect ethene to be more or less reactive than ethane?

EXERCISE 4.23

Answers on page 232

a Calculate $\Delta H^{\ominus}$ for the following reactions from bond energy terms. Draw energy level diagrams in your answer.
i) $H_2(g) + F_2(g) \rightarrow 2HF(g)$
ii) $H_2(g) + Cl_2(g) \rightarrow 2HCl(g)$
b Which of the two reactions above is energetically more feasible?
c What two factors contribute to the high negative value of $\Delta H^{\ominus}$ for the H_2/F_2 reaction?

EXERCISE 4.24

Answers on page 232

Figure 4.5 shows some bond energy terms for Group V elements. Use the data to answer the questions that follow.

Figure 4.5

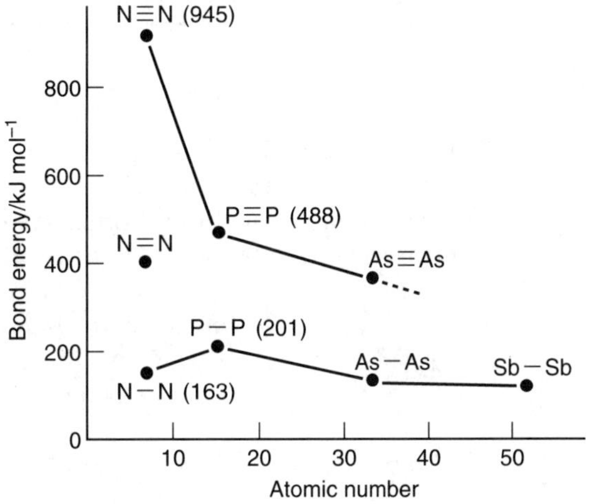

a State the type and number of bonds in a P_4 molecule. (The P_4 molecule has P atoms at the corners of a tetrahedron.)

b Calculate ΔH° for the following processes:

i) $4P(g) \rightarrow P_4(g)$

ii) $4N(g) \rightarrow N_4(g)$

iii) $4P(g) \rightarrow 2P{\equiv}P(g)$

iv) $4N(g) \rightarrow 2N{\equiv}N(g)$

c In view of your answer to **b**, suggest why nitrogen forms diatomic molecules whereas P, As and Sb in the vapour state all exist as X_4 molecules.

EXERCISE 4.25

Answers on page 232

Use the mean bond energies, $\bar{E}$, shown in Table 4.4 to answer the questions that follow.

Table 4.4

Bond	O=O	O—O	S=S	S—S
$\bar{E}$/kJ mol^{-1}	498	142	431	264

a Calculate the enthalpies of reaction for the following changes. Assume that S_2 and O_8 have the same structures as O_2 and S_8.

i) $S_8(g) \rightarrow 4S_2(g)$

ii) $O_8(g) \rightarrow 4O_2(g)$

b Why does oxygen consist of diatomic molecules whereas sulphur contains S_8 molecules?

So far your work has been confined to covalent compounds. You have seen how to use bond energy terms to calculate the enthalpy change of formation of a molecular compound from its constituent gaseous atoms – an important step in the calculation of ΔH° for a reaction. Now we look at energy changes in the formation of ionic compounds.

LATTICE ENERGY (LATTICE ENTHALPY)

It is often useful to know the enthalpy of formation of an ionic compound from its constituent gaseous ions. The energy value is called the 'lattice energy' of the compound, because ionic bonding always leads to a solid crystal lattice. It provides a measure of the strength of ionic bonds.

> $\Delta H^{\ominus}_{lat}$ **The lattice energy of an ionic compound** is the enthalpy change which occurs when one mole of it is formed, as a crystal lattice, from its constituent gaseous ions.

This definition of lattice energy always gives a negative sign, e.g.

$$Na^+(g) + Cl^-(g) \rightarrow Na^+Cl^-(s); \quad \Delta H^{\ominus}_{lat} = -781 \text{ kJ mol}^{-1}$$

If we were considering the energy required to separate the salt into its separate ions then the enthalpy change must have a positive sign:

$$Na^+Cl^-(s) \rightarrow Na^+(g) + Cl^-(g); \quad \Delta H^{\ominus}_{lat} = +781 \text{ kJ mol}^{-1}$$

You may find that some older books quote positive values of lattice energy, but you will not go wrong if you always relate $\Delta H^{\ominus}$ to the appropriate equation.

Since it is impossible to determine lattice energies directly by experiment we use an indirect method where we construct an energy diagram called a Born–Haber cycle.

Calculating lattice energy using a Born–Haber cycle

The Born–Haber cycle is yet another application of Hess' law but the alternative routes involve more steps than you have used so far.

To construct a Born–Haber cycle you need two enthalpy changes not previously mentioned in this volume, ionisation energy and electron affinity.

> $\Delta H^{\ominus}_{i}$ **The ionisation energy of an element** is the enthalpy change which occurs when one mole of its gaseous atoms loses one mole of electrons to form one mole of gaseous positive ions.

$$\text{e.g.} \quad Na(g) \rightarrow Na^+(g); \quad \Delta H^{\ominus}_{i} = +500 \text{ kJ mol}^{-1}$$

> $\Delta H^{\ominus}_{e}$ **The electron affinity of an element** is the enthalpy change which occurs when one mole of its gaseous atoms accepts one mole of electrons to form one mole of gaseous negative ions.

$$\text{e.g.} \quad Br(g) + e^- \rightarrow Br^-(g); \quad \Delta H^{\ominus}_{e} = -342 \text{ kJ mol}^{-1}$$

We start with a Revealing Exercise.

REVEALING EXERCISE

Below is a generalised Born–Haber cycle for an ionic compound.

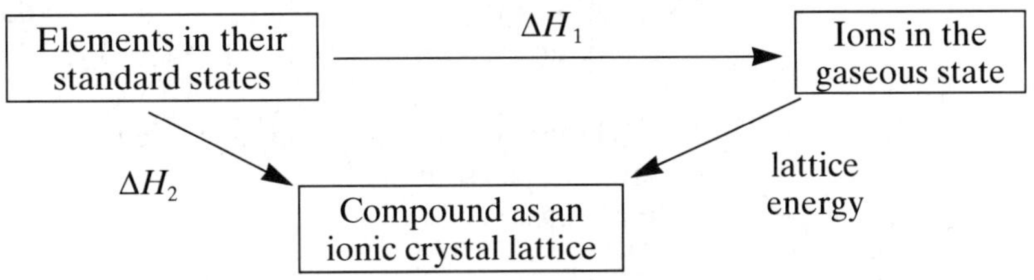

Q1 Redraw the cycle including the appropriate formulae for the formation of sodium chloride.

A1

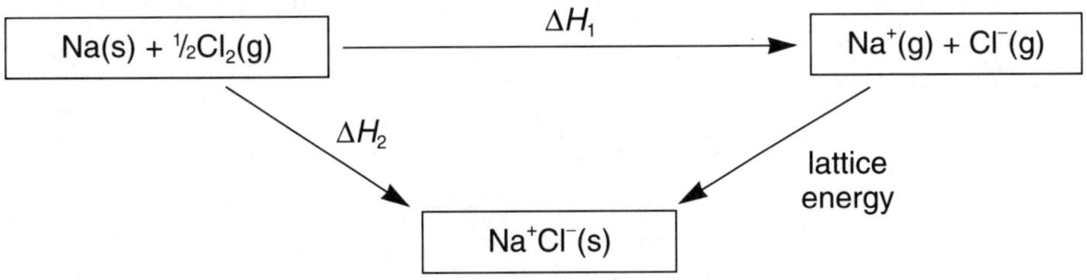

Q2 ΔH_1 is the sum of the energy change associated with four steps, two for sodium and two for chlorine. Write thermochemical equations for these four steps.

A2 $Na(s) \rightarrow Na(g)$; $\Delta H^{\ominus}_{at} = +108\ kJ\ mol^{-1}$
$Na(g) \rightarrow Na^+(g) + e^-$; $\Delta H^{\ominus}_{i} = +500\ kJ\ mol^{-1}$
$\frac{1}{2}Cl_2(g) \rightarrow Cl(g)$; $\Delta H^{\ominus}_{at} = +121\ kJ\ mol^{-1}$
$Cl(g) + e^- \rightarrow Cl^-(g)$; $\Delta H^{\ominus}_{e} = -364\ kJ\ mol^{-1}$

Q3 Calculate ΔH_1.

A3 $\Delta H_1 = \Delta H^{\ominus}_{at}(Na) + \Delta H^{\ominus}_{i}(Na) + \Delta H^{\ominus}_{at}(\frac{1}{2}Cl_2) + \Delta H^{\ominus}_{e}(Cl)$
So $\Delta H_1 = (+108 + 500 + 121 - 364)\ kJ\ mol^{-1} = +365\ kJ\ mol^{-1}$

Q4 What name is given to ΔH_2? Look up a value in your data book.

A4 ΔH_2 is the enthalpy change of formation of sodium chloride.
$\Delta H^{\ominus}_{f}[NaCl(s)] = -411\ kJ\ mol^{-1}$

Q5 Insert the values for ΔH_1 and ΔH_2 in the energy cycle.

A5

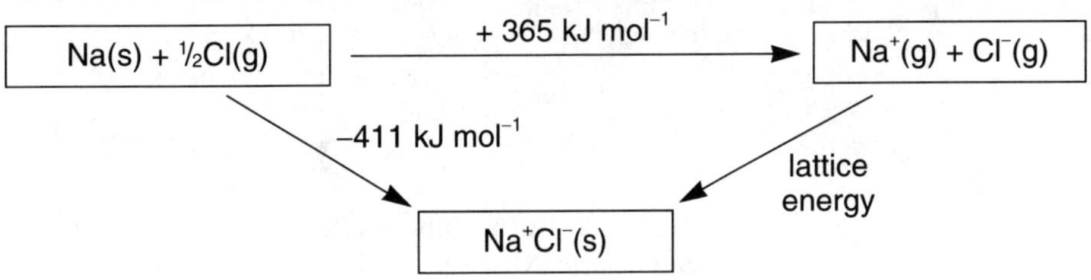

Q6 Use Hess' law to calculate the lattice energy for sodium chloride.

A6 365 kJ mol^{-1} + lattice energy = –411 kJ mol^{-1}
∴ lattice energy = –411 kJ mol^{-1} – 365 kJ mol^{-1} = **–776 kJ mol^{-1}**

Born–Haber cycles are often drawn as energy-level diagrams. Compare the following energy-level diagram with the cycle you drew in the Revealing Exercise and trace the various steps.

Figure 4.6 Born–Haber cycle for sodium chloride, drawn approximately to scale.

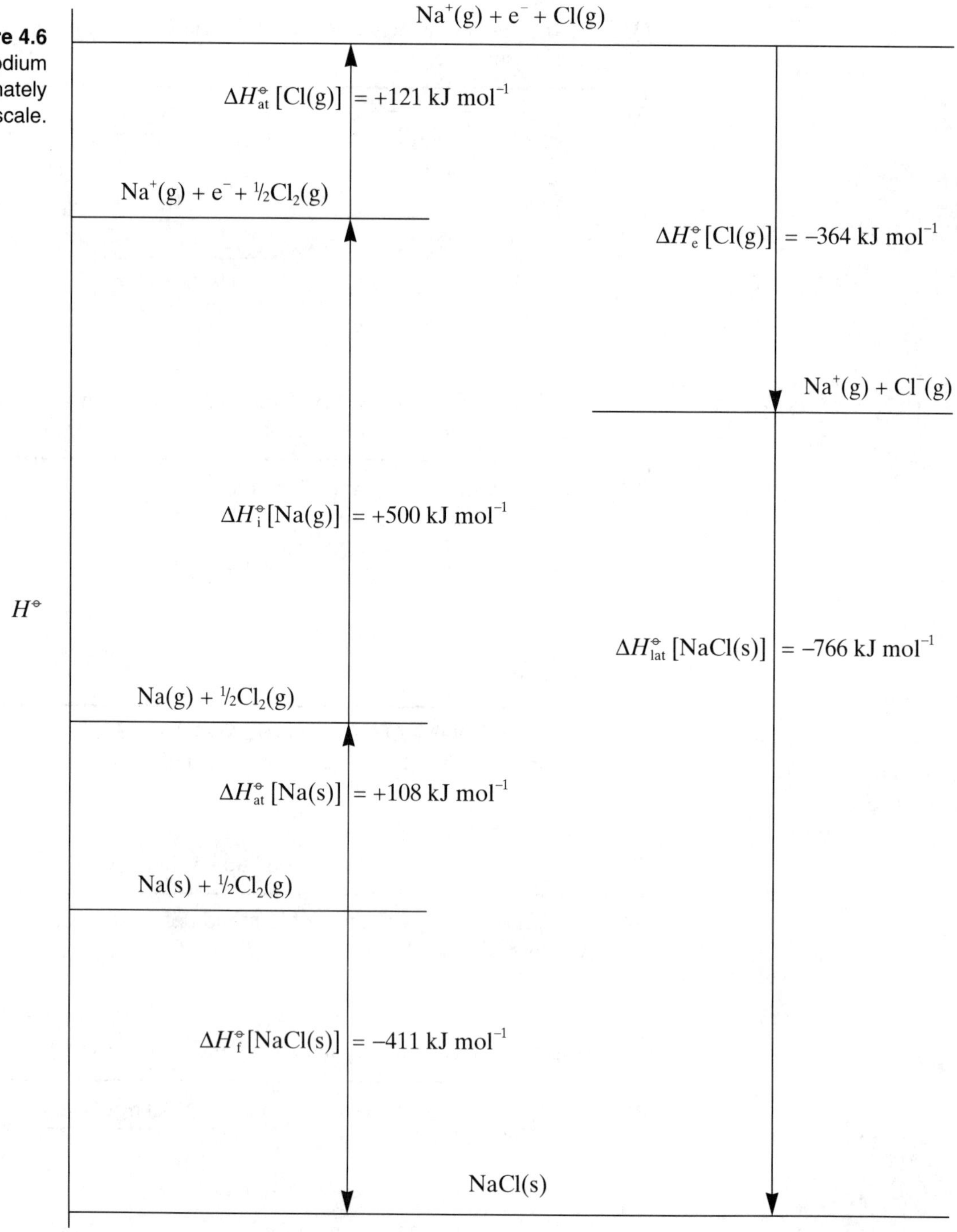

Note that Born–Haber cycles are frequently **not** drawn to scale, in order to save time and space. Also, the electrons are not always included in the ionisation steps.

Try drawing some Born–Haber cycles for yourself in the following Exercises, and use the cycles to calculate some lattice energies.

EXERCISE 4.26

Answers on page 233

Draw a Born–Haber cycle for each of the following ionic compounds, and then calculate their lattice energies. (Note that in sodium hydride, the hydrogen forms a negative ion.) The cycles should be drawn as energy-level diagrams but need not be drawn to scale.

Table 4.5

		Metal		**Non-metal**	
Compound	$\Delta H^\ominus_f$ /kJ mol^{-1}	$\Delta H^\ominus_{at}$ /kJ mol^{-1}	$\Delta H^\ominus_i$ /kJ mol^{-1}	$\Delta H^\ominus_{at}$ /kJ mol^{-1}	$\Delta H^\ominus_e$ /kJ mol^{-1}
KBr (K^+, Br^-)	−392	+89	+420	+112	−342
NaH (Na^+, H^-)	−57	+108	+500	+218	−72

EXERCISE 4.27

Answers on page 234

Complete Table 4.6 and draw Born–Haber cycles to obtain lattice energies. Note that in this example you may have to combine successive values for $\Delta H^\ominus_i$ and $\Delta H^\ominus_e$. Also, whereas $\Delta H^\ominus_i$ is **always** positive, $\Delta H^\ominus_e$ may be positive or negative.

Table 4.6

		Metal		**Non-metal**	
Compound	$\Delta H^\ominus_f$ /kJ mol^{-1}	$\Delta H^\ominus_{at}$ /kJ mol^{-1}	$\Delta H^\ominus_i$ /kJ mol^{-1}	$\Delta H^\ominus_{at}$ /kJ mol^{-1}	$\Delta H^\ominus_e$ /kJ mol^{-1}
$BaCl_2$ (Ba^{2+}, $2Cl^-$)	−860	+175	+500 +1000	+121	−364
SrO (Sr^{2+}, O^{2-})					

EXERCISE 4.28

Answers on page 235

Table 4.7

Process	$\Delta H^\ominus$**(298 K)/kJ mol**$^{-1}$
First electron affinity of oxygen	−140
Second electron affinity of oxygen	+790
Standard enthalpy of atomisation of oxygen	+250
Standard enthalpy of atomisation of boron	+590
Standard enthalpy of formation of the oxide of boron	−1270
First ionisation energy of boron	+800
Second ionisation energy of boron	+2400
Third ionisation energy of boron	+3700
Fourth ionisation energy of boron	+25000
Fifth ionisation energy of boron	+32800

a Draw a labelled Born–Haber cycle for the formation of the most likely oxide of boron (assumed to be ionic). Calculate a lattice energy for the oxide of boron.

b Would you expect your value for the lattice energy of the oxide of boron to be in good or poor agreement with a theoretically derived value? Give your reason(s).

Born–Haber cycles are not used exclusively to calculate lattice energies, as the next two Exercises illustrate. In particular, since it is difficult to obtain values of electron affinity by other methods, they are sometimes determined from Born–Haber cycles using values of lattice energy calculated from the laws of electrostatics.

EXERCISE 4.29

Answer on page 235

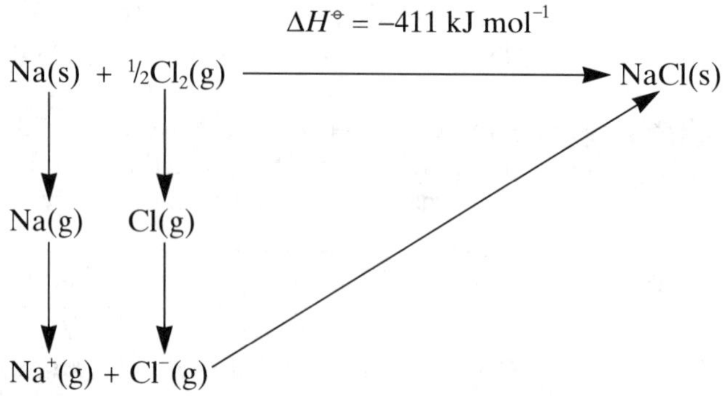

Use the above diagram and the following data to calculate $\Delta H^\ominus$ for the reaction:

$$Cl(g) + e^- \rightarrow Cl^-(g)$$

$Na(s) \rightarrow Na(g)$; $\Delta H^\ominus = 108\ kJ\ mol^{-1}$
$Na(g) \rightarrow Na^+(g) + e^-$; $\Delta H^\ominus = 500\ kJ\ mol^{-1}$
$\frac{1}{2}Cl_2(g) \rightarrow Cl(g)$; $\Delta H^\ominus = 121\ kJ\ mol^{-1}$
$NaCl(s) \rightarrow Na^+(g) + Cl^-(g)$; $\Delta H^\ominus = 776\ kJ\ mol^{-1}$

EXERCISE 4.30

Answers on page 235

The Born–Haber cycle for the formation of calcium chloride is given below:

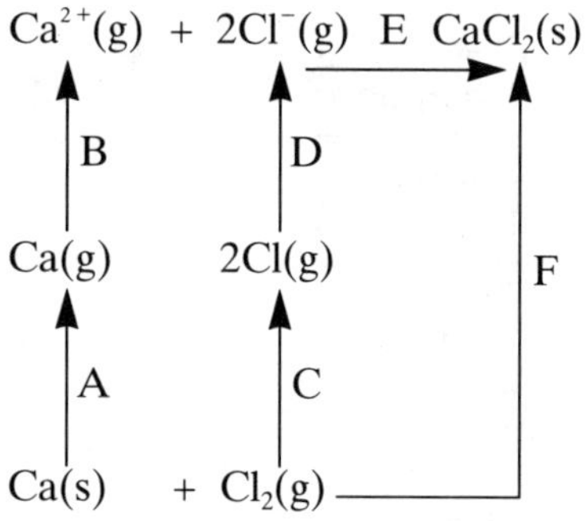

A = 177 kJ mol^{-1}, C = 242 kJ mol^{-1}, F = –795 kJ mol^{-1},
B = 1690 kJ mol^{-1}, E = –2197 kJ mol^{-1}.

a A is the enthalpy change of sublimation (atomisation) of solid calcium. Similarly, define the following:
i) B,
ii) C,
iii) E,
iv) F.

b Calculate the enthalpy change D.

Another use of Born–Haber cycles enables us to discuss the stoichiometry of compounds which might be formed by direct combination.

Lattice energy and stoichiometry

We now use enthalpy data and suitable Born–Haber cycles to predict the formula of a compound.

In the Exercise which follows, you construct Born–Haber cycles for MgCl, $MgCl_2$ and $MgCl_3$ to determine enthalpy changes of formation and then decide which would be the most likely formula for magnesium chloride.

EXERCISE 4.31
Answers on page 236

a Construct Born–Haber cycles for MgCl, $MgCl_2$ and $MgCl_3$, inserting all the values except $\Delta H^\ominus_f$. Since experimentally determined lattice energies for MgCl and $MgCl_3$ are not available, use the theoretically calculated values:

$$\Delta H^\ominus_{lat}\,[MgCl] = -753\ kJ\ mol^{-1}$$

$$\Delta H^\ominus_{lat}\,[MgCl_3] = -5440\ kJ\ mol^{-1}$$

b Use the cycles to obtain values for:
i) $\Delta H^\ominus_f[MgCl]$,
ii) $\Delta H^\ominus_f[MgCl_2]$,
iii) $\Delta H^\ominus_f[MgCl_3]$.

c Which of the three compounds MgCl, $MgCl_2$, $MgCl_3$ is/are energetically stable with respect to the elements?

d Calculate the enthalpy change for the hypothetical reaction:

$$2MgCl(s) \rightarrow MgCl_2(s) + Mg(s)$$

using the $\Delta H^\ominus_f$ values you calculated in part **b**.

e Discuss briefly the relative stability of MgCl and $MgCl_2$ in the light of your answer to **d**. Does this explain why MgCl is **not** known?

Born–Haber cycles are also used to determine another quantity which cannot easily be measured directly – enthalpy of hydration, which we consider in the next section.

ENTHALPY OF HYDRATION

Enthalpy of hydration is concerned with the bonding between dissolved ions and surrounding water molecules. It is closely related to two quantities you have already studied, lattice enthalpy and enthalpy of solution.

You may find it helpful to refer to the diagram below, which shows what happens when an ionic salt dissolves in water. Then you should be able to do the following Exercises.

Figure 4.7
(We have shown only one layer of attached water molecules surrounding each ion. There may be several layers, especially around small, highly charged ions.)

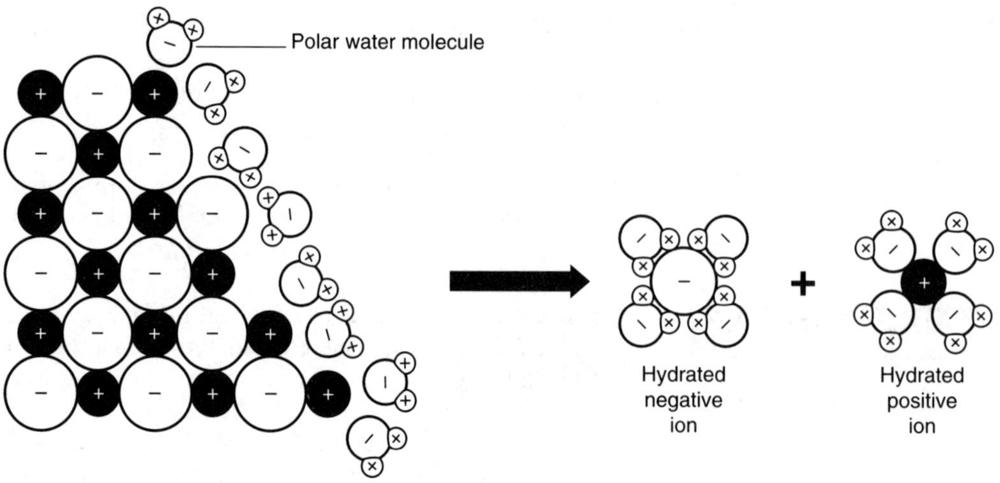

EXERCISE 4.32

Answers on page 237

a Write a thermochemical equation representing the complete separation of the ions in solid sodium chloride, including a value for ΔH° obtained from your data book.

b Write thermochemical equations representing the bonding of water molecules to the separated ions. Note that since we do not know precisely how many water molecules are attached, we write $Na^+(aq)$ rather than $Na(H_2O)_n^+$ for the hydrated ions, and put '+ aq' in the equation instead of '+ nH_2O'. ($\Delta H^\circ_{hyd}(Na^+) = -406$ kJ mol^{-1}; $\Delta H^\circ_{hyd}(Cl^-) = -364$ kJ mol^{-1}.)

c Add together the equations you have written (in the same way as you add algebraic equations). What process does the resulting equation represent, and what is the value for ΔH°?

In the next Exercise you look at a similar process using an energy-level diagram.

EXERCISE 4.33

Answers on page 237

The following is an energy-level diagram for calcium chloride.

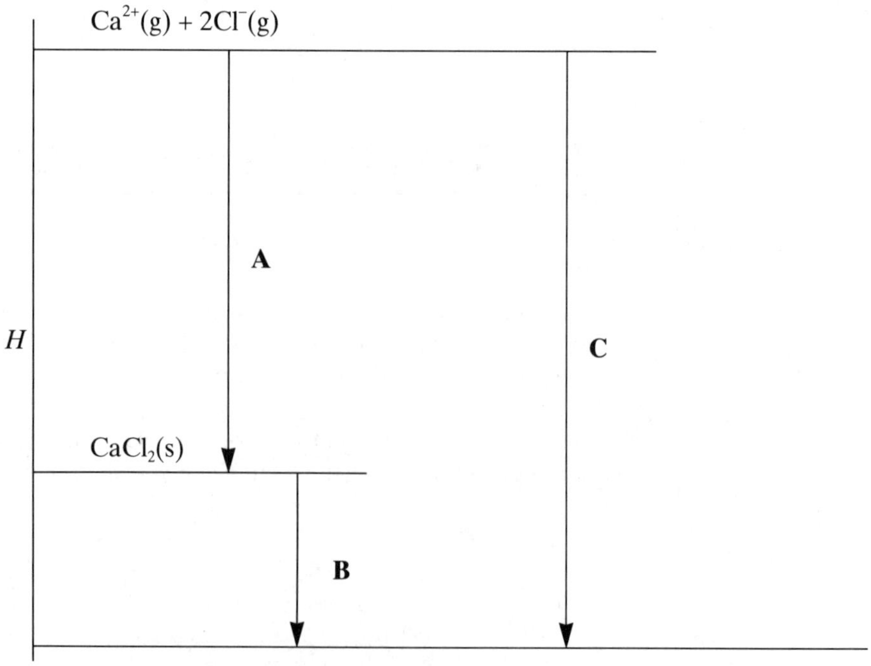

a What are the quantities denoted as **A**, **B** and **C**?

b Look up values of **A** and **C** in your data book and use them to calculate a value for **B**.

The preceding Exercises should have made the following relationship clear to you:

$$\Delta H_{solution} = \Delta H_{hydration} - \Delta H_{lattice}$$

Also, you should realise that enthalpy of solution, because it is the small difference between two much larger quantities, can change from positive to negative with only relatively small variations in either lattice enthalpy or hydration enthalpy. This, of course, has some bearing on the solubility of ionic compounds, because most spontaneous changes have negative $\Delta H^{\ominus}$ values. You already know that this rule is not altogether reliable, so you should not be surprised to find some soluble substances with small **positive** values for ΔH_{sol}. Nevertheless, the following statement is generally true:

'The more negative the enthalpy of solution, the more soluble the substance.'

Test the statement in the following Exercise.

EXERCISE 4.34
Answers on page 237

a Draw a Hess cycle to show how the enthalpy of solution of anhydrous magnesium chloride, $MgCl_2$, can be found by considering the change as taking place in two stages via separate gaseous ions. Calculate the enthalpy of solution, using the following data:
$\Delta H_{lat}(MgCl_2) = -2526$ kJ mol^{-1}; $\Delta H_{hyd}(Mg^{2+}) = -1920$ kJ mol^{-1};
$\Delta H_{hyd}(Cl^-) = -364$ kJ mol^{-1}.
b Represent the changes on an energy-level diagram.
c Would you expect $MgCl_2$ to be soluble in water? Explain your answer.

We now give you another method for determining experimentally the enthalpy change of a reaction, i.e. using the technique of thermometric titration.

THERMOMETRIC TITRATIONS

In thermometric titrations we make use of the fact that reactions in solution are accompanied by temperature changes, and thus it is possible to follow the course of a reaction using a thermometer.

EXERCISE 4.35
Answers on page 238

The data below are taken from an experiment used to determine the concentrations of two acids, hydrochloric acid, HCl, and ethanoic acid, CH_3CO_2H, by thermometric titration. The temperature of the mixture was measured at regular intervals during the titration. The enthalpy change for each reaction can be calculated from the maximum temperature rise.

Titration of 25.0 cm^3 of hydrochloric acid with 50.0 cm^3 of 1.00 M NaOH:

Table 4.8

Volume added/cm^3	0.0	5.0	10.0	15.0	20.0	25.0	30.0	35.0	40.0	45.0	50.0
Temperature/°C	22.2	24.4	26.4	28.4	30.1	31.1	30.4	29.9	29.2	28.8	28.2

Titration of 25.0 cm^3 of ethanoic acid with 50.0 cm^3 of 1.00 M NaOH:

Table 4.9

Volume added/cm^3	0.0	5.0	10.0	15.0	20.0	25.0	30.0	35.0	40.0	45.0	50.0
Temper-ature/°C	21.0	22.6	24.3	25.9	27.2	28.7	28.6	28.0	27.5	27.0	26.5

a Plot temperature (y-axis) against volume of acid added (x-axis) for each acid on the same graph.

b Extrapolate the curves as shown in Fig. 4.8. The point at which they meet corresponds to both the volume of acid required for neutralisation and to the maximum temperature.

Figure 4.8

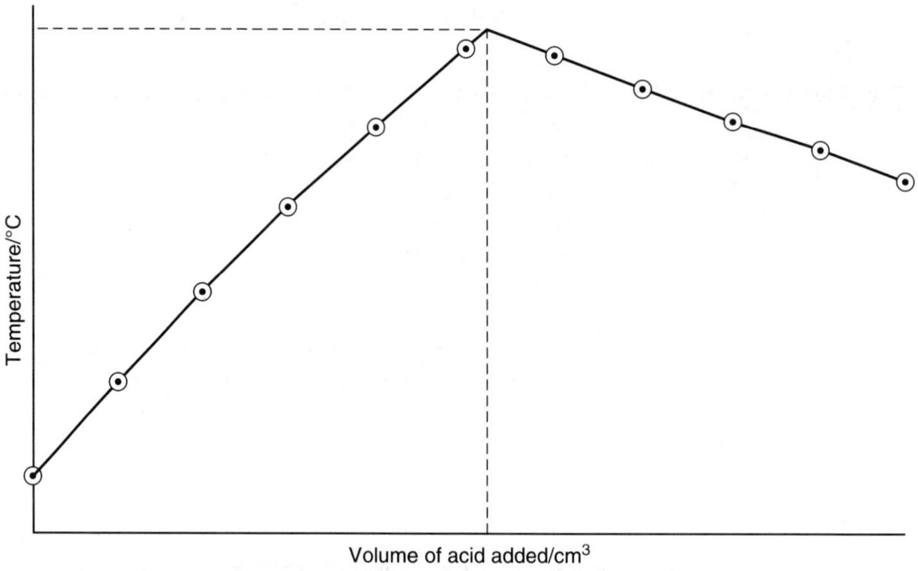

c Calculate the concentration of each of the acids.

d From the maximum temperature rise, determine the quantity of energy released in each titration. Assume that the specific heat capacity of the solutions is the same as that for water, 4.18 kJ kg^{-1} K^{-1} and that the heat capacity of the cup is zero.

e Calculate the standard enthalpy change of neutralisation for each reaction.

In the next section we explore further the unreliability of ΔH° as an indicator of the direction of change. This requires some understanding of free energy and entropy and may not, therefore, be appropriate for your syllabus.

THE DIRECTION OF CHANGE

You have already learned that the use of $\Delta H^\ominus$ to predict the stability of substances and the direction of change is useful but not completely reliable. It is $\Delta G^\ominus$ (standard free energy change) rather than $\Delta H^\ominus$ which is always negative for a reaction which goes to completion.

Fortunately, however, $\Delta G^\ominus$ is closely related to $\Delta H^\ominus$ and for most reactions (at 298 K) has similar values. Check this in your data book by looking up values of $\Delta G_f^\ominus$ and $\Delta H_f^\ominus$; they are usually tabulated in adjacent columns.

Provided $\Delta G_f^\ominus$ values are known, you can calculate $\Delta G^\ominus$ for a reaction just as you calculated $\Delta H^\ominus$. Use the relationship:

$$\Delta G^\ominus = \Sigma\Delta G_f^\ominus[\text{products}] - \Sigma\Delta G_f^\ominus[\text{reactants}]$$

(Free energy is often called 'Gibbs energy' or 'Gibbs free energy' after the famous American thermodynamicist, J. Willard Gibbs, who first introduced the term – hence the use of the letter G.)

EXERCISE 4.36
Answers on page 239

Calculate $\Delta G^\ominus$ for the following reactions and compare your answers with the values of $\Delta H^\ominus$ calculated in Exercise 4.9. Comment on any difference.

a $CH_3OH(l) + 1\frac{1}{2}O_2(g) \rightarrow CO_2(g) + 2H_2O(l)$
b $2CO(g) + O_2(g) \rightarrow 2CO_2(g)$
c $ZnCO_3(s) \rightarrow ZnO(s) + CO_2(g)$
d $2Al(s) + Fe_2O_3(s) \rightarrow 2Fe(s) + Al_2O_3(s)$

The last Exercise illustrates the fact that $\Delta H^\ominus$ is usually a good guide for predicting the relative stabilities of substances and the direction of change. You will often use $\Delta H^\ominus$ for this purpose, but you should remember that it is better to use $\Delta G^\ominus$, if values are available (frequently they are not!).

We now show you how $\Delta H^\ominus$ and $\Delta G^\ominus$ are related to each other.

Free energy and enthalpy, and also changes in these quantities, are related by the following equations:

$$\boldsymbol{G = H - TS}$$
$$\boldsymbol{\Delta G = \Delta H - T\Delta S}$$

where S is another measurable (or calculable) thermodynamic property called entropy and T is the absolute temperature. Before we discuss entropy, however, we want to be sure that you understand the use of the symbol $^\ominus$ in this context.

ΔH, ΔG and ΔS refer to **any** change in a system. For a chemical reaction, this change may involve any amounts of reactants under any conditions and may also be a **partial** change into a mixture of reactants and products.

$\Delta H^\ominus$, $\Delta G^\ominus$ and $\Delta S^\ominus$ refer only to a **complete** change in a system from reactants to products as shown in an equation, where reactants and products are at the **same standard conditions** of temperature, pressure and concentration.

The simple rule that a spontaneous change in a system can only occur if $\Delta G < 0$ is, unfortunately, not always easy to apply because we have to work with tabulated values of $\Delta G^\ominus$, not ΔG. It is not strictly true that spontaneous change can only occur if $\Delta G^\ominus < 0$: it has been shown that although a reaction with a small positive value of $\Delta G^\ominus$ cannot proceed to completion, it **can** proceed towards an equilibrium mixture of reactants and products, because **that** change has a negative value of ΔG.

The following diagram shows how the possibility of change is related to values of ΔG and $\Delta G^\ominus$.

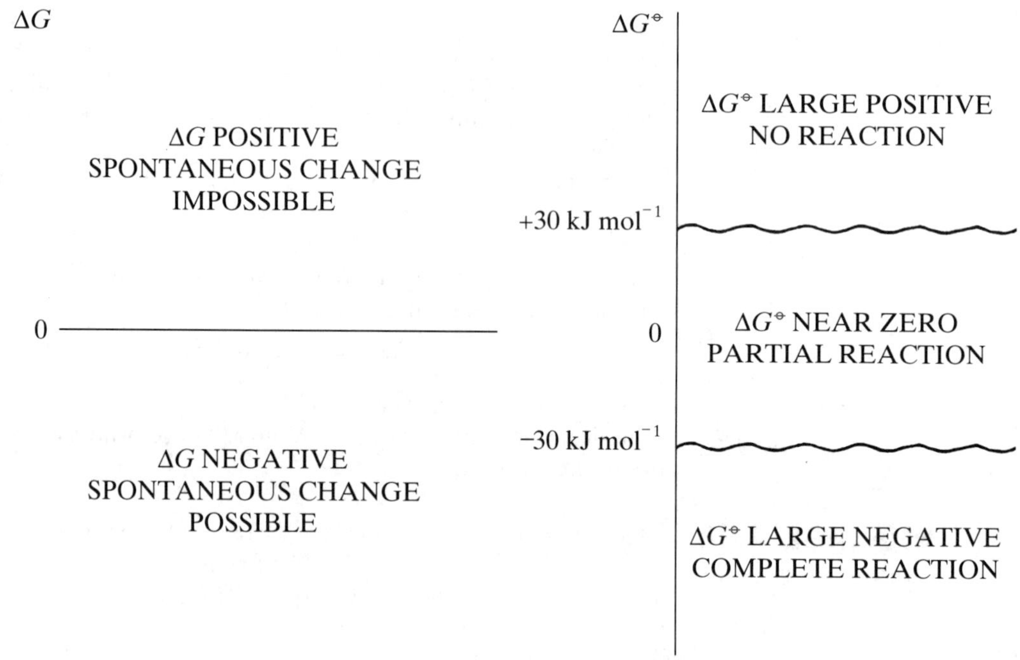

The boundary between partial reaction and complete reaction is, of course, not clearly defined. As $\Delta G^\ominus$ becomes more negative, reaction favours products more and more until, at values of $\Delta G^\ominus$ below about -30 kJ mol^{-1}, it can be regarded as complete.

Similarly, as $\Delta G^\ominus$ becomes more positive, reaction favours products less and less until, at values of $\Delta G^\ominus$ above about $+30$ kJ mol^{-1}, it can be regarded as not occurring at all.

You will learn more about incomplete changes, and the significance of the values ± 30 kJ mol^{-1}, in later chapters on equilibrium. At this stage you should simply try to remember the following expressions:

$\Delta G < 0$ spontaneous change possible

$\Delta G^\ominus < 0$ reaction possible; products favoured

$\Delta G^\ominus < \sim -30$ kJ mol^{-1} complete reaction possible

$\Delta G^\ominus = \Delta H^\ominus - T\Delta S^\ominus$

The following sections on entropy will help you to apply these expressions.

Entropy, *S*

The only completely satisfactory definition of entropy is a mathematical one, which you need not concern yourself with at A-level. However, the definition does allow the calculation of absolute values of entropy for any amount of any substance. Non-mathematical definitions are imprecise, but you may find the following statements help you (eventually!) to get some mental picture of entropy.

1. Entropy is an indication of the degree of disorder in a system.
2. Entropy is a measure of the extent to which energy is dispersed.
3. Entropy is a measure of 'sameness', i.e. substances with high entropy are more nearly alike than those with low entropy.

It is a fundamental law of experience that entropy tends to increase in natural processes. Applying this to the three statements gives:

1. Disorder tends to increase.
2. Energy tends to become dispersed.
3. Things that are different tend to become less distinguishable.

Clearly it is possible for us to create order, to localise energy and to separate mixtures, all of which **decrease** the entropy of a system, but we believe this can only be done at the expense of a greater increase in the entropy of the surroundings. This means that in the passage of any interval of time:

$$\Delta S(\text{total}) > 0$$

You may like to reflect on the philosophical significance of this statement as regards the nature of time and the future of our universe. (Does it mean that we are inexorably moving towards a state of maximum entropy where energy and atoms are totally and randomly dispersed – the so-called 'heat-death'?) However, it is of more relevance to your A-level studies to focus on entropy changes within a defined system. Since

$$\Delta S(\text{total}) = \Delta S(\text{system}) + \Delta S(\text{surroundings})$$

we can say that:

$$\Delta S(\text{system}) + \Delta S(\text{surroundings}) > 0$$

and therefore that, in many cases, $\Delta S(\text{system}) > 0$.

Thus, just as the enthalpy of a reacting system often (but not always) decreases ($\Delta H^\ominus$ negative), so the entropy often increases ($\Delta S^\ominus$ positive).

Whenever you use tables of entropy, you should bear in mind three points.

1. The tabulated entropy values give the entropy of one mole of substance in its standard state.
2. The unit is J K^{-1} mol^{-1} (not kJ K^{-1} mol^{-1}).
3. These are actual entropy values, not entropy changes.

If we can look up entropy values for reactants and products, it is a simple matter to calculate $\Delta S^\ominus$.

Calculating entropy changes

Because entropy is a function like enthalpy, we can calculate the entropy change for a process by the relation:

$$\Delta S^\ominus = S^\ominus[\text{products}] - S^\ominus[\text{reactants}]$$

We illustrate this by a Worked Example.

WORKED EXAMPLE Calculate the standard entropy change for the reaction:

$$2Cu(s) + O_2(g) \rightarrow 2CuO(s)$$

Solution 1. Look up the value of $S^\ominus$ for each substance, multiply it by the number of moles in the reaction, and put this value under the formula.

	2Cu(s) +	$O_2(g)$ →	2CuO(s)
$S^\ominus$/J K^{-1} mol^{-1}	(2×33.3)	(204.9)	(2×43.5)

2. Calculate the total entropy of the products.

$$S^\ominus[\text{products}] = 2 \times 43.5 \text{ J K}^{-1} \text{ mol}^{-1} = 87.0 \text{ J K}^{-1} \text{ mol}^{-1}$$

3. Calculate the total entropy of the reactants.

$$S^\ominus[\text{reactants}] = [(2 \times 33.3) + 204.9] \text{ J K}^{-1} \text{ mol}^{-1} = 271.5 \text{ J K}^{-1} \text{ mol}^{-1}$$

4. Calculate the entropy change by subtraction.

$$\Delta S^\ominus = S^\ominus[\text{products}] - S^\ominus[\text{reactants}] = (87.0 - 271.5) \text{ J K}^{-1} \text{ mol}^{-1}$$
$$= \mathbf{-184.5 \text{ J K}^{-1} \text{ mol}^{-1}}$$

Now work out some values of $\Delta S^\ominus$ for yourself.

EXERCISE 4.37
Answers on page 239

Calculate $\Delta S^\ominus$ for the following reactions.

a $H_2(g) + C_2H_4(g) \rightarrow C_2H_6(g)$
b $N_2(g) + 3H_2(g) \rightarrow 2NH_3(g)$
c $2NaNO_3(s) \rightarrow 2NaNO_2(s) + O_2(g)$

EXERCISE 4.38
Answers on page 239

Predict whether $\Delta S^\ominus$ will be positive or negative for each of the following reactions. Then check your prediction by calculation.

a $CO(g) + Cl_2(g) \rightarrow COCl_2(g)$
b $2H_2O_2(l) \rightarrow 2H_2O(l) + O_2(g)$
c $C_2H_4(g) + 3O_2(g) \rightarrow 2CO_2(g) + 2H_2O(l)$
d $2KClO_3(s) \rightarrow 2KCl(s) + 3O_2(g)$
e $2C(s) + O_2(g) \rightarrow 2CO(g)$

Now that you know how to calculate $\Delta S^\ominus$, you can use the value to calculate $\Delta G^\ominus$ in order to see whether or not a reaction is possible.

Using $\Delta S^\ominus$ to calculate $\Delta G^\ominus$

You already know that the standard free energy change, $\Delta G^\ominus$, is related to enthalpy and entropy changes by the equation:

$$\Delta G^\ominus = \Delta H^\ominus - T\Delta S^\ominus$$

and that, for a reaction to be possible, $\Delta G^\ominus$ must be substantially negative. We now show you how to predict the feasibility of a reaction by calculating $\Delta G^\ominus$. We start with a Worked Example.

WORKED EXAMPLE

Calculate the standard free energy change, $\Delta G^\ominus$, for the process:

$$CaO(s) + H_2O(l) \rightarrow Ca(OH)_2(s)$$

at 25 °C, using the following values of standard enthalpy of formation and entropy.

Table 4.10

	$\Delta H_f^\ominus$/kJ mol^{-1}	$S^\ominus$/J K^{-1} mol^{-1}
$CaO(s)$	–635.5	39.7
$Ca(OH)_2(s)$	–986.6	76.1
$H_2O(l)$	–285.9	70.0

Solution

1. Calculate the standard enthalpy change.
   $\Delta H^{\ominus} = \Delta H_f^{\ominus}[\text{products}] - \Delta H_f^{\ominus}[\text{reactants}]$
   $= -986.6 \text{ kJ mol}^{-1} - (-635.5 - 285.9) \text{ kJ mol}^{-1}$
   $= (-986.6 + 921.4) \text{ kJ mol}^{-1} = -65.2 \text{ kJ mol}^{-1}$
2. Calculate the standard entropy change.
   $\Delta S^{\ominus} = S^{\ominus}[\text{products}] - S^{\ominus}[\text{reactants}]$
   $= 76.1 \text{ J K}^{-1} \text{ mol}^{-1} - (39.7 + 70.0) \text{ J K}^{-1} \text{ mol}^{-1}$
   $= -33.6 \text{ J K}^{-1} \text{ mol}^{-1} = -0.0336 \text{ kJ K}^{-1} \text{mol}^{-1}$
   Note the change of unit from J to kJ. This is necessary for the next step.
3. Substitute into the expression for $\Delta G^{\ominus}$.
   $\Delta G^{\ominus} = \Delta H^{\ominus} - T\Delta S^{\ominus}$
   $= -65.2 \text{ kJ mol}^{-1} - (298 \text{ K} \times -0.0336 \text{ kJ K}^{-1} \text{ mol}^{-1})$
   $= (-65.2 + 10.0) \text{ kJ mol}^{-1} = \mathbf{-55.2\ kJ\ mol^{-1}}$
   The negative value of $\Delta G^{\ominus}$ indicates that at 25 °C and 1.00 atm, this reaction can proceed spontaneously (although it **may**, of course, be slow).

Now try some similar calculations for yourself.

EXERCISE 4.39
Answers on page 239

Calculate the standard free energy change, $\Delta G^{\ominus}$, accompanying each of the following processes at 25 °C.

a $2NO(g) + O_2(g) \rightarrow N_2O_4(g)$
b $NH_3(g) + HCl(g) \rightarrow NH_4Cl(s)$
c $H_2O(l) \rightarrow H_2O(g)$

You may have realised that it would have been simpler to calculate $\Delta G^{\ominus}$ in the last exercise using the expression you met earlier:

$$\Delta G^{\ominus} = \Sigma\Delta G_f^{\ominus}[\text{products}] - \Sigma\Delta G_f^{\ominus}[\text{reactants}]$$

However, the great advantage of using $\Delta H^{\ominus}$ and $\Delta S^{\ominus}$ is that you can calculate $\Delta G^{\ominus}$ at temperatures other than 298 K as we show in the next section.

Temperature and spontaneous processes

You would expect $\Delta G^{\ominus}$ to vary considerably with temperature because of the form of the equation:

$$\Delta G^{\ominus} = \Delta H^{\ominus} - T\Delta S^{\ominus}$$

This variation means that some processes which are not possible at low temperatures become feasible at higher temperatures, and vice versa. For an example, consider the familiar process:

$$H_2O(l) \rightarrow H_2O(g)$$

You know that this change does not occur to completion (in a closed system) at 25 °C but that it proceeds spontaneously at temperatures above 100 °C. This shows that $\Delta G^{\ominus}$ has very different values at different temperatures.

Note that the variation of $\Delta G^{\ominus}$ with temperature is quite separate from the fact that an increase in temperature increases the **rate** of reaction. Remember that thermodynamics tells us nothing at all about how fast a feasible reaction will occur. To calculate $\Delta G^{\ominus}$ at a temperature other than 25 °C we should, strictly speaking, use values of $\Delta H^{\ominus}$ and $\Delta S^{\ominus}$ at the appropriate temperature. Fortunately, the variation of $\Delta H^{\ominus}$ and $\Delta S^{\ominus}$ with temperature is small enough for us to ignore in this context.

EXERCISE 4.40
Answers on page 239

Calculate $\Delta G^{\ominus}$ (1000 K) for the following reactions and compare with given values of $\Delta G^{\ominus}$ (298 K). State whether or not the reactions are feasible at each temperature.

a $2NO(g) + O_2(g) \rightarrow N_2O_4(g)$; $\Delta G^{\ominus}$ (298 K) = -75.6 kJ mol^{-1}

b $NH_3(g) + HCl(g) \rightarrow NH_4Cl(s)$; $\Delta G^\ominus$ (298 K) = –92.3 kJ mol^{-1}
c $H_2O(l) \rightarrow H_2O(g)$; $\Delta G^\ominus$ (298 K) = +8.6 kJ mol^{-1}
d $CaCO_3(s) \rightarrow CaO(s) + CO_2(g)$; $\Delta G^\ominus$ (298 K) = +130.2 kJ mol^{-1}
(If your data book gives values for different forms of $CaCO_3$, use those for calcite.)

END-OF-CHAPTER QUESTIONS

Answers on page 240

A

4.1 The following are standard enthalpies of formation in kJ mol^{-1}:

$CO_2(g)$: –394 $H_2O(g)$: –242 $CO(g)$: –110

What is the standard enthalpy change, in kJ mol^{-1}, for the reaction:

$$CO(g) + H_2O(g) \rightarrow CO_2(g) + H_2(g)?$$

A

4.2 Consider the following thermochemical equations:

$H_2(g) + \frac{1}{2}O_2(g) \rightarrow H_2O(l)$; $\Delta H^\ominus$ = –286 kJ mol^{-1}

$H_2O(l) \rightarrow H_2O(g)$; $\Delta H^\ominus$ = +44 kJ mol^{-1}

Calculate the enthalpy change, in kJ mol^{-1}, for the reaction:

$2H_2(g) + O_2(g) \rightarrow 2H_2O(g)$.

A

4.3

Table 4.11

Compound	$CS_2(l)$	NOCl(g)	$CCl_4(l)$	$SO_2(g)$
$\Delta H_f^\ominus$ (298)/kJ mol^{-1}	88	53	–139	–296

From the data above, calculate the value of $\Delta H^\ominus$ (298) in kJ mol^{-1}, for the reaction:

$$CS_2(l) + 4NOCl(g) \rightarrow CCl_4(l) + 2SO_2(g) + 2N_2(g).$$

A P

4.4 Chemical companies manufacture containers filled with liquid butane for use by campers. The enthalpy change of combustion of butane is –3000 kJ mol^{-1}.
a Write an equation for the complete combustion of butane.
b A camper estimates that the liquid butane left in a container would give 1.2 dm^3 of butane gas (measured at ordinary temperature and pressure).
Calculate the mass of water at 20 °C that could be brought to the boiling point by burning this butane: use the following information.
Assume that 80% of the heat from the butane is absorbed by the water, the specific heat capacity of water is 4.2 J g^{-1} K^{-1}, and 1 mol of a gas occupies 24 dm^3 at ordinary temperatures and pressures.

A P

4.5 Calculate the standard enthalpy of formation of ethanoic acid, $CH_3CO_2H(l)$, from the following data:
$\Delta H_c^\ominus$[graphite] = –394 kJ mol^{-1}
$\Delta H_c^\ominus$[$H_2(g)$] = –286 kJ mol^{-1}
$\Delta H_c^\ominus$[$CH_3CO_2H(l)$] = –873.2 kJ mol^{-1}

A

4.6 Use the data in Table 4.12 to determine $\Delta H^\ominus$ and $\Delta G^\ominus$ (if you have studied it) for the following reactions:

a $N_2O_5(s) \rightarrow 2NO_2(g) + \frac{1}{2}O_2(g)$

b $N_2O_5(s) \rightarrow N_2O_4(g) + \frac{1}{2}O_2(g)$

Table 4.12

Compound	$\Delta H_f^\ominus$/kJ mol^{-1}	$\Delta G_f^\ominus$/kJ mol^{-1}
$N_2O_5(s)$	−43.1	+113.8
$NO_2(g)$	+33.2	+51.3
$N_2O_4(g)$	+9.2	+97.8

A

4.7 The standard enthalpy of formation of methane is given by:

$$\text{C (graphite)} + 2H_2(g) \rightarrow CH_4(g)$$

$$\Delta H^\ominus_{298} = -74 \text{ kJ (mol methane)}^{-1}$$

The enthalpies of atomisation of graphite and hydrogen are given by:

$$\text{C (graphite)} \rightarrow C(g); \qquad \Delta H^\ominus = +712 \text{ kJ mol}^{-1}$$

$$\tfrac{1}{2}H_2(g) \rightarrow H(g); \qquad \Delta H^\ominus = +215.5 \text{ kJ mol}^{-1}$$

Calculate the bond energy, in kJ mol^{-1}, of the C—H bond in methane.

4.8 Given that
$\Delta H_f^\ominus$ [$NH_3(g)$] = −46.2 kJ mol^{-1}
$\Delta H_{at}^\ominus$ [$N_2(g)$] = +473 kJ mol^{-1}
$\Delta H_{at}^\ominus$ [$H_2(g)$] = +218 kJ mol^{-1}
calculate the average N—H bond energy in ammonia.

A

4.9 Coke (carbon), ethanol and hydrogen can all be used as fuels.
a Write equations for the complete combustion of these fuels.
b Using the following standard enthalpies of formation,

$\Delta H_f^\ominus$/kJ mol^{-1}: $CO_2(g)$, −394; $H_2O(l)$, −286; $C_2H_5OH(l)$, −278

i) write down the standard enthalpy of combustion of coke,
ii) calculate the standard enthalpy of combustion of ethanol.

c Use the data given below to calculate the standard enthalpy of combustion of hydrogen.
Bond enthalpy term/kJ mol^{-1}: H—H, 436; O═O, 496; O—H, 463
$\Delta H_{vap}^\ominus$/kJ mol^{-1}: $H_2O(l)$, 41

d i) Use your answers in **b** and **c** to calculate the energy evolved, in kJ g^{-1}, when each fuel is burned.
ii) Comment on the advantages and disadvantages of hydrogen as a fuel.

A

4.10 The following table lists the enthalpy changes of combustion of several monohydric alcohols.

Table 4.13

Alcohol	$\Delta H_c^\ominus$/kJ mol^{-1}
Methanol	−715
Ethanol	−1367
Propan-1-ol	−2017
Butan-1-ol	−2675

a Draw a plot of $\Delta H_c^\ominus$ against the relative molecular mass of the alcohols.
b From your graph estimate a value for the enthalpy change of combustion of pentan-1-ol.

c Given $\Delta H_f^\circ[H_2O(l)] = -286$ kJ mol^{-1} and $\Delta H_f^\circ[CO_2(g)] = -394$ kJ mol^{-1}, calculate the standard enthalpy change of formation of ethanol.

d The enthalpy change of combustion of ethane-1,2-diol is –1180 kJ mol^{-1}. What would you expect the corresponding value for propane-1,3-diol to be?

4.11 Some energy data are tabulated below.

Table 4.14

Process	**ΔH° (298 K)/kJ mol^{-1}**
$Na(s) \rightarrow Na(g)$	+108
$\frac{1}{2}Cl_2(g) \rightarrow Cl(g)$	+121
$Na(g) \rightarrow Na^+(g) + e^-$	+496
$Cl(g) + e^- \rightarrow Cl^-$	−349
$Ca(g) \rightarrow Ca^{2+}(g) + 2e^-$	+1736
$Ca^{2+}(g) \rightarrow Ca^{3+}(g) + e^-$	+4941
$Ca^{2+}(g) + 2Cl^-(g) \rightarrow CaCl_2(s)$	−2220
$Ca^{3+}(g) + 3Cl^-(g) \rightarrow CaCl_3(s)$	−4800 (estimated)
$NaCl(s) \rightarrow Na^+(g) + Cl^-(g)$	+787
$NaCl(s) + water \rightarrow Na^+(aq) + Cl^-(aq)$	+4

Using this information,

a calculate the standard molar enthalpy change for the process

$$Na(s) + \tfrac{1}{2}Cl_2(g) \rightarrow Na^+(g) + Cl^-(g),$$

b calculate the standard enthalpy of formation of sodium chloride,

c explain why $CaCl_3$ does not exist, but $CaCl_2$ does. To show this by calculation, you will need an extra item of information:

$$Ca(s) \rightarrow Ca(g); \quad \Delta H^\circ = +178 \text{ kJ mol}^{-1}$$

4.12 In an experiment to determine the enthalpy of neutralisation of sodium hydroxide with sulphuric acid, 50 cm^3 of 0.40 M sodium hydroxide was titrated, thermometrically, with 0.50 M sulphuric acid. The results were plotted as follows:

Figure 4.9

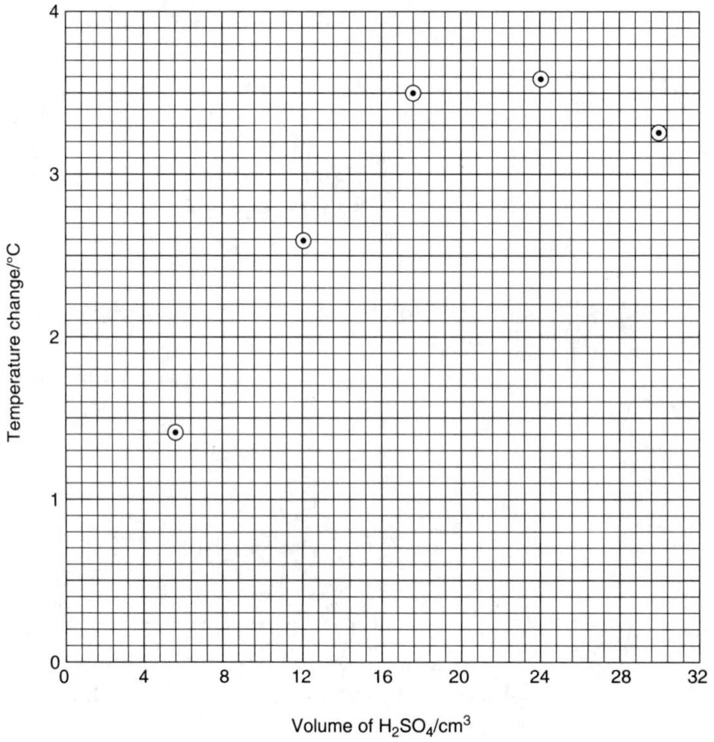

Calculate a value for the enthalpy of neutralisation of sodium hydroxide with sulphuric acid. (The specific heat capacity of water is 4.2 J K^{-1} g^{-1}.) What assumptions have you made in your calculations?

4.13 A student carefully measured out 45 cm^3 of the standard 2 mol dm^{-3} hydrochloric acid and placed it in a suitable vessel. The steady temperature of the solution was noted. 5 cm^3 of standard sodium hydroxide was measured into a small beaker and then this was tipped into the hydrochloric acid. The maximum temperature was recorded after stirring the solution well. The whole experiment was repeated using different volumes of acid and alkali. The results are plotted graphically below. (The specific heat capacities of the solutions were taken as 4.18 J g^{-1} °C^{-1} and the densities of the solutions were assumed to be 1 g cm^{-3}.)

Figure 4.10

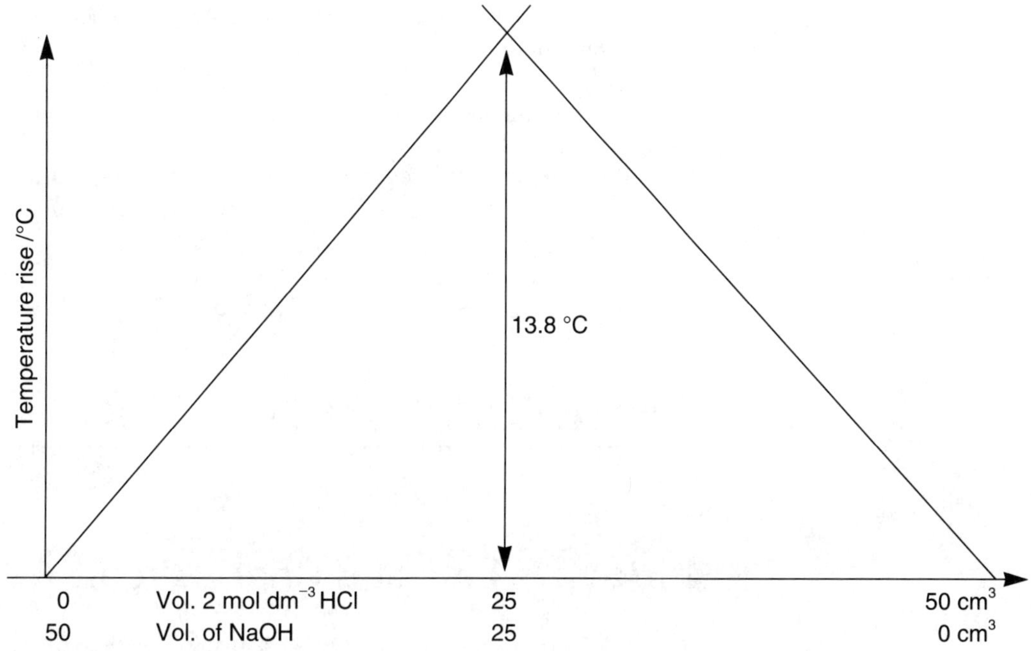

a Calculate a value for the heat evolved in the reaction.
b Calculate the enthalpy of neutralisation of hydrochloric acid and sodium hydroxide from these results.

4.14 **a** Calculate the standard entropy change and standard free energy change at 298 K accompanying the following process:

$$CO(g) + Cl_2(g) \rightarrow COCl_2(g); \Delta H^{\ominus} = -112 \text{ kJ mol}^{-1}$$

given the following table of standard entropy values.

Table 4.15

Substance	**$S^{\ominus}$/J mol^{-1} K^{-1}**
CO(g)	198
$COCl_2$(g)	289
Cl_2(g)	233

b Would you expect this reaction to be feasible i) at 298 K, and ii) at very high temperatures?

CHAPTER 5

GASES

INTRODUCTION AND PRE-KNOWLEDGE

The gaseous state is characterised by the fact that a fixed amount varies readily in shape and size and generally has low density and high compressibility relative to solids and liquids.

Throughout this chapter, we assume that you are familiar with the concepts of 'standard temperature and pressure' (s.t.p.) and 'absolute temperature' and that you have used a variety of different units for pressure. The pascal, symbol Pa, is the official SI unit for pressure and is equivalent to a force of one newton acting over an area of one square metre i.e. $1\ \text{Pa} = 1\ \text{N m}^{-2}$.

The pascal is a very small unit of pressure, so the kilopascal (kPa) is more often used. However, chemists still often use the units atmosphere (atm) and millimetres of mercury (mmHg). Partly because the pressure exerted by a gas is usually thought of in comparison with atmospheric pressure, and partly because barometers in schools are often graduated in millimetres of mercury, we have retained the use of the units atm and mmHg while also using Pa and kPa in some calculations.

Because the molecules of a gas are so far apart, the laws that describe their behaviour are remarkably simple.

We look in turn at several laws which describe quantitatively the behaviour of gases. These laws were discovered empirically (i.e. by studying the results of experiments) but were later shown to have the same theoretical basis.

THE GAS LAWS

Boyle's Law and Charles' Law

These laws were formulated by measuring how the volume of a sample of gas varies with temperature and pressure.

Boyle's Law states that the pressure of a fixed mass of gas at constant temperature is inversely proportional to its volume, i.e.

$$p \propto \frac{1}{V} \quad \text{or} \quad pV = \text{constant} \quad \text{or} \quad p_1V_1 = p_2V_2$$

Charles' Law states that the volume of a fixed mass of gas at constant pressure is directly proportional to the absolute temperature (K = °C + 273), i.e.

$$V \propto T \quad \text{or} \quad \frac{V}{T} = \text{constant} \quad \text{or} \quad \frac{V_1}{T_1} = \frac{V_2}{T_2}$$

Apply these laws in the three Exercises that follow.

EXERCISE 5.1
Answers on page 240

a A meteorological balloon is to be filled with helium at atmospheric pressure. What will the volume of the balloon be if it is to hold all the gas from a 25.0 dm^3 gas cylinder at 150 atm? Assume constant temperature.

b The balloon will burst if its volume exceeds 4000 dm^3. If the filling temperature is 15 °C, what is the maximum temperature the balloon can stand at atmospheric pressure?

EXERCISE 5.2
Answer on page 240

100 cm^3 of hydrogen, at 1.00 atm, is compressed to 37.0 cm^3 at constant temperature. What is the new pressure?

EXERCISE 5.3
Answer on page 240

73.0 cm^3 of nitrogen at 1456 mmHg is allowed to expand so that its pressure is 760 mmHg. What is its new volume at the same temperature?

(More problems on Boyle's Law and Charles' Law for practice and/or revision are in the End-of-chapter questions).

We show you how to combine Boyle's Law and Charles' Law in the following Worked Example.

WORKED EXAMPLE

A meteorological balloon has a volume of 6.15 m^3 when filled with helium at 14 °C and 762 mmHg. What will its volume be if the temperature rises to 19 °C and the pressure falls to 749 mmHg?

Solution

1. Calculate the effect on the volume of gas due to pressure change alone, using Boyle's Law:

$$pV = \text{constant} \quad \text{or} \quad p_1V_1 = p_2V_2$$

$$\therefore V_2 = V_1 \times \frac{p_1}{p_2} = 6.15\ \text{m}^3 \times \frac{762\ \text{mmHg}}{749\ \text{mmHg}}$$

Notice that because a **pressure decrease** gives a **volume increase**, we multiply the volume by a factor greater than one, i.e. 762/749 (and not by 749/762).

2. Calculate the effect on this new volume of gas due to a temperature change alone, using Charles' Law.

$$\frac{V}{T} = \text{constant} \quad \text{or} \quad \frac{V_2}{T_2} = \frac{V_3}{T_3}$$

$$\therefore V_3 = V_2 \times \frac{T_3}{T_2}$$ where V_2 is the volume calculated in step 1.

$$\therefore V_3 = \left(6.15\ \text{m}^3 \times \frac{762}{749}\right) \times \frac{(273 + 19)\,\text{K}}{(273 + 14)\,\text{K}}$$

Notice that because a **temperature increase** gives a **volume increase** we again multiply the volume by a factor greater than one, i.e. 292/287.

The final volume, V_3, is therefore given by:

$$V_3 = 6.15\ \text{m}^3 \times \frac{762}{749} \times \frac{292}{287} = \mathbf{6.37\ m^3}$$

Now try some Exercises. You will soon find that, with a little practice, you can do both steps in one:

new volume = original volume × pressure factor × temperature factor

or

$$V_2 = V_1 \times \frac{p_1}{p_2} \times \frac{T_2}{T_1}$$

Hint: always check whether the factors should be greater or less than one.

EXERCISE 5.4
Answer on page 240

What is the volume at s.t.p. of a sample of gas which occupies 38.2 cm^3 at 18 °C and 765 mmHg?

EXERCISE 5.5
Answer on page 240

79.0 cm^3 of hydrogen was collected from a reaction between zinc and sulphuric acid, at 21 °C and 756 mmHg. What volume would the gas occupy at s.t.p?

The combination of Boyle's Law and Charles' Law which you have been using is often quoted separately as the combined gas law.

The combined gas law

In the previous exercises you applied Boyle's Law and Charles' Law to problems where p, V and T all vary. The final relationship in the Worked Example

$$V_2 = V_1 \times \frac{p_1}{p_2} \times \frac{T_2}{T_1}$$

is a mathematical statement of the combined gas law, more commonly stated in the form:

$$\frac{p_1V_1}{T_1} = \frac{p_2V_2}{T_2}$$

You will use the combined gas law most often for calculating volume changes, but it can of course be used to calculate pressure or temperature changes in a very similar way. You will use it in the following Exercises.

EXERCISE 5.6
Answers on page 241

The table below shows a number of gas volumes and the conditions under which they were collected and measured. Calculate the volumes the gases would occupy if the conditions were changed to those indicated.

Table 5.1

	Volume of gas	Conditions of measurement	New conditions
a	29.2 cm^3	25 °C, 762 mmHg	s.t.p.
b	5.13 dm^3	62 °C, 1.02 atm	s.t.p.
c	132 cm^3	s.t.p.	25 °C, 758 mmHg
d	42.1 cm^3	31 °C, 752 mmHg	20 °C, 200 mmHg

EXERCISE 5.7
Answer on page 241

A sample of gas at 27 °C was heated so that both its pressure and its volume were doubled. What was the new temperature?

EXERCISE 5.8
Answer on page 241

A set of car tyres, each of volume 21.0 dm^3, were inflated until the pressure gauge read 28.5 lb in^{-2}. During a journey, the temperature of the tyres increased from 15 °C to 42 °C and their volume increased to 21.6 dm^3. What would the pressure gauge have read then? (The pressure of air in the tyre was the sum of the gauge pressure and atmospheric pressure, which may be taken as 14.7 lb in^{-2}.)

Having looked at some empirical laws, we now consider a very important theory which was developed to help explain them.

AVOGADRO'S THEORY AND ITS APPLICATIONS

Avogadro's theory (or hypothesis) states that equal volumes of gases, under the same conditions of temperature and pressure, contain equal numbers of molecules.

You have probably studied the way in which Avogadro's theory was used to explain Gay-Lussac's law of combining volumes and to help establish chemical formulae from measurements of reacting volumes of gases. However, we now often use the theory the other way round, i.e. to calculate reacting volumes of gases of known formulae.

■ Using Avogadro's theory to calculate reacting volumes

We show the procedure for this type of calculation in a Worked Example.

WORKED EXAMPLE What volume of oxygen would be required to burn completely 20 cm^3 of ethane, C_2H_6, and what volume of carbon dioxide would be produced? (Volumes measured at the same room temperature and pressure.)

Solution

1. Write the equation for the reaction:

$$C_2H_6(g) + 3\tfrac{1}{2}O_2(g) \rightarrow 2CO_2(g) + 3H_2O(l)$$

2. Write the amounts of each gas under the equation:

$$\begin{array}{llll} C_2H_6(g) + & 3\tfrac{1}{2}O_2(g) \rightarrow & 2CO_2(g) + & 3H_2O(l) \\ 1\text{ mol} & 3\tfrac{1}{2}\text{ mol} & 2\text{ mol} & \end{array}$$

3. Write under the amounts the relative volumes of each gas, by applying Avogadro's theory. Since equal volumes contain equal amounts, it follows that equal amounts occupy equal volumes.

$$\begin{array}{llll} C_2H_6(g) + & 3\tfrac{1}{2}O_2(g) \rightarrow & 2CO_2(g) + & 3H_2O(l) \\ 1\text{ mol} & 3\tfrac{1}{2}\text{ mol} & 2\text{ mol} & \\ 1\text{ volume} & 3\tfrac{1}{2}\text{ volumes} & 2\text{ volumes} & \\ V & 3\tfrac{1}{2} \times V & 2 \times V & \end{array}$$

4. Use the volume given in the question to calculate the other volumes.

Volume of ethane = $V = 20\ cm^3$
Volume of oxygen = $3\tfrac{1}{2} \times V = 3\tfrac{1}{2} \times 20\ cm^3 = \mathbf{70\ cm^3}$
Volume of carbon dioxide = $2 \times V = 2 \times 20\ cm^3 = \mathbf{40\ cm^3}$

Now try the following Exercise in the same way.

EXERCISE 5.9
Answers on page 241

In each of the following reactions the volume of some of the gases involved is given. Calculate the volumes of the gases which react and are produced. All gas volumes are measured at the same temperature and pressure.

a $2N_2O(g) \rightarrow \underset{15\ cm^3}{2N_2(g)} + O_2(g)$

b $4NH_3(g) + 3O_2(g) \rightarrow \underbrace{2N_2(g) + 6H_2O(g)}_{128\ cm^3}$

EXERCISE 5.10
Answer on page 241

What is the maximum volume of hydrogen chloride that can be obtained from 75 cm^3 of hydrogen at the same temperature and pressure?

EXERCISE 5.11
Answers on page 241

From the equation given, calculate the maximum volume of ammonia that can be oxidised by 15 cm^3 of oxygen at the same temperature and pressure. Also calculate the maximum volume of the products under the same conditions. What is the percentage change in volume during the reaction?

$$4NH_3(g) + 5O_2(g) \rightarrow 4NO(g) + 6H_2O(g)$$

EXERCISE 5.12
Answers on page 241

78.5 cm^3 of ethanol vapour, C_2H_5OH, at 110 °C was burnt in oxygen to carbon dioxide and water only. Calculate the volume of oxygen used and the volume of the products, all at the same temperature and pressure.

Another application of Avogadro's theory is in a method for finding the formula of a gaseous hydrocarbon by combustion.

Determining the formula of a gaseous hydrocarbon

We can easily measure the volumes of gases involved in the combustion of a hydrocarbon. From these volumes, and the equation for the reaction, we can obtain the formula of the hydrocarbon.

A Worked Example now shows you how to work out a formula from the results of a combustion experiment. Then you can try some problems for yourself.

WORKED EXAMPLE

10 cm^3 of a gaseous hydrocarbon at room temperature was mixed with 100 cm^3 of oxygen (an excess). After sparking the mixture and allowing it to cool to its original temperature the total volume was found to be 95 cm^3. This contracted to 75 cm^3 when in contact with a concentrated solution of sodium hydroxide. What is the formula of the hydrocarbon?

Solution

1. Let the formula of the hydrocarbon be C_xH_y and write an equation for the combustion of one mole of C_xH_y. Clearly, x moles of C will produce x moles of CO_2 and y moles of H will produce $\frac{y}{2}$ moles of H_2O. The amount of oxygen needed to do this will be $x + \frac{y}{4}$ moles of O_2.

$$\therefore C_xH_y(g) + (x + \tfrac{y}{4})O_2(g) \rightarrow xCO_2(g) + \tfrac{y}{2}H_2O(l)$$

2. Calculate the volume of carbon dioxide produced. This is equal to the contraction in volume when the sodium hydroxide absorbs carbon dioxide.

$$\text{Volume of } CO_2 \text{ formed} = 95\ cm^3 - 75\ cm^3 = 20\ cm^3$$

3. Calculate the volume of oxygen used up in the reaction. The final volume, 75 cm^3, must be the unused oxygen.

$$\text{Volume of oxygen used} = 100\ cm^3 - 75\ cm^3 = 25\ cm^3$$

4. Calculate the volume of water vapour (if any) produced. In this example the volumes are measured at room temperature and so the water is liquid and has negligible volume. (In other problems you may have to consider the volume of water vapour.)

5. Write down the volume of each gas under the equation and divide by the volume of hydrocarbon to obtain the relative volumes.

$C_xH_y(g)$ +	$(x + \frac{y}{4})O_2(g)$ →	$xCO_2(g)$ +	$\frac{y}{2}H_2O(l)$
10 cm^3	25 cm^3	20 cm^3	0
1 cm^3	2.5 cm^3	2 cm^3	0

6. Apply Avogadro's theory to write down the relative amount of each gas.

$C_xH_y(g)$ +	$(x + \frac{y}{4})O_2(g)$ →	$xCO_2(g)$ +	$\frac{y}{2}H_2O(l)$
10 cm^3	25 cm^3	20 cm^3	0
1 cm^3	2.5 cm^3	2 cm^3	0
1 mol	2.5 mol	2 mol	

7. Equate these amounts with the coefficients in the equation.

Thus $x = 2$ and $x + \frac{y}{4} = 2.5$
$\therefore y = 2$
∴ The formula is $\mathbf{C_2H_2}$

Alternatively the value of y can be found from the change in volume when combustion occurs (using the scaled down volumes).

$$\begin{aligned}
\text{Decrease in volume} &= \text{initial volume} - \text{final volume} \\
&= (10 + 100 - 95)\ \text{cm}^3/10 = 1.5\ \text{cm}^3 \\
\text{Decrease in amount of gas} &= (1 + x + \tfrac{y}{4}) - x = (1 + \tfrac{y}{4})\ \text{mol} \\
\therefore 1 + \tfrac{y}{4} &= 1.5 \\
\therefore y &= 2
\end{aligned}$$

Note that when volumes are measured at temperatures above 100 °C the volume of water vapour must be taken into account; in this case the decrease in amount of gas works out to be $(1 - \frac{y}{4})$ mol.

Now you should try some problems yourself.

EXERCISE 5.13
Answer on page 241

When 15 cm^3 of a gaseous hydrocarbon was exploded with 60 cm^3 of oxygen (an excess), the final volume was 45 cm^3. This decreased to 15 cm^3 on treatment with sodium hydroxide solution. What was the formula of the hydrocarbon? All measurements were made at the same room temperature and pressure.

EXERCISE 5.14
Answer on page 242

10 cm^3 of a gaseous organic compound X required 45 cm^3 of oxygen for complete combustion, both volumes measured at 120 °C. Which of the following compounds could X be?
A C_3H_8
B C_3H_6
C C_3H_7OH
D C_2H_5CHO

EXERCISE 5.15
Answer on page 242

20 cm^3 of a gaseous hydrocarbon was mixed with 200 cm^3 of oxygen (an excess) and exploded. The final volume, at the same room temperature and pressure, was 150 cm^3, but this was reduced to 70 cm^3 after treatment with a concentrated alkali. Calculate the formula of the hydrocarbon.

EXERCISE 5.16
Answers on page 242

$30\ cm^3$ of a gas, C_3H_x, was mixed with twice the minimum volume of oxygen needed for complete combustion and the mixture was exploded. After cooling to the same room temperature, and at the same pressure, the volume was $90\ cm^3$ less than before. Calculate the volume of oxygen used in the reaction, and the value of x.

In the next section we show how Avogadro's theory leads to an important and useful statement about the volume of one mole of gas.

Molar volume of gases

The converse of Avogadro's theory (i.e. putting it the other way round) is that if we have samples of different gases at the same temperature and pressure and containing the same number of molecules, then they must occupy equal volumes.

If the fixed number of molecules referred to above is the number in one mole, then the corresponding fixed volume is known as the molar volume, V_m. Molar volume, like other volumes, varies with temperature and pressure but its value at s.t.p. is most often used.

An expression which is very useful in calculations involving molar volume is:

$$\textbf{amount of gas} = \frac{\textbf{volume of gas}}{\textbf{molar volume}} \quad \text{or} \quad \boldsymbol{n = \frac{V}{V_m}}$$

where both volumes refer to the same conditions of temperature and pressure. This expression is very similar to the one you have already used often:

$$\text{amount of substance} = \frac{\text{mass}}{\text{molar mass}} \quad \text{or} \quad n = \frac{m}{M}$$

and is easily derived from it by dividing each mass by the appropriate density:

$$n = \frac{m/\rho}{M/\rho} = \frac{V}{V_m}$$

We use this expression in the following Worked Example, which shows you how to calculate molar volume from experimental data for a gas of known formula.

WORKED EXAMPLE

Calculate the molar volume at s.t.p. for carbon dioxide given that 2.50 g occupies $0.450\ dm^3$ at 3.00 atm and 16 °C.

Solution

1. Calculate the volume at s.t.p. as in previous examples using the combined gas law:

$$\frac{p_1V_1}{T_1} = \frac{p_2V_2}{T_2}$$

$$\text{or} \quad V_2 = V_1 \times \frac{p_1}{p_2} \times \frac{T_2}{T_1}$$

$$= 0.450\ dm^3 \times \frac{3.00\ atm}{1.00\ atm} \times \frac{273\ K}{(273 + 16)K} = 1.28\ dm^3$$

2. Calculate the amount, n, of CO_2 using the expression:

$$n = \frac{m}{M}$$

$$\therefore n = \frac{2.50\ g}{44.0\ g\ mol^{-1}} = 0.0568\ mol$$

3. Calculate the molar volume, V_m, using the expression:

$$n = \frac{V}{V_m}$$

$$\text{or} \quad V_m = \frac{V}{n} = \frac{1.28\ \text{dm}^3}{0.0568\ \text{mol}} = \mathbf{22.5\ dm^3\ mol^{-1}}$$

Now you should be able to calculate some molar volumes for yourself.

EXERCISE 5.17
Answers on page 242

Given the following experimental results, calculate the molar volume of each gas at s.t.p.
a 0.122 g of hydrogen, $H_2(g)$, occupies 0.211 dm^3 at 7.00 atm and 20.0 °C.
b 1.10 g of butane, $C_4H_{10}(g)$, occupies 34.4 dm^3 at 600 °C and 30.0 mmHg. (1 atm = 760 mmHg)
c 2.00 g of oxygen, $O_2(g)$, at 5.00 kPa* and 500 K occupies 51.9 dm^3. (1.00 atm = 100 kPa)

You will have noticed that the answers to the Exercise are not equal. Even with the most accurate experimental data, slight differences in molar volumes occur. This is because the gases do not observe the gas laws precisely – only a hypothetical ideal gas does this.

However, we very often assume that gases are ideal, and therefore assume that the molar volume of any gas is a constant. The value we use at s.t.p. is 22.4 $dm^3\ mol^{-1}$, a figure well worth remembering even though it is usually given in examinations when required.

$$\mathbf{V_m(s.t.p.) = 22.4\ dm^3\ mol^{-1}}$$

The molar volume is often used to calculate amounts of gas from volume measurements. In the following Exercises, you can use almost the same method as in the last Worked Example.

EXERCISE 5.18
Answer on page 242

Calculate the amount of gas contained in a 10.0 dm^3 globe at 27 °C and 350 mmHg.

EXERCISE 5.19
Answer on page 243

What is the volume of 8.00 g of oxygen at 23 °C and 0.200 atm?

We do not always take s.t.p. as our reference conditions. You may well have met the statement that the molar volume of any gas at ‘room temperature and pressure’ is 24.0 $dm^3\ mol^{-1}$. The next Exercise refers to this value.

EXERCISE 5.20
Answer on page 243

If the pressure is 1.00 atm, at what temperature is the molar volume of a gas 24.0 $dm^3\ mol^{-1}$?

EXERCISE 5.21
Answers on page 243

Calculate the molar volume of:
a ammonia at s.t.p. given that 0.0780 g occupies 86.1 cm^3 at 1.25 atm and 17 °C,
b methane, CH_4, at 25 °C and 1.00 atm, given that 0.0591 g occupies 90.2 cm^3 at s.t.p.

You should now be able to see the great value of Avogadro’s theory in a variety of useful calculations.

Next we consider a very important relationship, known as the ideal gas equation, which embodies both Avogadro’s theory and the combined gas law.

*Units for pressure are explained in the introduction to this chapter on page 82.

THE IDEAL GAS EQUATION

The ideal gas equation follows from the combined gas law and Avogadro's theory, as we now show. The combined gas law is written:

$$\frac{p_1V_1}{T_1} = \frac{p_2V_2}{T_2} \quad \text{or} \quad \frac{pV}{T} = \text{constant}$$

For one mole of gas, the volume is written as V_m, while the constant is written as R, and is known as the gas constant:

$$\text{i.e. } \frac{pV_m}{T} = R$$

We have already used Avogadro's theory to show that the molar volume, V_m, is the same for all gases under the same conditions. It follows, therefore, that R is the same for all gases. Rearranging the expression, we have:

$$pV_m = RT$$

This applies for one mole of gas, but for n moles it becomes:

$$\boldsymbol{pV = nRT} \quad \text{(the ideal gas equation)}$$

The gas constant, R

The gas constant, R, is a proportionality constant relating p, V, n and T. Its numerical value depends on the units chosen for the variables. T is always expressed in kelvin (K) and n in mol, but a variety of units are in use for p and V. Chemists often express p in atmospheres (atm) and V in dm^3; use these units in the next Exercise.

EXERCISE 5.22
Answer on page 243

Use the fact that one mole of ideal gas occupies 22.4 dm^3 at s.t.p. to calculate a value for R, the gas constant.

In the next Exercise, you calculate some other numerical values for R, using a variety of units. This involves using a conversion factor, as we show in a Worked Example.

WORKED EXAMPLE

R is given as 0.0821 atm dm^3 K^{-1} mol^{-1}. What value should be used if the volume of gas were given in m^3?

Solution

1. Rearrange the expression for R to isolate the unit to be changed.

$$R = 0.0821 \text{ atm } (\text{dm}^3) \text{ K}^{-1} \text{ mol}^{-1}$$

2. Make the appropriate substitution, given that:

$$1.00 \text{ m}^3 = 1.00 \times 10^3 \text{ dm}^3$$

Dividing by 10^3: $1.00 \times 10^{-3}\ m^3 = 1.00\ dm^3$

$$\therefore R = 0.0821\ atm\ (1.00 \times 10^{-3}\ m^3)\ K^{-1}\ mol^{-1}$$
$$= \mathbf{8.21 \times 10^{-5}\ atm\ m^3\ K^{-1}\ mol^{-1}}$$

Alternatively, the substitution can be made during the calculation using the ideal gas equation:

$$pV = nRT$$

$$\therefore R = \frac{pV}{nT} = \frac{1.00\ atm \times 22.4 \times 1.00 \times 10^{-3}\ m^3}{1.00\ mol \times 273\ K}$$

$$= \mathbf{8.21 \times 10^{-5}\ atm\ m^3\ K^{-1}\ mol^{-1}}$$

Note that when a quantity is expressed in a combination of units, those with positive indices are usually written first, in alphabetical order, followed by those with negative indices, also in alphabetical order. However, this is not a hard-and-fast rule.

EXERCISE 5.23

Answers on page 243

Obtain two more values of R, in different units.

a Substitute 1.00 atm = 760 mmHg in the expression:

$$R = 0.0821\ atm\ dm^3\ K^{-1}\ mol^{-1}$$

b Repeat the calculation of R as in Exercise 5.22, using the following conversion factors:

$$1.00\ atm = 1.00\ kPa = 100\ kN\ m^{-2}$$
$$1.00\ N\ m = 1.00\ J$$

The most widely used values for R are $0.0821\ atm\ dm^3\ K^{-1}\ mol^{-1}$* and $8.314\ J\ K^{-1}\ mol^{-1}$. It is most important in calculations that the value for R is consistent with the units of the data you are using. You must choose a value for R in the Exercises which follow, but first we give a Worked Example of a typical calculation involving the ideal gas equation.

WORKED EXAMPLE

Assuming ideal behaviour, calculate the volume occupied by 2.00 g of carbon monoxide at 20 °C under a pressure of $6250\ N\ m^{-2}$.

Solution

1. Calculate the molar mass of carbon monoxide:

$$M = (12.0 + 16.0)\ g\ mol^{-1} = 28.0\ g\ mol^{-1}$$

2. Identify the values to be substituted in the ideal gas equation, $pV = nRT$:

$$p = 6250\ N\ m^{-2}$$
$$n = 2.00\ g/28.0\ g\ mol^{-1} = 0.0714\ mol$$
$$T = (273 + 20)K = 293\ K$$
$$R = 8.314\ J\ K^{-1}\ mol^{-1}$$

(This value for R is appropriate because the pressure is in $N\ m^{-2}$, and 1 J = 1 N m.)

* The atmosphere has been redefined as 1.000 kPa (instead of 1.013 kPa). Thus R should be $0.0831\ atm\ dm^3\ K^{-1}\ mol^{-1}$, but you are still likely to meet the smaller value in this book and elsewhere.

3. Substitute the values in the equation $pV = nRT$ in the form:

$$V = \frac{nRT}{p}$$

$$V = \frac{0.0714 \text{ mol} \times 8.314 \text{ J K}^{-1} \text{ mol}^{-1} \times 293 \text{ K}}{6250 \text{ N m}^{-2}}$$

$$= \frac{0.0714 \times 8.314 \text{ J} \times 293}{6250 \text{ N m}^{-2}}$$

But 1.00 J = 1.00 N m

$$\therefore V = \frac{0.0714 \times 8.314 \text{ N m} \times 293}{6250 \text{ N m}^{-2}}$$

$$= \mathbf{0.0278 \ m^3} \text{ or } \mathbf{27.8 \ dm^3}$$

Note that the units used in the calculation determine the unit of the answer. It would be possible to do the calculation using $R = 0.0821$ atm dm^3 K^{-1} mol^{-1} but the answer would be expressed in a very odd unit of volume (atm dm^5 N^{-1})!

The next two Exercises are straightforward applications of the ideal gas equation.

EXERCISE 5.24

Answers on page 243

a What is the volume of 0.500 mol of sulphur dioxide, SO_2, at s.t.p.?
b What is the volume of 1.50 g of hydrogen, H_2, at 15 °C and a pressure of 750 mmHg?
c At what temperature will 4.71 g of nitrogen occupy 12.0 dm^3 at 760 mmHg?

In the next Exercise, you have to find reacting amounts from a chemical equation before applying the ideal gas equation.

EXERCISE 5.25

Answer on page 243

What mass of zinc is required to produce 2.00 dm^3 of hydrogen at 15 °C and 755 mmHg by reacting with acid?

In the next Exercise, you use the ideal gas equation to get an idea of the pressure and volume changes which occur in an explosion.

An explosion occurs when a chemical reaction produces large volumes of gas very rapidly. The gas cannot expand rapidly enough to avoid a very high pressure immediately around the explosive, and it is the spread of shock waves from this small region of high pressure which causes the noise and damage associated with explosions.

EXERCISE 5.26

Answers on page 244

TNT* is a widely used explosive which has a formula $C_7H_5N_3O_6$. This is mixed with a solid oxidant so that oxygen required for combustion can be supplied rapidly.

a Write an equation for the combustion of TNT. Assume that C and H atoms are completely oxidised and that N atoms emerge as nitrogen gas, N_2.
b What amount of gas is produced from 1.00 mol of TNT, and what volume would it occupy at 1.00 atm and 400 °C?
c Assuming that 1.00 mol of TNT mixed with oxidant occupies 0.500 dm^3, what is the increase in volume expressed as a percentage?
d Assume that the reaction occurs so fast that the gaseous products occupy only 2.00 dm^3 at 600 °C. What would be the resulting pressure?

These calculations could also have been done using a known value for molar volume, but the use of the ideal gas equation is more general.

Now try some further calculations.

*TNT is an abbreviation of the old name **tri**nitro**tol**uene. The modern systematic name is methyl-2,4,6-trinitrobenzene.

EXERCISE 5.27
Answers on page 244

What is the volume of
a 0.250 mol of methane at 17 °C and 750 mmHg?
b 0.521 g of ammonia at 50 °C and 762 mmHg?

EXERCISE 5.28
Answers on page 244

a What volume of carbon dioxide, at 25 °C and 764 mmHg, could be obtained from 7.31 g of calcium carbonate?
b 1.52 g of pure zinc was dissolved in dilute acid and the resulting gas, after drying, occupied 557 cm^3 at 1.01 atm. What was the temperature?
c Excess magnesium was added to 50.0 cm^3 of 0.102 M hydrochloric acid. What volume of dry gas would be produced at 14 °C and 751 mmHg?

EXERCISE 5.29
Answer on page 244

5.00 cm^3 of water is introduced into an evacuated tube of volume 150 cm^3. The tube is sealed and heated to 300 °C. What is the resulting pressure?

One of the most important applications of the ideal gas equation is in the determination of molar mass. This is the subject of the next section.

Determination of molar mass of a gas

The ideal gas equation can be transformed into an equation involving the mass of the gas. You can use this form of the equation to determine the molar masses of a gas and a volatile liquid as in the following Exercises.

EXERCISE 5.30
Answers on page 244

a Write down an expression for the amount, n, of a gas in terms of the mass, m, and the molar mass, M.
b Substitute your expression into the ideal gas equation to obtain an expression for the molar mass.

In the next Exercise, use the expression you have just derived.

EXERCISE 5.31
Answer on page 244

Calculate the molar mass of a substance, given that 3.72 g of its vapour occupies 2.00 dm^3 at 740 mmHg and 100 °C.

EXERCISE 5.32
Answer on page 244

A flask of 109 cm^3 capacity weighed 78.521 g when evacuated and 78.719 g when filled with a gas at 21 °C and 759 mmHg. What was the molar mass of the gas?

EXERCISE 5.33
Answers on page 245

Calculate the molar masses of gases A, B and C from the data given below:

Table 5.2

	Volume of gas/cm^3	Mass of gas/g	Temperature/°C	Pressure
A	257	0.672	25	1.00 atm
B	111	0.128	17	763 mmHg
C	526	1.60	21	103 kPa

(1.00 atm = 100 kPa)

In simple experiments to determine the molar mass of a gas, you are not likely to be able to weigh evacuated flasks. Instead, you deduct the mass of air from the mass of the 'empty' flask. Do this in the following Exercise, which is based on the results of an experiment in an A-level class.

EXERCISE 5.34
Answer on page 245

Calculate the molar mass of carbon dioxide from the experimental results tabulated below.

Table 5.3

Mass of flask filled with air	47.933 g
Mass of flask filled with CO_2	47.998 g
Mass of flask filled with water	152.8 g
Room temperature	26 °C
Atmospheric pressure	757 mmHg
Density of air under conditions of experiment	0.00118 g cm^{-3}

The method is equally applicable to volatile liquids. The experimental procedure is simpler because the mass of the vapour can be obtained by weighing the liquid before it is vaporised.

EXERCISE 5.35
Answer on page 245

0.25 cm^3 of a liquid of density 0.91 g cm^{-3} was injected into a syringe maintained at 160 °C. The liquid vaporised completely and occupied 81 cm^3 at 764 mmHg. What was the molar mass of the liquid?

EXERCISE 5.36
Answers on page 245

Calculate the molar masses of the volatile liquids D, E and F from the data given below:

Table 5.4

	Mass of liquid/g	Volume of vapour	Temperature/°C	Pressure
D	0.184	0.0792 dm^3	100	0.984 atm
E	0.295	121 cm^3	140	747 mmHg
F	0.163	65.0 cm^3	101	105 kPa

(1.00 atm = 100 kPa)

Another empirical gas law that provides us with an alternative method for determining the molar mass of a gas is Graham's Law of effusion.

Graham's Law of effusion

Graham's Law states that the rate of effusion (and the rate of diffusion) of a gas is inversely proportional to the square root of its density. Thus, if the rates of effusion of two different gases at the same conditions of temperature and pressure are compared, we have:

$$\frac{\text{rate}_1}{\text{rate}_2} = \sqrt{\frac{\text{density}_2}{\text{density}_1}}$$

Since the rate of effusion is inversely proportional to the time, t, taken for a fixed volume of gas to escape through a small hole, and the density, ρ, is directly proportional to the molar mass, M, other forms of the law are often used:

$$\text{e.g.} \quad \frac{t_1}{t_2} = \sqrt{\frac{\rho_1}{\rho_2}} = \sqrt{\frac{M_1}{M_2}} \quad \textit{or} \quad \left(\frac{t_1}{t_2}\right)^2 = \frac{\rho_1}{\rho_2} = \frac{M_1}{M_2}$$

Apply the appropriate form of the law in the following Exercises.

EXERCISE 5.37
Answers on page 245

A volumetric flask was filled with hydrogen, H_2, and another with sulphur dioxide, SO_2, for an experiment. The flasks were stoppered, but not perfectly sealed, and put aside for use later.

a Use molar masses to calculate the ratio of the densities of hydrogen and sulphur dioxide under the same conditions.

b Assuming the flasks and the gaps between stoppers and flasks were identical, how much more rapidly would the hydrogen escape (to be replaced by air) than the sulphur dioxide?

EXERCISE 5.38
Answer on page 245

A gas syringe, set up vertically, was filled first with hydrogen and then with carbon monoxide, and the gases were allowed to effuse through a tiny pin-hole under the weight of the piston.

The time taken for 75 cm^3 of hydrogen to escape was 25 seconds, compared with 93 seconds for carbon monoxide. Given the molar mass of hydrogen, calculate the molar mass of carbon monoxide.

EXERCISE 5.39
Answer on page 246

The isotopes of uranium ^{235}U and ^{238}U can be separated by using the different rates of diffusion of the gaseous fluorides UF_6. What is the ratio of these rates of diffusion?

EXERCISE 5.40
Answers on page 246

Oxygen was allowed to escape from a syringe through a small hole under the weight of the piston. It took 47 seconds for 50 cm^3 of oxygen to escape. When the experiment was repeated for another gas, 50 cm^3 took 17 seconds to escape.

a What was the molar mass of the gas?

b How long would the same volume of ammonia take to escape?

EXERCISE 5.41
Answer on page 246

The times taken for the effusion of equal volumes of carbon dioxide and a mixture of carbon dioxide with carbon monoxide were 28 seconds and 24 seconds respectively. Calculate the apparent molar mass of the mixture and the mole fraction of each gas.

■ Dalton's Law of partial pressures

The gas laws you have studied so far relate mainly to pure gases. Now we consider a law which enables us to consider mixtures of gases more fully.

You have learned that the pressure exerted by a gas is due to the combined effect of many collisions between individual molecules and the containing wall. Dalton invented the term 'partial pressure' to refer to the contribution to the total pressure made by the molecules of one particular sort in a mixture of gases.

Dalton's Law of partial pressures states that, in a mixture of gases, the total pressure is the sum of the partial pressures of the component gases. The partial pressure of a component gas is equal to the pressure which it would exert if it *alone* occupied the total volume.

The first Exercise in this section simply tests your understanding of the law.

EXERCISE 5.42
Answer on page 246

A globe contains oxygen at a pressure of 0.30 atm. Hydrogen is admitted until the total pressure is 0.80 atm, and then nitrogen is admitted until the total pressure is 0.90 atm. What is the partial pressure of each gas?

The last Exercise was very simple. You are more likely to meet problems where you have to calculate partial pressures from the amounts of gas present.

To do this, it is convenient to introduce a quantity known as 'mole fraction'. In any mixture of substances (whether gaseous or not) X_A, the mole fraction of A, is given by the expressions:

$$\textbf{mole fraction of A} = \frac{\textbf{amount of A}}{\textbf{total amount}} \quad \text{or} \quad X_A = \frac{n_A}{n}$$

We now derive a simple and useful expression relating partial pressure to mole fraction. Dalton's law tells us that the partial pressure of a gas in a mixture is the same as if that gas alone occupied the entire volume. The ideal gas equation can then be applied to each gas in turn.

For example, in a mixture of three gases, A, B and C, with partial pressures p_A, p_B and p_C, we can write for the gas A:

$$p_A V = n_A RT \quad \text{or} \quad p_A = n_A \times \frac{RT}{V}$$

We can express the quantity $\frac{RT}{V}$ in terms of the total pressure, p, and the total amount of gas, n:

$$pV = nRT \quad \text{or} \quad \frac{RT}{V} = \frac{p}{n}$$

Substituting into the expression for p_A:

$$p_A = n_A \times \frac{RT}{V} = n_A \times \frac{p}{n} = p \times \frac{n_A}{n}$$

i.e. $\mathbf{p_A = p \times X_A}$

In the same way, $p_B = p \times X_B$ and $p_C = p \times X_C$ or:

partial pressure = total pressure × mole fraction

By calculating mole fractions from the equation for the reaction you should be able to do the next Exercise.

EXERCISE 5.43

Answers on page 246

Some ammonia in a syringe is completely decomposed to nitrogen and hydrogen by passing it over heated iron wool. If the total pressure is then 760 mmHg, calculate the partial pressure of each gas.

If you can calculate partial pressures from amounts, then you should also be able to calculate amounts from partial pressures in the next Exercise.

EXERCISE 5.44

Answers on page 246

If the amount of oxygen in the globe referred to in Exercise 5.42 were 0.25 mol, how much of the other gases would be present?

An application of Avogadro's theory enables us to calculate mole fractions and partial pressures from the volumes of gases which make up a mixture.

Another way of stating Avogadro's theory is to say that the volume of a gas is proportional to its amount. Therefore, in a mixture of gases, A, B and C, we can express the mole fraction of A by:

$$X_A = \frac{\text{amount of A}}{\text{total amount}} = \frac{\text{volume of A}}{\text{total volume}}$$

We have already shown that amount is proportional to partial pressure. You will find it helpful therefore to remember these symmetrical relationships.

$$\frac{\textbf{partial pressure of A}}{\textbf{total pressure}} = \frac{\textbf{amount of A}}{\textbf{total amount}} = \frac{\textbf{volume of A}}{\textbf{total volume}}$$

Now try the following Exercises.

EXERCISE 5.45
Answer on page 246

85.0 cm^3 of moist air, at 1.00 atm, was dried carefully and its volume, measured at the same temperature and pressure, fell to 82.0 cm^3. What was the partial pressure of water vapour in the moist air?

EXERCISE 5.46
Answer on page 247
A

A diver should always breathe oxygen at a partial pressure of 2.0×10^4 N m^{-2}. What percentage of oxygen by volume should his breathing mixture contain when he dives to a depth of 40 m and the pressure on him rises to 5.0×10^5 N m^{-2}?

EXERCISE 5.47
Answers on page 247

75 cm^3 of hydrogen, 15 cm^3 of nitrogen, and 90 cm^3 of carbon dioxide, all measured at the same temperature and pressure, were mixed in a vessel. The total pressure was 2.4 atm. Calculate the partial pressure of each gas in the mixture.

EXERCISE 5.48
Answers on page 247

A 500 cm^3 globe contains oxygen at 1.00 atm. 300 cm^3 of nitrogen, measured at the same temperature and pressure, are added under pressure, and then carbon dioxide is added until the total pressure is 3.10 atm. Calculate the partial pressure of each gas in the mixture, and the volume of carbon dioxide used.

EXERCISE 5.49
Answers on page 247

A flask contains 200 cm^3 of oxygen at 1.00 atm, and another flask contains 500 cm^3 of helium at 2.00 atm. The two flasks are connected to allow the gases to mix. What is the partial pressure of each gas in the mixture, and the total pressure?

EXERCISE 5.50
Answers on page 247

A 600 cm^3 vessel contains 0.0232 g of nitrogen and 0.0417 g of carbon monoxide at 23 °C. Calculate the total pressure and the partial pressure of each gas.

EXERCISE 5.51
Answers on page 247

A mixture of 82.2 g of helium and 27.4 g of oxygen was contained in a cylinder at 25.0 atm and 11 °C. What was the volume of the cylinder, and the partial pressure of each gas?

EXERCISE 5.52
Answers on page 247

15 cm^3 of a mixture of carbon monoxide and methane was mixed with excess oxygen and exploded. There was a contraction in volume of 21 cm^3 at the same room temperature and 1.0 atm pressure. Calculate the mole fraction of each gas in the mixture and their partial pressures.

END-OF-CHAPTER QUESTIONS

Answers on page 248
A

5.1 A sample of gas occupied 400 cm^3 at s.t.p. Its volume, in cm^3, at 45 °C and 700 mmHg pressure would be:

A $400 \times \frac{700}{760} \times \frac{273}{318}$

B $400 \times \frac{700}{760} \times \frac{298}{318}$

C $400 \times \frac{760}{700} \times \frac{298}{318}$

D $400 \times \frac{760}{700} \times \frac{318}{298}$

E $400 \times \frac{760}{700} \times \frac{318}{273}$

A

5.2 Which sample of gas contains the **smallest** number of molecules?
A 25 litres of N_2 at 1 atm and 600 K. (1 litre = 1 dm^3)
B 15 litres of CO_2 at 1 atm and 300 K.
C 10 litres of H_2 at 2 atm and 300 K.
D 5 litres of CH_4 at 2 atm and 150 K.
E 5 litres of HCl at 4 atm and 450 K.

A

5.3 The following information refers to 5.60 g of a gas at a constant temperature of 0 °C.

Table 5.5

Pressure /atm	2.25	1.90	1.60	1.12	0.75
Volume /dm^3	2.00	2.35	2.85	4.00	6.00

a Plot a graph of pressure against $\frac{1}{\text{volume}}$.
b Is the gas behaving ideally? Give your reasons.
c Name and state the law which the information illustrates.
d Determine the gradient of the graph and write an expression relating the gradient, temperature and molar gas constant R.
e Hence, given that the molar volume is 22.4 $dm^3\ mol^{-1}$ at s.t.p., calculate the relative molecular mass of the gas.
f Calculate the density of the gas at 13 °C and 0.800 atm.

5.4 0.25 cm^3 of a liquid of density 0.91 g cm^{-3} was injected into a syringe maintained at 160 °C. The liquid vaporised completely and occupied 81 cm^3 at 764 mmHg. What was the molar mass of the liquid?
(1 atm = 760 mmHg, R = 0.0821 atm $dm^3\ K^{-1}\ mol^{-1}$.)

A(P)

5.5 In an experiment to determine the composition of a mixture of the gases propene, C_3H_6, and butene, C_4H_8, a student attempted to find the average relative molecular mass of the hydrocarbon mixture by finding the mass of a certain volume of it. The apparatus used was a standard volumetric flask, inverted and securely clamped in a well ventilated fume cupboard.
The results of the experiment were:

Mass of hydrocarbon mixture = 0.546 g
Volume of hydrocarbon mixture = 300 cm^3
Pressure = 1.00 atmosphere
Temperature = 27 °C

(Molar volume at 0 °C and 1 atm = 22.4 $dm^3\ mol^{-1}$.)

a Calculate the volume of the hydrocarbon mixture at 0 °C and 1.00 atm pressure.
b Calculate the average mass of one mole of the particles in the hydrocarbon mixture.
c From your calculation state which hydrocarbon is more abundant in the mixture.

A(P)

5.6 Using the ideal gas equation, calculate the number of molecules of neon in a bulb of capacity 2.48 dm^3, evacuated to a pressure of 0.001 mmHg at 27 °C.
1 mmHg = $13.60 \times 980.7 \times 10^{-2}$ N m^{-2} $\quad$ R = 8.31 J $K^{-1}\ mol^{-1}$
J = kg $m^2\ s^{-2}$ = N m $\quad$ $L = 6.02 \times 10^{23}\ mol^{-1}$

5.7 20 cm^3 of a gaseous hydrocarbon require 90 cm^3 of oxygen for complete combustion, both volumes being measured under the same conditions. Which one of the following is the hydrocarbon?
A CH_4
B C_2H_2
C C_2H_4
D C_3H_6
E C_3H_8

5.8 50 cm^3 of carbon dioxide at 760 mmHg are mixed with 150 cm^3 of hydrogen at the same pressure. If the resultant pressure of the mixture is also 760 mmHg, which one of the following is the partial pressure of the carbon dioxide in mmHg?
A 760
B 407
C 380
D 253
E 190

5.9 **a** When 0.32 g of liquid bromine is volatilised, the volume of vapour formed, measured at s.t.p., is 45 cm^3. Calculate the relative molecular mass of bromine. S.t.p. is 273 K and 1 atm (i.e. 101 kN m^{-2}), $R = 0.082$ atm $dm^3\ K^{-1}\ mol^{-1}$ (i.e. 8.31 J $K^{-1}\ mol^{-1}$).
b Does the relative molecular mass you have calculated refer to the substance in the liquid or the gaseous state? Explain.

5.10 Gas X has a density of 0.714 g dm^{-3} at 273 K and 101.3 kPa, and diffuses twice as quickly as gas Y under identical conditions.
a Calculate the relative molecular mass of gas X.
b Calculate the relative molecular mass of Y.
(The gas constant $R = 8.31$ J $K^{-1}\ mol^{-1}$.)

5.11 When liquid trichloromethane ($CHCl_3$) is introduced into an evacuated vessel at 50 °C it will evaporate until a vapour pressure of 70.0 kPa is reached.
Calculate the mass of liquid trichloromethane remaining when 4.00 g of liquid trichloromethane is introduced into an evacuated vessel of 1.00 dm^3 capacity at 50.0 °C. Assume that the gaseous state is ideal and that the volume of liquid remaining is very small.
(The gas constant $R = 8.31$ J $K^{-1}\ mol^{-1}$.)

5.12 A sample of gas occupied 400 cm^3 at s.t.p. What would be its volume at 45 °C and 700 mmHg pressure?

5.13 A 25.0 dm^3 cylinder of compressed helium is sufficient to fill a balloon of volume 6.00×10^3 dm^3 at 1.00 atm at the same temperature. What is the pressure in the cylinder?

5.14 32.0 cm^3 of carbon dioxide from a chemical reaction is collected in a syringe at 27 °C and 1.00 atm. What would be its volume at s.t.p.?

5.15 15 cm^3 of a gaseous hydrocarbon at 100 °C was mixed with an excess of oxygen and exploded by sparking. Cooling the products caused a reduction in volume equivalent to 30 cm^3 at the original temperature and pressure. A further reduction of 30 cm^3 occurred on exposure to concentrated alkali. What was the formula of the hydrocarbon?

CHAPTER 6

EQUILIBRIUM I – PRINCIPLES

INTRODUCTION AND PRE-KNOWLEDGE

Virtually all chemical reactions may be regarded as reversible and, in a closed system, a state of equilibrium will be reached in which both forward and reverse reactions are occurring simultaneously, but at equal rates.

Despite the dynamic nature of the equilibrium state, no change can be observed in macroscopic properties, particularly the concentrations of reactants and products. If the concentration of reactants is vanishingly small compared with the products, we say that the reaction has 'gone to completion', whereas if the situation is the opposite we say that the reaction 'does not occur'.

However, for any particular reaction, the relative concentrations of reactants and products may be changed by altering the conditions of temperature and pressure and by the addition of other materials. The qualitative effects of such changes in conditions are described by *Le Chatelier's principle*:

When a system at equilibrium is subjected to a change in conditions, the equilibrium position will shift in such a way as to relieve the immediate effect of the change, if that is possible.

We assume that you have observed some equilibrium shifts experimentally and discussed them qualitatively in terms of Le Chatelier's principle. In order to discuss them quantitatively, we now introduce the equilibrium law.

THE EQUILIBRIUM LAW

The equilibrium law applies to all equilibrium systems and we shall use it as a tool to make precise mathematical predictions and calculations. It was put forward as a result of careful experimental work, but has a firm basis in theoretical thermodynamics.

For any equilibrium system, which we can generalise by the equation

$$aA + bB \rightleftharpoons cC + dD$$

the concentrations of reactants and products are related by the expression:

$$K_c = \frac{[C]^c_{eqm}\,[D]^d_{eqm}}{[A]^a_{eqm}\,[B]^b_{eqm}}$$

K_c is constant at constant temperature.

For reactions involving gases, there is a similar expression for the equilibrium constant K_p (see page 113).

What will probably not be mentioned in your textbook is the fact that although the equilibrium constants K_c and K_p may express the equilibrium condition, they are not truly constant!

For any reversible reaction the **thermodynamic** equilibrium constant K, which **is** constant at constant temperature, is expressed in terms of 'relative activities' instead of concentration or partial pressures.

The 'relative activity' of a substance is a measure of the 'effective concentration' of that substance. In other words, although a substance A in a mixture may have a concentration [A], it may behave as if its concentration were somewhat less than [A], owing to molecular or ionic interactions.

If we restrict ourselves to systems where these interactions are small, such as gases at low pressure, covalent liquids or, in the case of ionic substances, solutions in which ionic concentration is low, then K_p and K_c will be reasonably constant (at constant temperature). Thus for A-level work we may use K_p and K_c as expressions of the equilibrium condition, but, in the light of the above, we should rarely be justified in calculating values to more than two, or at most three, significant figures.

We now show you, in a Worked Example, how to write an equilibrium law expression for a particular equilibrium system. We illustrate the conventions used in writing equilibrium law expressions and show you how to work out the unit of K_c.

WORKED EXAMPLE

a Write an expression for K_c for the reaction:

$$N_2O_4(g) \rightleftharpoons 2NO_2(g)$$

b What is the unit of K_c in this case? Assume the unit of concentration is mol dm^{-3}.

Solution

a By convention, the equilibrium law is written with the products in the numerator and the reactants in the denominator; the index (exponent) of the concentration term is the same as the stoichiometric coefficient in the balanced equation:

$$K_c = \frac{[NO_2(g)]^2_{eqm}}{[N_2O_4(g)]_{eqm}}$$

In future, we assume that the concentrations used in equilibrium law expressions are equilibrium ones and we leave out the subscript 'eqm'. This saves space and looks neater on the page. So the above expression becomes:

$$K_c = \frac{[NO_2(g)]^2}{[N_2O_4(g)]}$$

Warning note!

You would be wise to include the subscripts 'eqm' in your answers to examination questions. One examination board, for example, will award full marks only if such subscripts are included in the answers.

b To work out the unit, assume that:

$$[NO_2(g)] = x \text{ mol dm}^{-3} \quad \text{and} \quad [N_2O_4(g)] = y \text{ mol dm}^{-3}$$

Substitution into the equilibrium law gives:

$$K_c = \frac{(x \text{ mol dm}^{-3})^2}{y \text{ mol dm}^{-3}} = \frac{x^2}{y}\frac{(\text{mol dm}^{-3})^2}{\text{mol dm}^{-3}} = \frac{x^2}{y} \text{ mol dm}^{-3}$$

Now try writing some equilibrium law expressions by doing the next Exercise.

EXERCISE 6.1

Answers on page 248

a For each of the following reactions write an expression for K_c. Work out units for K_c and include them in your answers, assuming that concentrations are measured in mol dm^{-3}.

i) $2HBr(g) \rightleftharpoons H_2(g) + Br_2(g)$
ii) $2SO_2(g) + O_2(g) \rightleftharpoons 2SO_3(g)$
iii) $Cu(NH_3)_4^{2+}(aq) \rightleftharpoons Cu^{2+}(aq) + 4NH_3(aq)$
iv) $2NO(g) + O_2(g) \rightleftharpoons 2NO_2(g)$
v) $4PF_5(g) \rightleftharpoons P_4(g) + 10F_2(g)$
vi) $2NO(g) \rightleftharpoons N_2(g) + O_2(g)$
vii) $C_2H_5OH(l) + CH_3CO_2H(l) \rightleftharpoons CH_3CO_2C_2H_5(l) + H_2O(l)$

b Look at the examples in which K_c has no unit. What do all these reactions have in common?

As soon as you come to consider **numerical values** of K_c, you must remember two important points.

1. A particular value of K_c refers only to one particular temperature.
2. The form of the equilibrium law expression (and therefore the value of K_c) depends on the way the equation is written.

We illustrate these ideas in the following Exercises.

EXERCISE 6.2
Answers on page 248

The same equilibrium system may be represented by two different equations:

$$COCl_2(g) \rightleftharpoons CO(g) + Cl_2(g)$$

$$CO(g) + Cl_2(g) \rightleftharpoons COCl_2(g)$$

a Write expressions for two equilibrium constants, K_c and K_c'.
b What is the mathematical relation between K_c and K_c'?

EXERCISE 6.3
Answers on page 248

The equilibrium between dinitrogen tetroxide and nitrogen dioxide may be represented equally well by two different equations:

$$\tfrac{1}{2}N_2O_4(g) \rightleftharpoons NO_2(g)$$
$$N_2O_4(g) \rightleftharpoons 2NO_2(g)$$

a Write expressions for two equilibrium constants, K_c and K_c'.
b At 100 °C, $K_c' = 0.490$ mol dm^{-3} and at 200 °C, $K_c' = 18.6$ mol dm^{-3}. What are the values of K_c at these temperatures?

Calculating the value of an equilibrium constant

We illustrate the method of calculation by a Worked Example.

WORKED EXAMPLE

Equilibrium was established at 308 K for the system

$$CO(g) + Br_2(g) \rightleftharpoons COBr_2(g)$$

Analysis of the mixture gave the following values of concentration:

$$[CO(g)] = 8.78 \times 10^{-3}\ \text{mol dm}^{-3}$$

$$[Br_2(g)] = 4.90 \times 10^{-3}\ \text{mol dm}^{-3}$$

$$[COBr_2(g)] = 3.40 \times 10^{-3}\ \text{mol dm}^{-3}$$

Calculate the value of the equilibrium constant.

Solution

1. Write the equilibrium law expression for this reaction in terms of concentration:

$$K_c = \frac{[COBr_2(g)]}{[CO\ (g)][Br_2(g)]}$$

2. Substitute the equilibrium concentrations:

$$K_c = \frac{3.40 \times 10^{-3}\ \text{mol dm}^{-3}}{8.78 \times 10^{-3}\ \text{mol dm}^{-3} \times 4.90 \times 10^{-3}\ \text{mol dm}^{-3}}$$

3. Do the arithmetic; cancel units where appropriate.

$$K_c = 79.0\ \text{dm}^3\ \text{mol}^{-1}$$

Now try some similar problems.

EXERCISE 6.4
Answer on page 248

The equilibrium:

$$N_2O_4 \rightleftharpoons 2NO_2$$

can be established in an inert solvent at 298 K. Analysis of an equilibrium mixture gave the concentration of N_2O_4 as 0.021 mol dm^{-3} and the concentration of NO_2 as 0.010 mol dm^{-3}. Calculate the value of the equilibrium constant at 298 K.

EXERCISE 6.5
Answer on page 248

At 250 °C, equilibrium for the following system was established:

$$PCl_5(g) \rightleftharpoons PCl_3(g) + Cl_2(g)$$

Analysis of the mixture showed that:

$$[PCl_3(g)] = 1.50 \times 10^{-2}\ \text{mol dm}^{-3},\ [Cl_2(g)] = 1.50 \times 10^{-2}\ \text{mol dm}^{-3}$$
$$\text{and } [PCl_5(g)] = 1.18\ \text{mol dm}^{-3}.$$

Calculate the value of K_c at this temperature.

EXERCISE 6.6
Answer on page 249

Analysis of the equilibrium system

$$2SO_2(g) + O_2(g) \rightleftharpoons 2SO_3(g)$$

showed that:

$$[SO_2(g)] = 0.23\ \text{mol dm}^{-3},\ [O_2(g)] = 1.37\ \text{mol dm}^{-3}$$
$$[SO_3(g)] = 0.92\ \text{mol dm}^{-3}.$$

Calculate the value of K_c at this temperature.

In some problems, you may have to calculate concentrations from amounts and total volume before you apply the equilibrium law. Try this in the next Exercise.

EXERCISE 6.7
Answers on page 249

Some phosphorus pentachloride was heated at 250 °C in a sealed container until equilibrium was reached according to the equation:

$$PCl_5(g) \rightleftharpoons PCl_3(g) + Cl_2(g)$$

Analysis of the mixture showed that it contained 0.0042 mol of PCl_5, 0.040 mol of PCl_3 and 0.040 mol of Cl_2. The total volume was 2.0 dm^3.

a Calculate the concentration of each component and hence determine the equilibrium constant, K_c.

b The value of K_c for this system is much greater than the one you calculated in Exercise 6.4, and smaller than the one you calculated in Exercise 6.6. What can you say about the relative concentrations of reactants and products when
 i) K_c is very large,
 ii) K_c is very small?

Sometimes you can simplify a calculation by omitting the volume from the equilibrium law expression, as we now show.

When volume can be omitted from an equilibrium law expression

In the next Exercise you show that, for an equilibrium system where the equilibrium constant has no unit, amounts can be used directly in the equilibrium law expression without using the total volume.

EXERCISE 6.8

Answers on page 249

The table below shows the composition of two equilibrium mixtures at 485 °C.

Table 6.1

	Amount of H_2/mol	Amount of I_2/mol	Amount of HI/mol
1	0.02265	0.02840	0.1715
2	0.01699	0.04057	0.1779

a Write the equation for the formation of hydrogen iodide from hydrogen and iodine.
b Write an expression for the equilibrium constant, K_c.
c Calculate a value of K_c for each mixture, assuming the volume of the equilibrium mixture is 1.00 dm^3. Include units in your working.
d For mixture 1, calculate a value for the equilibrium constant, assuming the volume is 2.00 dm^3.
e For mixture 2, calculate a value for the equilibrium constant, assuming the volume is V dm^3. Show that the volume cancels.

Exercise 6.8 should convince you that the volume of the reaction mixture made no difference to the value of the equilibrium constant. We can generalise this for gaseous reactions and reactions in solution: if the equation shows equal numbers of molecules on both sides, then the equilibrium law expression is independent of the volume, and K_c is unitless.

In these circumstances, to calculate an equilibrium constant you can use equilibrium amounts rather than concentrations; or you can include V as an unknown value which cancels as in Exercise 6.8**e**.

In the next Exercise you obtain an equilibrium constant from the results of an experiment in which measured quantities of ethyl ethanoate and water were mixed in the presence of a catalyst, hydrochloric acid, and allowed to reach equilibrium:

$$CH_3CO_2C_2H_5(l) + H_2O(l) \rightleftharpoons C_2H_5OH(l) + CH_3CO_2H(l)$$

(Note that the state symbol (l) is not strictly appropriate since there is only one liquid phase, not four separate ones. Nor is (aq) appropriate, since the liquid phase is not primarily water.)

At equilibrium, the total amount of acid present (CH_3CO_2H + HCl) was determined by titration with a standard alkali. The equilibrium concentrations of all four components of the mixture may then be calculated, leading to a value of K_c. Each step in the calculation is very simple, but you may find the flow chart in Fig. 6.1 helps you to see how the steps fit together.

The quantities given directly from experimental measurements are listed on the left-hand side and those required for the equilibrium constant are on the right. Intermediate calculations are indicated by the arrows. All the steps shown are necessary but you can choose, to some extent, the order in which you do them.

Figure 6.1

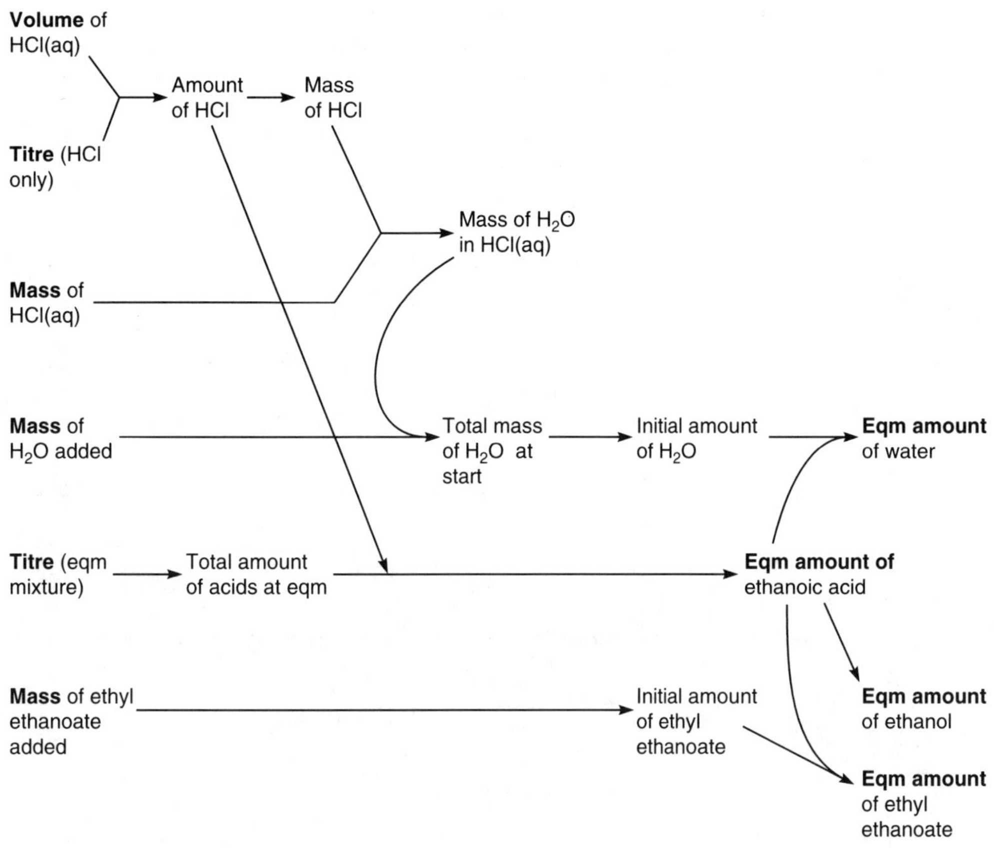

EXERCISE 6.9

Answer on page 249

The following substances, in the quantities shown, were mixed in a specimen tube and allowed to reach equilibrium.

3.64 g of ethyl ethanoate, $CH_3CO_2C_2H_5$
0.99 g of water, H_2O
5.17 g of 2.00 M hydrochloric acid, HCl(aq)

Precisely the same mass of the hydrochloric acid was found to require 10.50 cm^3 of 0.974 M NaOH for neutralisation. The equilibrium mixture required 39.20 cm^3 of the same alkali for neutralisation. Calculate the equilibrium constant, K_c.

■ Various numerical problems involving equilibrium constants

We illustrate different types of problem by means of Worked Examples and Exercises.

WORKED EXAMPLE

For the equilibrium:

$$PCl_5(g) \rightleftharpoons PCl_3(g) + Cl_2(g)$$

$$K_c = 0.19 \text{ mol dm}^{-3} \text{ at } 250\ °C.$$

One equilibrium mixture at this temperature contains PCl_5 at a concentration of 0.20 mol dm^{-3} and PCl_3 at a concentration of 0.010 mol dm^{-3}.

Calculate the concentration of Cl_2 in this mixture.

Solution 1. Start by writing the equation (even if it is given in the question). Leave space where the equilibrium concentrations can be tabulated under the formulae of the compounds:

$$PCl_5(g) \rightleftharpoons PCl_3(g) + Cl_2(g)$$

Equilibrium
concn./mol dm^{-3}

2. Indicate the equilibrium concentrations under the equation. Let the concentration of chlorine, $[Cl_2(g)]$, be x mol dm^{-3}.

	$PCl_5(g)$	$\rightleftharpoons$ $PCl_3(g)$	+ $Cl_2(g)$
Equilibrium concn./mol dm^{3}	0.20	0.010	x

3. Write the equilibrium law expression in terms of concentration:

$$K_c = \frac{[PCl_3(g)]\,[Cl_2(g)]}{[PCl_5(g)]}$$

4. Substitute all values into the expression (remember to include units):

$$0.19 \text{ mol dm}^{-3} = \frac{0.010 \text{ mol dm}^{-3} \times x \text{ mol dm}^{-3}}{0.20 \text{ mol dm}^{-3}}$$

5. Solve the equation for x:

$$x = \frac{0.19 \times 0.20}{0.010} = 3.8$$

$$\therefore\ [Cl_2(g)] = 3.8 \text{ mol dm}^{-3}$$

Now do the following Exercises.

EXERCISE 6.10
Answer on page 249

Some N_2O_4 dissolved in trichloromethane was allowed to reach equilibrium at a known temperature:

$$N_2O_4 \rightleftharpoons 2NO_2$$

At this point the concentration of NO_2 was 1.85×10^{-3} mol dm^{-3}.
What was the equilibrium concentration of N_2O_4? $K_c = 1.06 \times 10^{-5}$ mol dm^{-3} at this temperature.

EXERCISE 6.11
Answer on page 249

At 1400 K, $K_c = 2.25 \times 10^{-4}$ mol dm^{-3} for the equilibrium:

$$2H_2S(g) \rightleftharpoons 2H_2(g) + S_2(g)$$

In an equilibrium mixture, $[H_2S(g)] = 4.84 \times 10^{-3}$ mol dm^{-3} and $[S_2(g)] = 2.33 \times 10^{-3}$ mol dm^{-3}. Calculate the equilibrium concentration of H_2.

EXERCISE 6.12
Answer on page 250

In the following equilibrium:

$$H_2(g) + I_2(g) \rightleftharpoons 2HI(g)$$

$K_c = 54.1$, at a particular temperature. The equilibrium mixture was found to contain H_2 at a concentration of 0.48×10^{-3} mol dm^{-3} and HI at a concentration of 3.53×10^{-3} mol dm^{-3}.What is the equilibrium concentration of I_2?

EXERCISE 6.13
Answer on page 250

For the equilibrium:

$$C_5H_{10}(l) + CH_3CO_2H(l) \rightleftharpoons CH_3CO_2C_5H_{11}(l)$$
(pentene) (ethanoic acid) (pentyl ethanoate)

$K_c = 540\ dm^3\ mol^{-1}$ at a certain temperature. An equilibrium mixture at this temperature contains 5.66×10^{-3} mol dm^{-3} of pentene and 2.55×10^{-3} mol dm^{-3} of ethanoic acid. Calculate the concentration of pentyl ethanoate in the mixture.

EXERCISE 6.14
Answer on page 250

This question concerns the equilibrium system:

$$CH_3CO_2H(l) + C_2H_5OH(l) \rightleftharpoons CH_3CO_2C_2H_5(l) + H_2O(l);\ K_c = 4.0 \text{ at } 25\ °C$$

In a particular experiment, 0.33 mol of CH_3CO_2H, 0.66 mol of $CH_3CO_2C_2H_5$ and 0.66 mol of H_2O are found to be present. What amount of C_2H_5OH is present?

The next Exercise is similar, but requires an extra step at the start; i.e. calculating concentrations from amounts and total volume.

EXERCISE 6.15
Answers on page 250

In another equilibrium mixture of the reaction:

$$PCl_5(g) \rightleftharpoons PCl_3(g) + Cl_2(g)$$

at 250 °C in a 2.0 dm^3 vessel, there is 0.15 mol of PCl_3 and 0.090 mol of Cl_2.
$K_c = 0.19$ mol dm^{-3} at 250 °C.
a Calculate the amount of PCl_5 present at equilibrium.
b Calculate the mass of PCl_5 present at equilibrium.

Don't forget that if the total number of molecules does not change during a homogeneous reaction, the volume doesn't make any difference to the calculation and you need not include it (see page 104).

Now we show you how to apply the equilibrium law to a slightly different type of problem.

WORKED EXAMPLE

6.75 g of SO_2Cl_2 was put into a 2.00 dm^3 vessel, the vessel was sealed and its temperature raised to 375 °C. At equilibrium, the vessel contained 0.0345 mol of Cl_2. Calculate the equilibrium constant for the reaction:

$$SO_2Cl_2(g) \rightleftharpoons SO_2(g) + Cl_2(g)$$

Solution

1. First, note that the number of molecules **does** change, so you will have to calculate concentrations from amounts.
2. Write the balanced equation, leaving room for initial concentrations above and equilibrium concentrations below:

Initial
concn./mol dm^{-3}

$$SO_2Cl_2(g) \rightleftharpoons SO_2(g) + Cl_2(g)$$

Equilibrium
concn./mol dm^{-3}

3. Calculate the initial concentrations and put the values above the appropriate formulae in the equation.

$$\text{The amount, } n\text{, of } SO_2Cl_2 = \frac{m}{M} = \frac{6.75\text{ g}}{135\text{ g mol}^{-1}} = 0.0500\text{ mol}$$

$$c = \frac{n}{V}$$

$$\therefore\ [SO_2Cl_2(g)] = \frac{0.0500\text{ mol}}{2.00\text{ dm}^3} = 0.0250\text{ mol dm}^{-3}$$

The initial concentrations of SO_2 and Cl_2 are both zero. Write down the initial concentrations:

	$SO_2Cl_2(g)$	$\rightleftharpoons$	$SO_2(g)$	+	$Cl_2(g)$
Initial concn./mol dm^{-3}	0.025		0		0
Equilibrium concn./mol dm^{-3}					

4. Now work out the equilibrium concentrations from the data and balanced equation.
   a There is 0.0345 mol of Cl_2 present at equilibrium so there must also be 0.0345 mol of SO_2 present and 0.0345 mol of SO_2Cl_2 must have been used up. The concentrations are:

$$[Cl_2(g)] = \frac{0.0345\text{ mol}}{2.00\text{ dm}^3} = 0.0173\text{ mol dm}^{-3}$$

$$[SO_2(g)] = \frac{0.0345\text{ mol}}{2.00\text{ dm}^3} = 0.0173\text{ mol dm}^{-3}$$

   b The **amount** of SO_2Cl_2 left = initial amount – amount reacted

$$= (0.0500 - 0.0345)\text{ mol} = 0.0155\text{ mol}$$

$$\therefore\ [SO_2Cl_2(g)] = \frac{0.0155\text{ mol}}{2.00\text{ dm}^3} = 0.00775\text{ mol dm}^{-3}$$

   Write these equilibrium concentrations under the equation:

	$SO_2Cl_2(g)$	$\rightleftharpoons$	$SO_2(g)$	+	$Cl_2(g)$
Initial concn./mol dm^{-3}	0.025		0		0
Equilibrium concn./mol dm^{-3}	0.00775		0.0173		0.0173

5. From here proceed as in the Worked Example on page 102. Write the equilibrium law expression and substitute the equilibrium concentrations:

$$K_c = \frac{[SO_2(g)]\,[Cl_2(g)]}{[SO_2Cl_2(g)]}$$

$$= \frac{0.0173\text{ mol dm}^{-3} \times 0.0173\text{ mol dm}^{-3}}{0.00775\text{ mol dm}^{-3}} = \mathbf{0.0386\ mol\ dm^{-3}}$$

Now try the following six Exercises, two of which are from A-level questions.

EXERCISE 6.16
Answer on page 250

Ethanoic acid, CH_3CO_2H, and pentene, C_5H_{10}, react to produce pentyl ethanoate in an inert solvent. A solution was prepared containing 0.020 mol of pentene and 0.010 mol of ethanoic acid in 600 cm^3 of solution. At equilibrium there was 9.0×10^{-3} mol of pentyl ethanoate. Calculate the value of K_c from these data.

$$CH_3CO_2H + C_5H_{10} \rightleftharpoons CH_3CO_2C_5H_{11}$$

EXERCISE 6.17
Answer on page 250

A mixture of 1.90 mol of hydrogen and 1.90 mol of iodine was allowed to reach equilibrium at 710 K. The equilibrium mixture was found to contain 3.00 mol of hydrogen iodide. Calculate the equilibrium constant at 710 K for the reaction:

$$H_2(g) + I_2(g) \rightleftharpoons 2HI(g)$$

EXERCISE 6.18
Answer on page 250

If a mixture of 6.0 g of ethanoic acid and 6.9 g of ethanol is allowed to reach equilibrium, 7.0 g of ethyl ethanoate is formed. Calculate K_c.

$$CH_3CO_2H + C_2H_5OH \rightleftharpoons CH_3CO_2C_2H_5 + H_2O$$

EXERCISE 6.19
Answer on page 251

In the following equilibrium system:

$$H_2(g) + I_2(g) \rightleftharpoons 2HI(g)$$

20.57 mol of hydrogen and 5.22 mol of iodine were allowed to reach equilibrium at 450 °C. At this point, the mixture contained 10.22 mol of hydrogen iodide. Calculate the value of K_c at this temperature.

EXERCISE 6.20
Answer on page 251

The equilibrium:

$$N_2O_4(l) \rightleftharpoons 2NO_2(l)$$

was established in a solvent at 10 °C starting with 0.1307 mol dm^{-3} of N_2O_4 ; the equilibrium mixture was found to contain 0.0014 mol dm^{-3} of NO_2. Calculate the value of K_c at this temperature.

EXERCISE 6.21
Answer on page 251

In the following equilibrium:

$$C_2H_5OH(l) + CH_3CO_2H(l) \rightleftharpoons CH_3CO_2C_2H_5(l) + H_2O(l)$$

2.0 mol of ethanol and 1.0 mol of ethanoic acid were allowed to react to equilibrium at 25 °C. At equilibrium the mixture contained 0.845 mol of ethyl ethanoate and the total volume was 30 cm^3. Calculate the value of K_c at this temperature.

The type of problem we present in the next section often requires the solution of a quadratic equation, although it is not always possible to tell just by reading the question. The chemistry is usually straightforward, but if you find the mathematics difficult, remember that in an examination you will secure a good proportion of the marks simply by applying the correct principle without actually solving the equation at the last step. In any case, some examining boards rarely set problems which require the solution of quadratic equations.

Problems involving quadratic equations

The general form of a quadratic equation is:

$$ax^2 + bx + c = 0$$

where a, b and c are constants. The solution of this equation is given by the formula:

$$x = \frac{-b \pm \sqrt{b^2 - 4ac}}{2a}$$

and it is easy to solve this using a calculator.

WORKED EXAMPLE Carbon monoxide and chlorine react to form phosgene, $COCl_2$. A mixture was prepared containing 0.20 mol of CO and 0.10 mol of Cl_2 in a 3.0 dm^3 vessel. At the temperature of the experiment $K_c = 0.410\ dm^3\ mol^{-1}$. Calculate the concentration of $COCl_2$ at equilibrium.

Solution

1. As in almost all these calculations, write the equation, leaving space above and below for initial and equilibrium conditions:

Initial amount/mol	
	$CO(g) + Cl_2(g) \rightleftharpoons COCl_2(g)$
Equilibrium amount/mol	

2. In this case it is convenient to work first in terms of initial and equilibrium amounts, and then divide them by the volume later when substituting concentrations into the equilibrium law. Indicate the initial amounts:

Initial amount/mol	0.20	0.10	0
	$CO(g)$ +	$Cl_2(g)$ $\rightleftharpoons$	$COCl_2(g)$
Equilibrium amount/mol			

3. Calculate the amounts at equilibrium in terms of one unknown amount.
   a Let the amount of $COCl_2$ formed be x mol.
   b The balanced equation tells us that at equilibrium:

$$\text{amount of CO} = \text{initial amount} - \text{amount reacted}$$
$$= (0.20 - x)\ \text{mol}$$

   c By similar reasoning, at equilibrium:

$$\text{amount of } Cl_2 = (0.10 - x)\ \text{mol}$$

4. Put these equilibrium amounts in the appropriate spaces under the equation:

Initial amount/mol	0.20	0.10	0
	$CO(g)$ +	$Cl_2(g)$ $\rightleftharpoons$	$COCl_2(g)$
Equilibrium amount/mol	$(0.20 - x)$	$(0.10 - x)$	x

5. Write the equilibrium law in terms of concentrations:

$$K_c = \frac{[COCl_2(g)]}{[CO(g)]\,[Cl_2(g)]}$$

6. Put the equilibrium amounts in terms of concentrations by dividing by the volume; substitute these into the equilibrium law:

$$[COCl_2(g)] = \frac{x}{3.0}\ \text{mol dm}^{-3}$$

$$[CO(g)] = \frac{0.20 - x}{3.0}\ \text{mol dm}^{-3}$$

$$[Cl_2(g)] = \frac{0.10 - x}{3.0}\ \text{mol dm}^{-3}$$

$$0.410\ \text{dm}^3\ \text{mol}^{-1} = \frac{\frac{x}{3.0}\ \text{mol dm}^{-3}}{\frac{0.20 - x}{3.0}\ \text{mol dm}^{-3} \times \frac{0.10 - x}{3.0}\ \text{mol dm}^{-3}}$$

7. The equation becomes:

$$0.410 = \frac{\frac{x}{3.0}}{\frac{0.20 - x}{3.0} \times \frac{0.10 - x}{3.0}}$$

Simplify this equation:

$$0.410 = \frac{3x}{(0.20 - x) \times (0.10 - x)} = \frac{3x}{0.020 - 0.30x + x^2}$$

$$\text{or } 0.0082 - 0.123x + 0.410x^2 = 3x$$

Put this quadratic equation in the standard form:

$$0.410x^2 - 3.123x + 0.0082 = 0$$

8. Solve this quadratic equation. The solutions of this equation are:

$$x = \frac{-b \pm \sqrt{b^2 - 4ac}}{2a}$$

for which $a = 0.410$, $b = -3.123$, $c = 0.0082$.

$$x = \frac{3.123 \pm \sqrt{(3.123)^2 - (4 \times 0.410 \times 0.0082)}}{(2 \times 0.410)}$$

$$= \frac{3.123 \pm \sqrt{9.7531 - 0.0134}}{0.820} = \frac{3.123 \pm \sqrt{9.7397}}{0.820} = \frac{3.123 \pm 3.121}{0.820}$$

$$\therefore x = 7.6 \text{ or } x = 2.4 \times 10^{-3}$$

Solving a quadratic equation produces two roots. Because the system began with 0.20 mol of CO and 0.10 mol of Cl_2, it is impossible to produce 7.6 mol of $COCl_2$. Therefore, the correct root is $x = 2.4 \times 10^{-3}$.

9. The amount of $COCl_2$ at equilibrium is 2.4×10^{-3} mol.

$$\text{Now, concentration} = \frac{\text{amount}}{\text{volume}}$$

$$\therefore\ [COCl_2(g)] = \frac{2.4 \times 10^{-3}\ \text{mol}}{3.0\ \text{dm}^3} = 8.0 \times 10^{-4}\ \text{mol dm}^{-3}$$

Now try the following Exercises, some of which require the solution of a quadratic equation.

EXERCISE 6.22

Answer on page 251

Carbon monoxide will react with steam under the appropriate conditions according to the following reversible reaction:

$$CO(g) + H_2O(g) \rightleftharpoons CO_2(g) + H_2(g);\quad \Delta H^\circ = -40\ \text{kJ mol}^{-1}$$

Calculate the number of moles of hydrogen in the equilibrium mixture when three moles of carbon monoxide and three moles of steam are placed in a reaction vessel of constant volume and maintained at a temperature at which the equilibrium constant has a numerical value of 4.00.

EXERCISE 6.23

Answers on page 251

For the equilibrium:

$$PCl_5(g) \rightleftharpoons PCl_3(g) + Cl_2(g)$$

$K_c = 0.19$ mol dm^{-3} at 250 °C. 2.085 g of PCl_5 was heated to 250 °C in a sealed vessel of 500 cm^3 capacity and maintained at this temperature until equilibrium was established. Calculate the concentrations of PCl_5, PCl_3 and Cl_2 at equilibrium.

EXERCISE 6.24

Answer on page 252

For the reaction:

$$H_2(g) + I_2(g) \rightleftharpoons 2HI(g)$$

the equilibrium constant = 49.0 at 444 °C. If 2.00 mol of hydrogen and 2.00 mol of iodine are heated in a closed vessel at 444 °C until equilibrium is attained, calculate the composition in moles of the equilibrium mixture.

EXERCISE 6.25

Answer on page 252

In the following equilibrium:

$$CH_3CO_2H(l) + C_2H_5OH(l) \rightleftharpoons CH_3CO_2C_2H_5(l) + H_2O(l)$$

8.0 mol of ethanoic acid and 6.0 mol of ethanol were placed in a 2.00 dm^3 vessel. What is the equilibrium amount of water? K_c at this temperature = 4.5.

EXERCISE 6.26

Answers on page 252

In the following equilibrium:

$$2HI(g) \rightleftharpoons H_2(g) + I_2(g)$$

1.0 mol of I_2 and 2.0 mol of H_2 are allowed to react in a 1.0 dm^3 vessel at 440 °C. What are the equilibrium concentrations of HI, H_2 and I_2 at this temperature, given that $K_c = 0.02$ at 440 °C?

EXERCISE 6.27
Answer on page 252

4.0 g of phosphorus pentachloride is allowed to reach equilibrium at 250 °C in a 750 cm^3 vessel. $K_c = 0.19$ mol dm^{-3} for the following reaction at 250 °C:

$$PCl_5(g) \rightleftharpoons PCl_3(g) + Cl_2(g)$$

Calculate the amount of phosphorus pentachloride at equilibrium.

So far in this chapter we have expressed the equilibrium law only in terms of the equilibrium **concentrations** of reactants and products. However, for gaseous systems, it is often convenient to use partial pressures rather than concentrations. The equilibrium law still applies, but gives a different equilibrium constant, K_p, which we now consider.

THE EQUILIBRIUM CONSTANT, K_p

For equilibrium involving gases, the equilibrium law can be expressed in terms of partial pressures because the partial pressure of each gas (i.e. its contribution to the total pressure) is proportional to its concentration. The equilibrium constant is then given the symbol K_p.

EXERCISE 6.28
Answers on page 253

Write an expression for K_p for each of the following equilibrium systems. Assume all pressures are measured in atmospheres (atm), and include units for K_p.

a $2NH_3(g) \rightleftharpoons N_2(g) + 3H_2(g)$
b $2SO_2(g) + O_2(g) \rightleftharpoons 2SO_3(g)$
c $C(s) + CO_2(g) \rightleftharpoons 2CO(g)$
d $CaCO_3(s) \rightleftharpoons CaO(s) + CO_2(g)$
e $NH_4HS(s) \rightleftharpoons NH_3(g) + H_2S(g)$

To calculate K_p, we simply substitute equilibrium partial pressures into the equilibrium law expression. It is just like substituting equilibrium concentrations to calculate K_c. To see if you can do this, try the following Exercises.

EXERCISE 6.29
Answer on page 253

In the equilibrium system:

$$2SO_2(g) + O_2(g) \rightleftharpoons 2SO_3(g)$$

at 700 K, the partial pressures of the gases in an equilibrium mixture are:

$$p_{SO_2} = 0.090 \text{ atm}, \ p_{SO_3} = 4.5 \text{ atm}, \ p_{O_2} = 0.083 \text{ atm}.$$

Calculate K_p for this system.

EXERCISE 6.30
Answer on page 253

In the equilibrium system:

$$N_2O_4(g) \rightleftharpoons 2NO_2(g)$$

at 55 °C, the partial pressures of the gases are $p_{N_2O_4} = 0.33$ atm and $p_{NO_2} = 0.67$ atm. Calculate K_p for this system.

EXERCISE 6.31
Answer on page 253

In the following equilibrium system:

$$2SO_2(g) + O_2(g) \rightleftharpoons 2SO_3(g)$$

at a certain temperature, the partial pressures of the gases are $p_{SO_2} = 2.3$ atm, $p_{O_2} = 4.5$ atm, $p_{SO_3} = 2.3$ atm. Calculate K_p for this system.

EXERCISE 6.32
Answer on page 253

Consider the following reaction:

$$H_2(g) + I_2(g) \rightleftharpoons 2HI(g)$$

At a certain temperature, analysis of the equilibrium mixture of the gases yielded the following results: $p_{H_2} = 0.25$ atm, $p_{I_2} = 0.16$ atm, $p_{HI} = 0.40$ atm. Calculate K_p for this reaction at the same temperature.

In the next Exercise, you have to calculate the partial pressures before substituting in the equilibrium law expression. Remember that:

partial pressure = mole fraction × total pressure

$$\textbf{mole fraction} = \frac{\textbf{amount of one component}}{\textbf{total amount of all components}}$$

EXERCISE 6.33
Answers on page 253

Analysis of the equilibrium system:

$$N_2(g) + 3H_2(g) \rightleftharpoons 2NH_3(g)$$

showed 25.1 g of NH_3, 12.8 g of H_2 and 59.6 g of N_2.

a Calculate the mole fraction of each gas.

b The total pressure of the system was 10.0 atm. Calculate the partial pressure of each gas.

c Calculate K_p for the system.

The following Exercises are similar.

EXERCISE 6.34
Answer on page 253

Phosphorus pentachloride dissociates on heating according to the equation:

$$PCl_5(g) \rightleftharpoons PCl_3(g) + Cl_2(g)$$

At a pressure of 10.0 atm and a temperature of 250 °C the amount of each gas present at equilibrium was: 0.33 mol of PCl_5, 0.67 mol of PCl_3 and 0.67 mol of Cl_2. Calculate the value of K_p.

EXERCISE 6.35
Answer on page 253

Ammonium aminomethanoate, $NH_2CO_2NH_4$, decomposes according to the equation:

$$NH_2CO_2NH_4(s) \rightleftharpoons 2NH_3(g) + CO_2(g)$$

In a particular system at 293 K, there was 0.224 mol of CO_2 and 0.142 mol of NH_3 at a total pressure of 1.83 atm. Calculate K_p for this system.

EXERCISE 6.36
Answer on page 254

In the following reaction:

$$CO_2(g) + H_2(g) \rightleftharpoons CO(g) + H_2O(g)$$

at a temperature of 100 °C and a pressure of 2.0 atm, the amounts of each gas present at equilibrium were 6.2×10^{-3} mol of CO, 6.2×10^{-3} mol of H_2O, 0.994 mol of CO_2 and 0.994 mol of H_2. Calculate the value of K_p.

EXERCISE 6.37
Answer on page 254

An equilibrium mixture contained 0.40 mol of I atoms and 0.60 mol of I_2 molecules. The total pressure was 1.0 atm. Calculate the value of K_p at this temperature.

$$I_2(g) \rightleftharpoons 2I(g)$$

EXERCISE 6.38
Answer on page 254

In the following reaction:

$$H_2(g) + I_2(g) \rightleftharpoons 2HI(g)$$

at a temperature of 485 °C and a pressure of 2.00 atm, the amount of each gas present at equilibrium was 0.56 mol of H_2, 0.060 mol of I_2 and 1.27 mol of HI. Calculate K_p at this temperature.

K_p and K_c generally have different values and units (look, for example, at Exercises 6.7 and 6.34), but the two quantities are related because partial pressure is proportional to concentration. Ask your teacher if your syllabus requires you to know this relationship; if not, skip the next section.

The relationship between K_p and K_c

You will derive the relationship by means of a Revealing Exercise which concerns the equilibrium system:

$$N_2O_4(g) \rightleftharpoons 2NO_2(g)$$

REVEALING EXERCISE

Q1 Write expressions for the amounts of the two gases, n_{NO_2} and $n_{N_2O_4}$, in terms of $[NO_2(g)]$, $[N_2O_4(g)]$ and the total volume V.

A1 $n_{NO_2} = [NO_2(g)]\ V$, $n_{N_2O_4} = [N_2O_4(g)]\ V$.

Q2 Use the ideal gas equation to relate the amounts, n_{NO_2} and $n_{N_2O_4}$, to partial pressures.

A2 $p_{NO_2}V = n_{NO_2}RT$, $p_{N_2O_4}V = n_{N_2O_4}RT$.

Q3 Combine **A1** and **A2** to write expressions for p_{NO_2} and $p_{N_2O_4}$ in terms of $[NO_2(g)]$ and $[N_2O_4(g)]$.

A3 $p_{NO_2} = [NO_2(g)]\ RT$, $p_{N_2O_4} = [N_2O_4(g)]\ RT$.

Q4 Write an expression for K_p and substitute partial pressures from **A3**.

A4 $$K_p = \frac{p^2_{NO_2}}{p_{N_2O_4}} = \frac{[NO_2(g)]^2\ (RT)^2}{[N_2O_4(g)]\ RT}$$

Q5 Use **A4** to write an expression for K_p in terms of K_c.

A5 $K_p = K_cRT$.

In general,

$$K_p = K_c(RT^{\Delta n})$$

where Δn = number of moles of gaseous products – number of moles of gaseous reactants.

Use this expression to do the following Exercises. (Choose the value of R which has the most appropriate units.)

EXERCISE 6.39
Answer on page 254

For the equilibrium system:

$$PCl_5(g) \rightleftharpoons PCl_3(g) + Cl_2(g)$$

K_p = 0.811 atm at 523 K. Calculate K_c at this temperature.

EXERCISE 6.40
Answer on page 254

In the equilibrium system:

$$N_2(g) + 3H_2(g) \rightleftharpoons 2NH_3(g)$$

K_c = 2.0 dm^6 mol^{-2} at 620 K. Calculate K_p at this temperature.

EXERCISE 6.41
Answer on page 254

For the system:

$$H_2(g) + I_2(g) \rightleftharpoons 2HI(g)$$

show that $K_p = K_c$ at any temperature.

Now that you can write an expression for K_p, we show you how to use it to consider the effect on a system of changing the total pressure.

The effect on an equilibrium system of changing pressure

We assume that you can apply Le Chatelier's principle to determine the **direction** of any shift in equilibrium position caused by changing the total pressure. Now you can apply the equilibrium law to show why Le Chatelier's principle works in relation to pressure changes. You can also calculate the pressure change required to alter the equilibrium concentrations to a given extent.

We show you how to achieve these objectives by means of a Revealing Exercise, followed by a Worked Example, both of which concern the equilibrium system:

$$2SO_2(g) + O_2(g) \rightleftharpoons 2SO_3(g)$$

REVEALING EXERCISE

Q1 Write the equilibrium law expression for the system $2SO_2(g) + O_2(g) \rightleftharpoons 2SO_3(g)$.

A1

$$K_p = \frac{p^2_{SO_3}}{p^2_{SO_2} \times p_{O_2}}$$

Q2 Write an expression for the partial pressure of each gas in terms of its mole fraction X and the total pressure p_T.

A2 $p_{SO_3} = X_{SO_3}p_T$, $p_{SO_2} = X_{SO_2}p_T$, $p_{O_2} = X_{O_2}p_T$

Q3 Substitute your answers to **Q2** into the equilibrium law expression for K_p and simplify the equation.

A3

$$K_p = \frac{(X_{SO_3}p_T)^2}{(X_{SO_2}p_T)^2 \times (X_{O_2}p_T)} = \frac{1}{p_T} \times \frac{X^2_{SO_3}}{X^2_{SO_2} \times X_{O_2}}$$

Q4 If the pressure of the system is doubled (at constant temperature), which mole fractions must increase and which must decrease? (The new values must give the same value of K_p as before.)

A4 The mole fraction of SO_3 must increase and the mole fractions of SO_2 and O_2 must decrease.

Q5 In which direction must the equilibrium position shift in order to change the mole fractions? Is this in accordance with Le Chatelier's principle?

A5 The equilibrium shifts to the right, as predicted by Le Chatelier's principle.

The same method could be applied to any gaseous equilibrium system. We now show you, in a Worked Example, how to use the expression you derived in **A3** of the Revealing Exercise.

WORKED EXAMPLE At 1100 K, $K_p = 0.13$ atm^{-1} for the system:

$$2SO_2(g) + O_2(g) \rightleftharpoons 2SO_3(g)$$

If 2.00 mol of SO_2 and 2.00 mol of O_2 are mixed and allowed to react, what must the total pressure be to give a 90% yield of SO_3?

Solution

1. Write the balanced equation, leaving room above and below to put in initial amounts and equilibrium amounts:

Initial amount/mol	
	$2SO_2(g) + O_2(g) \rightleftharpoons 2SO_3(g)$
Equilibrium amount/mol	

2. Put in the initial amounts:

Initial amount/mol	2.00	2.00	0
	$2SO_2(g)$ +	$O_2(g)$ ⇌	$2SO_3(g)$
Equilibrium amount/mol			

3. Calculate the amounts at equilibrium:
   a The amount of SO_3 produced = 2.00 mol × (90/100) = 1.80 mol.
   b The amount of SO_2 left at equilibrium
   = initial amount – amount reacted
   = (2.00 – 1.80) mol = 0.20 mol.
   c Amount of oxygen reacting = ½ × amount of SO_2 reacting
   = ½ × 1.80 mol = 0.900 mol.
   The amount of oxygen remaining at equilibrium
   = initial amount – amount reacting
   = (2.00 – 0.900) mol = 1.10 mol.

4. Tabulate this equilibrium information.

Initial amount/mol	2.0	2.0	0
	$2SO_2(g)$ +	$O_2(g) \rightleftharpoons$	$2SO_3(g)$
Equilibrium amount/mol	0.20	1.10	1.80

5. Write an expression for the partial pressure of each gas. The total amount of gas present = (0.20 + 1.10 + 1.80) mol = 3.10 mol.

$$p_{SO_3} = X_{SO_3} p_T = \frac{1.80}{3.10} \times p_T$$

Similarly for the other two partial pressures:

$$p_{SO_2} = \frac{0.20}{3.10} \times p_T; \quad p_{O_2} = \frac{1.10}{3.10} \times p_T$$

6. Write an expression for K_p and substitute the partial pressures:

$$K_p = \frac{p^2_{SO_3}}{p^2_{SO_2} \times p_{O_2}}$$

$$= \frac{1}{p_T} \times \frac{X^2_{SO_3}}{X^2_{SO_2} \times X_{O_2}}$$

$$\therefore\ 0.13\ \text{atm}^{-1} = \frac{1}{p_T} \frac{\left(\frac{1.80}{3.10}\right)^2}{\left(\frac{0.20}{3.10}\right)^2 \times \left(\frac{1.10}{3.10}\right)}$$

7. Solve this equation for p_T:

$$\therefore 0.13\ \text{atm}^{-1} = \frac{1}{p_T} \frac{0.337}{0.00416 \times 0.355}$$

$$p_T = \frac{0.337}{0.13\ \text{atm}^{-1} \times 0.00416 \times 0.355} = 1.76 \times 10^3\ \text{atm}$$

The pressure calculated in the Worked Example is far too high for economic production. In the next Exercise, you calculate the pressure required for a more modest conversion.

EXERCISE 6.42
Answer on page 254

For the system described in the Worked Example, calculate the pressure necessary for 20% conversion of the SO_2.

Now try some similar Exercises.

EXERCISE 6.43
Answer on page 254

Sulphur dioxide and oxygen in the ratio 2 moles : 1 mole were mixed at a constant temperature of 1110 K and a constant pressure of 9 atm in the presence of a catalyst. At equilibrium, one third of the sulphur dioxide had been converted into sulphur trioxide:

$$2SO_2(g) + O_2(g) \rightleftharpoons 2SO_3(g)$$

Calculate the equilibrium constant (K_p) for this reaction under these conditions.

EXERCISE 6.44
Answer on page 255

In the following reaction:

$$2NO_2(g) \rightleftharpoons 2NO(g) + O_2(g)$$

at a temperature of 700 K, the amount of each gas present at equilibrium is 0.96 mol of NO_2, 0.04 mol of NO, and 0.02 mol of O_2. If $K_p = 6.8 \times 10^{-6}$ atm what must the total pressure have been to achieve this particular equilibrium mixture?

EXERCISE 6.45
Answer on page 255

K_p = 3.72 atm at 1000 K for the system:

$$C(s) + H_2O(g) \rightleftharpoons CO(g) + H_2(g)$$

What pressure must be applied to obtain a mixture containing 2.0 mol of CO, 1.0 mol of H_2 and 4.0 mol of H_2O?

EXERCISE 6.46
Answer on page 255

A sample of dinitrogen tetroxide is 66% dissociated at a pressure of 98.3 kPa (747 mmHg) and a temperature of 60 °C. (Standard pressure = 100 kPa (760 mmHg).) Calculate the value of K_p for the equilibrium at 60 °C, stating the units.

When considering mixtures of gases, a useful concept is 'average molar mass'. We deal with this in the next section.

Average molar mass of a gaseous mixture

You might be asked to calculate the average molar mass of a mixture of gases from the amounts present or, alternatively, to calculate the relative amounts present from the average molar mass of a mixture.

We show you how to do this calculation in a Worked Example. You may like to try to solve the problem yourself before you look at our solution.

WORKED EXAMPLE

Calculate the average molar mass of a mixture which contains 0.20 mol of SO_2, 1.10 mol of O_2 and 1.80 mol of SO_3.

Solution

Each component of the mixture contributes to the average molar mass, $\overline{M}$, in proportion to its mole fraction, X.

$$\overline{M} = X_{SO_2} M_{SO_2} + X_{O_2} M_{O_2} + X_{SO_3} M_{SO_3}$$

$$= \frac{0.20}{3.10} \times 64.1 \text{ g mol}^{-1} + \frac{1.10}{3.10} \times 32.0 \text{ g mol}^{-1} + \frac{1.80}{3.10} \times 80.1 \text{ g mol}^{-1}$$

$$= (4.1 + 11.4 + 46.5) \text{ g mol}^{-1} = \mathbf{62.0 \text{ g mol}^{-1}}$$

Now do the following Exercises.

EXERCISE 6.47
Answers on page 255

$$2NO_2(g) \rightleftharpoons 2NO(g) + O_2(g).$$

For this system, a particular equilibrium mixture has the composition 0.96 mol of $NO_2(g)$, 0.040 mol of NO(g), 0.020 mol of $O_2(g)$ at 700 K and 0.20 atm.

a Calculate the equilibrium constant, K_p, for this reaction under the stated conditions.
b Calculate the average molar mass of the mixture under the stated conditions.

EXERCISE 6.48
Answers on page 255

0.20 mol of carbon dioxide was heated with excess carbon in a closed vessel until the following equilibrium was attained:

$$CO_2(g) + C(s) \rightleftharpoons 2CO(g)$$

It was found that the average molar mass of the gaseous equilibrium mixture was 36 g mol^{-1}.

a Calculate the mole fraction of carbon monoxide in the equilibrium gaseous mixture.
b The pressure at equilibrium in the vessel was 12 atm. Calculate K_p for the equilibrium at the temperature of the experiment.
c Calculate the mole fraction of carbon monoxide which would be present in the equilibrium mixture if the pressure were reduced to 2.0 atm at the same temperature.

EXERCISE 6.49
Answers on page 256

For the equilibrium:

$$N_2(g) + 3H_2(g) \rightleftharpoons 2NH_3(g)$$

at a particular temperature and a total pressure of 2.0 atm, an equilibrium mixture has the composition of 1.0 mol NH_3, 3.6 mol H_2 and 13.5 mol N_2.

a Calculate the equilibrium constant, K_p, for this reaction under the stated conditions.
b Calculate the average molar mass of the mixture under the stated conditions.

EXERCISE 6.50
Answers on page 256

For the equilibrium:

$$PCl_5(g) \rightleftharpoons PCl_3(g) + Cl_2(g)$$

a particular equilibrium mixture contains 0.20 mol PCl_5, 0.010 mol PCl_3 and 3.8 mol Cl_2, at a particular temperature and a total pressure of 3.0 atm.

a Calculate the equilibrium constant, K_p, for this reaction under the stated conditions.
b Calculate the average molar mass of the mixture under the stated conditions.

Now we consider the effect on an equilibrium system of a change in temperature.

The variation of equilibrium constant with temperature

Your syllabus may not include this topic, so you should ask your teacher if it is necessary for you to study any of this section.

You should already know that, by applying Le Chatelier's principle the **direction** of a shift in equilibrium brought about by a temperature change is determined by the sign of the standard enthalpy change, ΔH°. You should not be surprised to find that the **extent** of the shift depends on the **value** of ΔH°.

Since a change in temperature does not, of itself, change the concentration of a substance, it follows that any shift in equilibrium position must be due to a change in the equilibrium constant.

The effect on equilibrium of changing temperature is therefore different from the effects of changing concentration and pressure. Remember that:

equilibrium constants vary with temperature;
they do not vary with pressure or concentration;

if ΔH° is positive, K increases with temperature (more products);
if ΔH° is negative, K decreases with temperature (more reactants).

You also know, from Chapter 4 (*Chemical Energetics*), that the extent to which a reaction 'goes' depends on the value of the standard free energy change, ΔG°. You might therefore expect ΔG° and K to be related.

We were unable to explain fully at that stage why it is that although ΔG (not ΔG°) must be negative for a spontaneous change, reactions where ΔG° has a small **positive** value may proceed partially, towards an equilibrium position. The reason is that ΔG° and K are related mathematically by an equation, which we now introduce without derivation.

$$\Delta G^\circ = -RT \ln K \quad \text{or} \quad \Delta G^\circ = -2.30\, RT \log K$$

We use 'ln' to mean 'logarithm to the base *e*' and 'log' to mean 'logarithm to the base 10'.

Note that for reactions involving a change in the amount of gas present, K in the equation above must be K_p; in other cases we may use K_c.

The equation shows that

if $\Delta G^\ominus$ is zero, log K is zero and $K = 1$

if $\Delta G^\ominus$ is negative, log K is positive and $K > 1$

if $\Delta G^\ominus$ is positive, log K is negative and $K < 1$

We can combine the above equation with an equation that you have already met in Chapter 4:

$$\Delta G^\ominus = \Delta H^\ominus - T\Delta S^\ominus$$

Combining these two equations gives

$$-RT \ln K = \Delta H^\ominus - T\Delta S^\ominus$$

or

$$\ln K = -\frac{\Delta H^\ominus}{RT} + \frac{\Delta S^\ominus}{R}$$

$\Delta H^\ominus$ and $\Delta S^\ominus$ vary a little with temperature but can be treated as constants over a limited temperature range. A plot of ln K against $1/T$ should therefore be a straight line, because the equation is of the form

$$y = mx + c$$

where the slope

$$m = -\frac{\Delta H^\ominus}{R}$$

and the intercept

$$c = \frac{\Delta S^\ominus}{R}$$

We can therefore use a series of measurements of equilibrium constants at different temperatures in two important ways.

1. We can calculate a value for $\Delta H^\ominus$. This method is often applied when calorimetry cannot be used.
2. We can obtain values of K for temperatures other than those at which measurements were taken.

We illustrate these two uses in the following Exercises.

EXERCISE 6.51
Answers on page 256

Use the information in the table below for the reaction:

$$N_2O_4(g) \rightleftharpoons 2NO_2(g)$$

in answering the questions which follow.

Table 6.2

T/K	$(1/T)$/K^{-1}	K_p/atm	log (K_p/atm)
350	2.86×10^{-3}	3.89	+0.59
400	2.50×10^{-3}	4.79×10^{1}	+1.68
450	2.22×10^{-3}	3.47×10^{2}	+2.54
500	2.00×10^{-3}	1.70×10^{3}	+3.23

a Plot a graph of log K_p (y-axis) against $1/T$ (x-axis).
b Measure the slope of the graph and so calculate $\Delta H^\ominus$ for this reaction. You may assume that $\Delta H^\ominus$ is constant over this temperature range.
c What is the value of log K_p for this reaction at:
i) 375 K,
ii) 475 K,
iii) 550 K?
What assumption must you make to determine log K_p at 550 K?
d Use the graph to determine the effect of an increase in temperature on the position of equilibrium for this reaction.

EXERCISE 6.52

Answers on page 257

Use the information in the table below for the reaction:

$$N_2(g) + 3H_2(g) \rightleftharpoons 2NH_3(g)$$

and answer the questions which follow.

Table 6.3

T/K	$(1/T)$/K^{-1}	K_p/atm^{-2}	log (K_p/atm^{-2})
345	2.9×10^{-3}	1×10^{3}	+3.0
385	2.6×10^{-3}	3.16×10^{1}	+1.5
500	2.0×10^{-3}	3.55×10^{-2}	−1.45
700	1.43×10^{-3}	7.76×10^{-5}	−4.11

a Calculate $\Delta H^\ominus$ for the reaction by a graphical method.
b Use the graph to determine the effect of an increase in temperature on the position of equilibrium for this reaction.
c i) What do you notice about the gradients of the graphs drawn in this Exercise and in the last?
ii) What determines whether the gradient will be positive or negative?

It is not always necessary to plot a graph of log K against $1/T$ in order to estimate K at some other temperature. We can derive an expression which enables you to calculate K directly.

Let K_1 = equilibrium constant at temperature T_1 and K_2 = equilibrium constant at temperature T_2. Then

$$\ln K_1 = -\frac{\Delta H^\ominus}{RT_1} + \frac{\Delta S^\ominus}{R}$$

and

$$\ln K_2 = -\frac{\Delta H^\ominus}{RT_2} + \frac{\Delta S^\ominus}{R}$$

Subtracting

$$\ln K_2 - \ln K_1 = -\frac{\Delta H^\ominus}{RT_2} + \frac{\Delta H^\ominus}{RT_1}$$

or

$$\ln\left(\frac{K_2}{K_1}\right) = \frac{\Delta H^\ominus}{R}\left(\frac{1}{T_1} - \frac{1}{T_2}\right)$$

If we use logarithms to the base 10, the expression becomes

$$\log\left(\frac{K_2}{K_1}\right) = \frac{\Delta H^\ominus}{2.30R}\left(\frac{1}{T_1} - \frac{1}{T_2}\right)$$

Thus, if four of the quantities $\Delta H^\ominus$, K_1, K_2, T_1, and T_2 are known, the fifth may be calculated.

Use the expression derived above in the following Exercises.

EXERCISE 6.53

Answer on page 257

At 1065 °C, $K_p = 0.0118$ atm for the reaction:

$$2H_2S(g) \rightleftharpoons 2H_2(g) + S_2(g);\ \Delta H^\ominus = 177.3\ \text{kJ mol}^{-1}$$

Calculate the equilibrium constant for the reaction at 1200 °C.

EXERCISE 6.54

Answer on page 257

A particular reaction has a value of $K_p = 2.44$ atm at 1000 K and 3.74 atm at 1200 K. Calculate $\Delta H^\ominus$ for this reaction.

EXERCISE 6.55

Answers on page 258

The table below gives information about the values at different temperatures of the equilibrium constant for the reaction:

$$N_2(g) + O_2(g) \rightleftharpoons 2NO(g)$$

It also gives the partial pressures of NO in equilibrium with two different mixtures of nitrogen and oxygen at the given temperatures.

Table 6.4

		Partial pressure of NO/atm × 10^2	
Temperature/K	$10^4 K_p$	$p_{N_2} = 0.8$ atm, $p_{O_2} = 0.2$ atm	$p_{N_2} = 0.8$ atm, $p_{O_2} = 0.05$ atm
1800	1.21	0.44	0.22
2000	4.08	0.81	0.40
2200	11.00	1.33	0.67
2400	25.10	2.00	1.00
2600	50.30	2.84	1.42

a Write an expression for the equilibrium constant, K_p, for the given reaction.

b Use the expression you have written to explain why the values in the fourth column are half those in the third.

c What condition of temperature and pressure should be used to obtain the best yield of NO? Justify your answer.

d Is the reaction exothermic or endothermic? Justify your answer.

EXERCISE 6.56 Calculate $\Delta H^\ominus$ from the data given in Exercise 6.55.

Answer on page 258

Having considered some homogeneous equilibrium systems, we now apply the equilibrium law to an important group of heterogeneous systems – sparingly soluble salts in contact with their saturated solutions.

SOLUBILITY PRODUCTS

Equilibria involving sparingly soluble salts deserve special attention because of their importance in chemical analysis. A simple application of the equilibrium law helps us to understand how precipitation can be controlled.

The solubility product, K_s, for a solution of a sparingly soluble salt in equilibrium with the solid is simply related to K_c, as shown in the following example:

$$Ag_3PO_4(s) \rightleftharpoons 3Ag^+(aq) + PO_4^{3-}(aq)$$

$$K_c = \frac{[Ag^+(aq)^3][PO_4^{3-}(aq)]}{[Ag_3PO_4(s)]}$$

$[Ag_3PO_4(s)]$ is constant and is combined with K_c to form a new constant, K_s.

$$K_c \times [Ag_3PO_4(s)] = K_s = [Ag^+(aq)]^3[PO_4^{3-}(aq)]$$

Numerical problems are often concerned with the relationship between solubility product and solubility, as we now show by a Worked Example.

Calculating solubility product from solubility

WORKED EXAMPLE The solubility of silver sulphide, Ag_2S, is 2.48×10^{-15} mol dm^{-3}. Calculate its solubility product.

Solution

1. Write the equation representing the dissolving of silver sulphide, leaving space above and below the equation for initial and equilibrium concentrations:

Initial concn./mol dm^{-3}		0	0
	$Ag_2S(s) \rightleftharpoons$	$2Ag^+(aq)$ +	$S^{2-}(aq)$
Equilibrium concn./mol dm^{-3}			

2. Calculate the concentrations of sulphide ions and silver ions.
   The solubility indicates the amount of silver sulphide in a saturated solution. The balanced equation shows that every mole of Ag_2S which dissolves produces one mole of S^{2-} and two moles of Ag^+. Hence, at this temperature, in a saturated solution of Ag_2S

$$[S^{2-}(aq)] = 2.48 \times 10^{-15} \text{ mol dm}^{-3}$$

and

$$[Ag^+(aq)] = 2 \times 2.48 \times 10^{-15} \text{ mol dm}^{-3} = 4.96 \times 10^{-15} \text{ mol dm}^{-3}$$

Put these values below the equation:

Initial concn./mol dm^{-3}			0		0
	$Ag_2S(s)$	$\rightleftharpoons$	$2Ag^{+}(aq)$	+	$S^{2-}(aq)$
Equilibrium concn./mol dm^{-3}			4.96×10^{-15}		2.48×10^{-15}

3. Write the expression for K_s, substitute the equilibrium concentrations and do the arithmetic.
$K_s = [Ag^{+}(aq)]^2[S^{2-}(aq)]$
$= (4.96 \times 10^{-15} \text{ mol dm}^{-3})^2 \times (2.48 \times 10^{-15} \text{ mol dm}^{-3})$
$= \mathbf{6.10 \times 10^{-44} \text{ mol}^3 \text{ dm}^{-9}}$

In this type of calculation some students wonder why a concentration term is both doubled and squared. The explanation is as follows: firstly, the stoichiometry tells us that $[Ag^{+}(aq)]$ is twice the solubility; secondly, the form of the equilibrium law tells us that the silver ion concentration is squared.

Now try the following Exercises.

EXERCISE 6.57
Answers on page 258

Calculate the solubility products of the following substances from their solubilities.

Table 6.5

Substances	**Solubility/mol dm^{-3}**
a $CdCO_3$	1.58×10^{-7}
b CaF_2	2.15×10^{-4}
c $Cr(OH)_3$	1.39×10^{-8}

EXERCISE 6.58
Answers on page 258

In an experiment to determine the solubility product of calcium hydroxide, $Ca(OH)_2$, 25.0 cm^3 of a saturated solution required 11.45 cm^3 of 0.100 M HCl for neutralisation.

Calculate the concentration of hydroxide ions and, hence, the solubility and solubility product of calcium hydroxide at the temperature of the experiment.

Now that you know how to calculate solubility product from solubility, we show you how to do the reverse process.

Calculating solubility from solubility product

WORKED EXAMPLE

Calculate the solubility, in mol dm^{-3}, of calcium sulphate. K_s for calcium sulphate $= 8.64 \times 10^{-8} \text{ mol}^2 \text{ dm}^{-6}$.

Solution

1. Write the equation for the solution of the salt, leaving room above and below for concentrations. Initially there is no $Ca^{2+}(aq)$ or $SO_4^{2-}(aq)$.

Initial concn./mol dm^{-3}			0		0
	$CaSO_4(s)$	$\rightleftharpoons$	$Ca^{2+}(aq)$	+	$SO_4^{2-}(aq)$
Equilibrium concn./mol dm^{-3}					

2. From the balanced equation we know that the equilibrium concentration of $Ca^{2+}(aq)$ is equal to that of $SO_4^{2-}(aq)$. Let this be x mol dm^{-3}.

	$CaSO_4(s) \rightleftharpoons$	$Ca^{2+}(aq)$ +	$SO_4^{2-}(aq)$
Initial concn./mol dm^{-3}		0	0
Equilibrium concn./mol dm^{-3}		x	x

3. Write the expression for K_s:

$$K_s = [Ca^{2+}(aq)][SO_4^{2-}(aq)]$$

4. Substitute K_s and the equilibrium concentrations.

$$8.64 \times 10^{-8}\ \text{mol}^2\ \text{dm}^{-6} = x\ \text{mol dm}^{-3} \times x\ \text{mol dm}^{-3}$$
$$x^2 = 8.64 \times 10^{-8}$$
$$x = 2.94 \times 10^{-4}$$
$$\therefore \text{solubility} = \mathbf{2.94 \times 10^{-4}\ mol\ dm^{-3}}$$

Try the next Exercise, using a similar method.

EXERCISE 6.59
Answers on page 258

Calculate the solubilities of the following slightly soluble substances from their solubility products.

Table 6.6

Substance	Solubility product
a CuS	6.3×10^{-36} mol^2 dm^{-6}
b $Fe(OH)_2$	6.0×10^{-15} mol^3 dm^{-9}
c Ag_3PO_4	1.25×10^{-20} mol^4 dm^{-12}

The idea of solubility product strictly applies only to slightly soluble electrolytes, i.e. solutions in which the concentration of ions is very low. Usually the total ion concentration is not more than 0.01 mol dm^{-3}.

At higher concentrations, the ions interact with one another so that their effective concentrations differ from their actual concentrations. In this respect, very dilute solutions may be regarded as ‘ideal solutions’ which obey simple laws, just as gases at low pressure observe the ideal gas law.

Calcium hydroxide is a border-line case; strictly, it is too soluble for its solution to be regarded as ideal, but the error is small enough for us to use calcium hydroxide as an example.

So far in this chapter we have considered only pure solutions. Now we deal with mixtures of solutions with a common ion. The effect is often demonstrated in concentrated solutions; calculations are valid only for dilute solutions.

The common ion effect

Provided that the total ion concentration is no more than about 0.1 mol dm^{-3}, you can use solubility products in calculations involving mixtures of solutions. We illustrate this in a Worked Example.

WORKED EXAMPLE

Calculate the solubility of silver chloride, AgCl, in
a pure water,
b 0.10 M NaCl.

$$K_s(AgCl) = 2.0 \times 10^{-10}\ \text{mol}^2\ \text{dm}^{-6}.$$

Solution **a** This is not a new type of calculation but you need the result for comparison. Refer to the Worked Example on page 125 if necessary.

$$\text{Solubility of AgCl} = \sqrt{K_s} = \sqrt{2.0 \times 10^{-10}\ \text{mol}^2\ \text{dm}^{-6}}$$
$$= \mathbf{1.4 \times 10^{-5}\ mol\ dm^{-3}}$$

b i) Write the equation and indicate the concentrations present initially and at equilibrium.

Initial concn./mol dm^{-3}		0	0.10
	$AgCl(s) \rightleftharpoons$	$Ag^+(aq)$ +	$Cl^-(aq)$
Equilibrium concn./mol dm^{-3}		x	$(x + 0.10)$

If x mol dm^{-3} of AgCl dissolves, then x mol dm^{-3} of $Ag^+(aq)$ and x mol dm^{-3} of $Cl^-(aq)$ are produced. The total concentration of chloride ions equals that produced by the AgCl dissolving plus the 0.10 mol dm^{-3} initially present.

ii) Write the equilibrium law expression:

$$K_s = [Ag^+(aq)]\,[Cl^-(aq)]$$

iii) Substitute the equilibrium concentrations and the solubility product:

$$2.0 \times 10^{-10}\,\text{mol}^2\ \text{dm}^{-6} = x \times (x + 0.10)\ \text{mol}^2\ \text{dm}^{-6}$$

iv) Now you can make a very useful approximation. In otherwise pure water the concentration of $Ag^+(aq)$ from dissolved AgCl is 1.4×10^{-5} mol dm^{-3}. Applying Le Chatelier's principle, addition of Cl^- will drive the equilibrium to the left thereby reducing the silver ion concentration still further. So you can assume that the concentration of chloride ions from the AgCl is negligible compared to the 0.10 mol dm^{-3} solution of Cl^- present initially. Expressed mathematically,

$$(x + 0.10)\ \text{mol dm}^{-3} \approx 0.10\ \text{mol dm}^{-3}.$$

(You can test this approximation after the calculation.)

v) The equilibrium law expression now becomes:

$$2.0 \times 10^{-10} = x \times 0.10$$

$$\therefore\ x = \frac{2.0 \times 10^{-10}}{0.10} = 2.0 \times 10^{-9}$$

vi) The solubility in 0.10 M NaCl = $[Ag^+(aq)]$
$= x$ mol dm^{-3} = $\mathbf{2.0 \times 10^{-9}\ mol\ dm^{-3}}$

Note the dramatic effect of the common ion; this solubility is less than one thousandth of the solubility in water!

You can quite simply check that the approximation in step iv) is valid:

$$x + 0.10 = (2.0 \times 10^{-9}) + 0.10 = 0.100000002 \approx 0.10$$

A better, but more tedious, check is to calculate x from the quadratic equation in step iii):

$$2.0 \times 10^{-10} = x \times (x + 0.10)$$

$$\therefore\ x^2 + 0.10x - 2.0 \times 10^{-10} = 0$$

$$\therefore\ x = \frac{-0.10 \pm \sqrt{0.010 + 8.0 \times 10^{-10}}}{2} = 2.0 \times 10^{-9}$$

You will find it very useful to make such approximations because they often make calculations much simpler, as you can see in the following Exercises.

EXERCISE 6.60
Answers on page 259

A

At 20 °C, the solubility product of strontium sulphate is 4.0×10^{-7} mol^2 dm^{-6}, and that of magnesium fluoride is 7.2×10^{-9} mol^3 dm^{-9}. Estimate, to two significant figures, the solubility at 20 °C in mol dm^{-3} of

a strontium sulphate in a 0.1 M solution of sodium sulphate;
b magnesium fluoride in a 0.2 M solution of sodium fluoride.

EXERCISE 6.61
Answer on page 259

A saturated solution of strontium carbonate was filtered. When 50 cm^3 of the filtrate was added to 50 cm^3 of 1.0 M sodium carbonate solution, some strontium carbonate was precipitated. Calculate the concentration of strontium ions remaining in the solution. All the work was done at 25 °C.

We now come to the last application of the solubility product principle that we cover in this chapter.

Will precipitation occur?

To predict whether precipitation will occur when solutions are mixed, you need to recognise that solubility product is a value which the product of ion concentrations in solution **can never exceed** at equilibrium.

We show you how to apply this idea in a Worked Example.

WORKED EXAMPLE

If 50.0 cm^3 of 0.050 M $AgNO_3$ is mixed with 50.0 cm^3 of 0.010 M $KBrO_3$, will a precipitate of $AgBrO_3$ form? $K_s(AgBrO_3) = 6 \times 10^{-5}$ mol^2 dm^{-6}.

Solution

1. Work out what the concentrations of Ag^+ and BrO_3^- would be after mixing but before any reaction:

   a Amount of Ag^+ = 0.050 dm^3 × 0.050 mol dm^{-3}

$$[Ag^+(aq)] = \frac{0.050\ \text{dm}^3 \times 0.050\ \text{mol dm}^{-3}}{0.100\ \text{dm}^3} = 0.025\ \text{mol dm}^{-3}$$

   (The volume is doubled, so the concentration is halved.)

   b Similarly, $[BrO_3^-(aq)] = \frac{1}{2} \times 0.010$ mol dm^{-3} = 0.0050 mol dm^{-3}.

2. Calculate the product of the ion concentrations (called the ion product).
   Ion product = $[Ag^+(aq)][BrO_3^-(aq)]$
   = 0.025 mol dm^{-3} × 0.0050 mol dm^{-3} = 1.25×10^{-4} mol^2 dm^{-6}.

3. Compare the value of the ion product with the solubility product.

   If the ion product is greater than K_s, precipitation will occur.
   If the ion product is less than K_s, precipitation will not occur.

∴ a **silver bromate precipitate will appear** because the ion product is greater than the solubility product.

To make sure that you can do this kind of problem, try the following two Exercises.

EXERCISE 6.62
Answer on page 259

Will a precipitate of BaF_2 form if 150 cm^3 of 0.1 M $Ba(NO_3)_2$ is mixed with 50.0 cm^3 of 0.05 M KF? $K_s = 1.7 \times 10^{-6}$ mol^3 dm^{-9}.

EXERCISE 6.63
Answers on page 259

You are given the following **numerical values** only for the solubility products of various salts at 25 °C.

K_s [silver chloride] $= 2 \times 10^{-10}$
K_s [lead(II) bromide] $= 3.9 \times 10^{-5}$
K_s [silver bromate(V)] $= 6.0 \times 10^{-5}$
K_s [magnesium hydroxide] $= 2.0 \times 10^{-11}$

a State the units for each of the above solubility products.

b Which of the following pairs of solutions (all of concentration 1.0×10^{-3} mol dm^{-3}) will form a precipitate when equal volumes are mixed at 25 °C? Give reasons for your answers.

i) Silver nitrate and sodium chloride.
ii) Lead(II) nitrate and sodium bromide.
iii) Silver nitrate and potassium bromate(V).
iv) Magnesium sulphate and sodium hydroxide.

We now deal with the last application of the equilibrium law in this chapter, namely the distribution of a solute between two immiscible solvents.

DISTRIBUTION EQUILIBRIUM

When a solute is distributed between two immiscible solvents, there is a dynamic equilibrium at the interface, where the solute is transferred from one solvent to the other and back at equal rates. For example, the distribution of ammonia between water and 1,1,1-trichloroethane (tce) may be represented:

$$NH_3(tce) \rightleftharpoons NH_3(aq)$$

Application of the equilibrium law gives an expression for the equilibrium constant which, in this situation, is known as the distribution coefficient, K_d, or the partition coefficient.

$$K_d = \frac{[NH_3(aq)]}{[NH_3(tce)]}$$

We illustrate this by means of a Worked Example.

WORKED EXAMPLE

Aqueous ammonia and Volasil 244 were shaken together and the two layers were then allowed to separate. (Volasil 244 is a cyclic silicone, that is a non-toxic and 'environmentally friendly' solvent). Samples of each layer were titrated with standard hydrochloric acid. 10 cm^3 of the aqueous layer required 17.0 cm^3 of 0.500 M HCl for neutralisation, whilst 10 cm^3 of the organic layer required 6.0 cm^3 of 0.010 M HCl.

Calculate the value of the distribution coefficient for this system.

Solution 1. Calculate the concentration of ammonia in the aqueous layer. Substituting in the expression:

$$\frac{c_A V_A}{c_B V_B} = \frac{a}{b} \text{ where A refers to HCl and B to } NH_3$$

gives
$$\frac{0.50 \text{ mol dm}^{-3} \times 17.0 \text{ cm}^3}{c_B \times 10.0 \text{ cm}^3} = \frac{1}{1}$$

$$\therefore c_B = 0.50 \text{ mol dm}^{-3} \times \frac{17.0}{10.0} = \mathbf{0.85 \text{ mol dm}^{-3}}$$

2. Calculate the concentration of ammonia in the organic layer. By the same method as in 1:

$$c_B = 0.010 \text{ mol dm}^{-3} \times \frac{6.0}{10.0} = \mathbf{6.0 \times 10^{-3} \text{ mol dm}^{-3}}$$

3. $$K_d = \frac{[NH_3(aq)]}{[NH_3(Vol)]}$$

$$= \frac{0.85 \text{ mol dm}^{-3}}{6.0 \times 10^{-3} \text{ mol dm}^{-3}} = \mathbf{142}$$

Use similar expressions in the following Exercises.

EXERCISE 6.64

Answer on page 259

Calculate the value of the distribution coefficient for the distribution equilibrium of butanedioic acid between water and ether from the following data. All experiments were performed at the same temperature.

Table 6.7

Concn. of acid in water /mol dm^{-3}	0.0759	0.108	0.158	0.300
Concn. of acid in ether /mol dm^{-3}	0.0114	0.0162	0.0237	0.0451

The distribution of solute between two solvents is sometimes used in extraction and purification procedures. For instance, you could remove most of the ammonia from a solution in Volasil 244 by shaking it with water in several portions, as you discover in the next Exercise.

These next two Exercises involve the use of solvents which were widely used up to January 1995, but because they are now known to be ozone-damaging chemicals, they are no longer available.

EXERCISE 6.65

Answers on page 260

This Exercise concerns the removal of ammonia from a solution in 1,1,1-trichloroethane (0.10 mol dm^{-3}) by shaking with water. $K_d = 290$.

How much ammonia remains in the organic layer after shaking 100 cm^3 of the solution with:

a 100 cm^3 of water,
b four successive 25 cm^3 portions of water?

In the distribution equilibria you have considered so far, the solute is in the same molecular form in both solvents. For example, ammonia in 1,1,1-trichloroethane exists entirely as molecules of NH_3, and in water almost entirely as molecules of NH_3. There is a slight reaction in water:

$$NH_3(aq) + H_2O(l) \rightleftharpoons NH_4^+(aq) + OH^-(aq)$$

but K_c for this is about 10^{-7}, so virtually all the ammonia remains as molecules.

By contrast, ethanoic acid exists in aqueous solution almost entirely as CH_3CO_2H molecules and in 1,1,1-trichloroethane almost entirely as $(CH_3CO_2H)_2$ molecules, i.e. as dimers. You need not concern yourself with the derivation of the expression for K_D, but the simple result is that a squared term appears:

$$K_d = \frac{[CH_3CO_2H(tce)]}{[CH_3CO_2H(aq)]^2}$$

The next Exercise concerns a similar equilibrium system.

EXERCISE 6.66
Answer on page 260

Trichloromethane was added to a series of aqueous solutions of ethanoic acid (acetic acid) and the mixtures were shaken at laboratory temperature. By titration the following concentrations of ethanoic acid were found in the two layers:

Table 6.8

Trichloromethane/g dm^{-3}	17.5	43.5	84.6
Water/g dm^{-3}	292	479	642

By neglecting any dissociation of ethanoic acid in water, deduce its molecular formula in trichloromethane.

You may use graph paper if you wish.

END-OF-CHAPTER QUESTIONS

Answers on page 261

6.1 At 700 K, analysis of the system:

$$N_2(g) + 3H_2(g) \rightleftharpoons 2NH_3(g)$$

showed that $[N_2(g)] = 13.6\ \text{mol dm}^{-3}$, $[H_2(g)] = 1.0\ \text{mol dm}^{-3}$, $[NH_3(g)] = 1.5\ \text{mol dm}^{-3}$. Calculate the value of K_c at this temperature.

6.2 At a certain temperature, the equilibrium constant for the reaction:

$$CH_3CO_2H(l) + C_2H_5OH(l) \rightleftharpoons CH_3CO_2C_2H_5(l) + H_2O(l)$$

is 4. If one mole of ethanoic acid (CH_3CO_2H) is added to one mole of ethanol and the mixture is allowed to reach equilibrium, how many moles of ethanoic acid will be present in the equilibrium mixture?

6.3 For the equilibrium:

$$C_2H_5OH(l) + C_2H_5CO_2H(l) \rightleftharpoons C_2H_5CO_2C_2H_5(l) + H_2O(l)$$

propanoic acid ethyl propanoate

$K_c = 7.5$ at 50 °C. If 50.0 g of C_2H_5OH is mixed with 50.0 g of $C_2H_5CO_2H$, what mass of ethyl propanoate will be formed at equilibrium?

6.4 At 1000 K, K_p = 1.9 atm for the system:

$$C(s) + CO_2(g) \rightleftharpoons 2CO(g)$$

What must the total pressure be to have an equilibrium mixture containing 0.013 mol of CO_2 and 0.024 mol of CO?

6.5 For the equilibrium:

$$N_2O_4(g) \rightleftharpoons 2NO_2(g)$$

1.00 mol of N_2O_4 was introduced into a vessel and allowed to attain equilibrium at 308 K. It was found that the average molar mass of the mixture was 72.4 g mol^{-1}.

a Calculate the mole fraction of NO_2 in the equilibrium mixture.

b The pressure at equilibrium was 1.00 atm. Calculate K_p for the system at the temperature of the experiment.

c Calculate the mole fraction of NO_2 which would be present in the equilibrium mixture if the pressure were increased to 6.00 atm at the same temperature.

A

6.6 25 g of a solid X were dissolved in 100 g of water. The solution was shaken with 20 cm^3 of tetrachloromethane, and, when equilibrium was attained, 10 g of X were found to be dissolved in the tetrachloromethane. The molecular state of X was the same in both solvents. What is the value of the partition coefficient, $[X]_{CCl_4}/[X]_{H_2O}$?

A 0.67
B 1.5
C 2.5
D 3.33
E 7.5

A

6.7 If a solution which is 0.0001 M with respect to carbonate ions, CO_3^{2-}, is mixed with an equal volume of a 0.0001 M solution of ions of a Group II metal, which of the following carbonates would be precipitated?

Table 6.9

	K_s (298 K)/$mol^2\ dm^{-6}$
1 $MgCO_3$	1.1×10^{-5}
2 $CaCO_3$	5.0×10^{-9}
3 $SrCO_3$	1.1×10^{-10}

A(P)

6.8 The equation for the reaction of ethanol and ethanoic acid is given below:

$$CH_3CO_2H(l) + C_2H_5OH(l) \rightleftharpoons CH_3CO_2C_2H_5(l) + H_2O(l)$$

3.0 g of ethanoic acid and 2.3 g of ethanol were equilibrated at 100 °C for about one hour and then quickly cooled in an ice-bath. 50 cm^3 of aqueous sodium hydroxide of concentration 1.0 mol dm^{-3} were then added and the mixture titrated with hydrochloric acid of the same concentration. 33.3 cm^3 of acid were required.

Give an expression for K_c and calculate its value using the data provided.

A(P)

6.9 The equilibrium constant K_p for the reaction:

$$PCl_5(g) \rightleftharpoons PCl_3(g) + Cl_2(g)$$

is 1.06×10^6 N m^{-2} at 250 °C.

A sample of PCl_5 at an initial pressure of 1.01×10^6 N m^{-2} dissociates in the vapour phase at 250 °C in a vessel of fixed volume.

a Calculate the partial pressure of each species present at equilibrium and the final total pressure.

b Will the same partial pressures be obtained if equal amounts (i.e. equal numbers of moles) of PCl_3 and Cl_2 are mixed at 250 °C and the **final** pressure is the same as in part **a**? Explain your reasoning.

[The solution to a quadratic equation of the general form $ax^2 + bx + c = 0$ is $x = \{-b \pm \sqrt{(b^2 - 4ac)}\}/2a$.]

6.10 **a** Write an expression for the solubility product of lead(II) chloride.

b The solubility product of lead(II) chloride is 1.6×10^{-5} $mol^3\,dm^{-9}$ at a given temperature.

i) What is the solubility in mol dm^{-3} of lead(II) chloride in water at the same temperature?

ii) How many moles of chloride ion must be added to a 1.0 M solution of lead(II) nitrate at the same temperature in order just to cause a precipitate of lead(II) chloride? Assume that no change in volume occurs on adding the chloride ion.

6.11 When lead sulphate is added to an aqueous solution of sodium iodide, the following equilibrium is obtained:

$$PbSO_4(s) + 2I^-(aq) \rightleftharpoons PbI_2(s) + SO_4^{2-}(aq)$$

The equilibrium constant for this reaction may be determined by adding an excess of lead sulphate to a known volume of a standard solution of sodium iodide and allowing the mixture to equilibrate in a water-bath thermostatically controlled at the desired temperature. Cold water is then added to the reaction mixture to 'freeze' the equilibrium and the mixture is then titrated with standard silver nitrate solution. In a typical experiment using 50.0 cm^3 of 0.1 M sodium iodide, a titre of 31.0 cm^3 of 0.1 M silver nitrate was obtained.

a Give an expression for the equilibrium constant, K, of the reaction.

b Why is it not necessary to know the mass of lead sulphate used in the experiment?

c From the data given above, calculate:

i) the concentration of iodide ions present initially,

ii) the concentration of iodide ions present at equilibrium,

iii) the concentration of iodide ions which have reacted,

iv) the concentration of sulphate ions formed,

v) a value for K.

CHAPTER 7

EQUILIBRIUM II – ACIDS AND BASES

INTRODUCTION AND PRE-KNOWLEDGE

In this chapter we extend your study of equilibrium by applying the principles developed in Chapter 6 to equilibria involving weak acids and weak bases, including buffer solutions and acid–base indicators.

We assume that you are already familiar with Le Chatelier's principle and can apply the equilibrium law in calculations involving equilibrium constants. You have already met several types of equilibrium constant (K_c, K_p, K_s, and K_d): in this chapter we introduce some more, the first of which concerns the ionisation of water.

IONISATION OF WATER

Water is a special case of an amphiprotic substance; it can react with itself:

$$H_2O(l) + H_2O(l) \rightleftharpoons H_3O^+(aq) + OH^-(aq)$$

The evidence that this process takes place in even the purest water is that it has a slight electrical conductivity. A few ions must be present to carry a current and these must have come from the ionisation of the water molecules themselves. A simpler way of writing the equation is:

$$H_2O(l) \rightleftharpoons H^+(aq) + OH^-(aq)$$

Application of the equilibrium law gives:

$$K_c = \frac{[H^+(aq)][OH^-(aq)]}{[H_2O(l)]}$$

Use the above expression in the following Exercise.

EXERCISE 7.1
Answer on page 261

In pure water at 25 °C, the equilibrium concentration of un-ionised water molecules is 55.6 mol dm^{-3}. Given that K_c is 1.80×10^{-16} mol dm^{-3}, calculate the fraction of water molecules ionised at this temperature. (Hint: calculate $[H^+(aq)]$ and $[OH^-(aq)]$ first.)

So few water molecules are ionised that their concentration remains effectively constant (proportional to the density of water). We can therefore include $[H_2O(l)]$ with the constant K_c:

$$K_c \times [H_2O(l)] = [H^+(aq)][OH^-(aq)]$$

Let $K_c \times [H_2O(l)]$ be another constant, K_w:

$$\therefore \mathbf{K_w = [H^+(aq)][OH^-(aq)]}$$

This expression is known as the **ionic product of water**. In your reading you may also see it written in the form:

$$K_w = [H_3O^+(aq)][OH^-(aq)]$$

This means the same and can be used in exactly the same way as the above expression; its only difference is that it has been derived using the 'full' equation for the ionisation of water:

$$H_2O(l) + H_2O(l) \rightleftharpoons H_3O^+(aq) + OH^-(aq)$$

The value of the ionic product has been measured experimentally and found to be 1.0×10^{-14} mol^2 dm^{-6} at 25 °C. Since it is directly related to the equilibrium constant for the self-ionisation of water, this value is, of course, temperature dependent.

The aim of the following Revealing Exercise is to examine what happens when the equilibrium in pure water is upset by the addition of hydrogen ions from an acid or hydroxide ions from an alkali. Also, since K_w is constant, we derive a definition of acidic, neutral and basic solutions in terms of the concentration of hydrogen ions.

REVEALING EXERCISE

Consider the equilibrium:

$$H_2O(l) \rightleftharpoons H^+(aq) + OH^-(aq) \qquad K_w = 1.0 \times 10^{-14} \text{ mol}^2 \text{ dm}^{-6}$$

Q1 Calculate $[H^+(aq)]$ and $[OH^-(aq)]$ in pure water.

A1 $K_w = [H^+(aq)][OH^-(aq)] = 1.0 \times 10^{-14}$ mol^2 dm^{-6}
Since $[H^+(aq)] = [OH^-(aq)]$, then

$[H^+(aq)]^2 = 1.0 \times 10^{-14}$ mol^2 dm^{-6}
$\therefore [H^+(aq)] = \sqrt{1.0 \times 10^{-14} \text{ mol}^2 \text{ dm}^{-6}} = 1.0 \times 10^{-7}$ mol dm^{-3}
and $[OH^-(aq)] = 1.0 \times 10^{-7}$ mol dm^{-3}

Q2 Acid is added to the solution to increase the concentration of hydrogen ions. How does the equilibrium shift?

A2 It shifts to the left.

Q3 Is the concentration of hydroxide ions less than or greater than 1.0×10^{-7} mol dm^{-3} now?

A3 It is less than 1.0×10^{-7} mol dm^{-3}.

Q4 Does the concentration of hydroxide ions ever decrease to zero? Explain.

A4 No. The ionisation of water is an equilibrium system. The product of concentrations of hydrogen ions and hydroxide ions must always be 1.0×10^{-14} mol^2 dm^{-6}. The only way for one value to drop to zero would be for the other to be infinity, which is impossible.

Q5 What is the effect of adding alkali to the equilibrium system existing in pure water?

A5 It shifts the equilibrium to the left.

Q6 Is the concentration of hydrogen ions greater than or less than 1.0×10^{-7} mol dm^{-3} now?

A6 It is less than 1.0×10^{-7} mol dm^{-3}.

Q7 Does the hydrogen ion concentration ever decrease to zero? Explain.

A7 No, for similar reasons to those given in **A4**.

To summarise, in a neutral solution:

$$[H^+(aq)] = [OH^-(aq)]$$
$$[H^+(aq)] = [OH^-(aq)] = 1.0 \times 10^{-7}\ \text{mol dm}^{-3}\ (\text{at } 25\ °\text{C})$$

In an acidic solution:

$$[H^+(aq)] > [OH^-(aq)]$$

and $$[H^+(aq)] > 1.0 \times 10^{-7}\ \text{mol dm}^{-3}$$

In a basic, or alkaline, solution:

$$[H^+(aq)] < [OH^-(aq)]$$

and $$[H^+(aq)] < 1.0 \times 10^{-7}\ \text{mol dm}^{-3}$$

In the next section you learn how to use the expression for the ionic product of water to calculate hydroxide ion concentrations.

Using the expression for the ionic product of water

Since the product of concentrations of hydrogen ions and hydroxide ions is always constant for any dilute solution at a given temperature, we can calculate the concentration of one of these ions given the concentration of the other. We show you how to do this in a Worked Example.

WORKED EXAMPLE Calculate the hydroxide ion concentration in an acid solution which has a concentration of hydrogen ions of 0.10 mol dm^{-3} given that:

$$K_w = 1.0 \times 10^{-14}\ \text{mol}^2\ \text{dm}^{-6}\ \text{at } 25\ °\text{C}$$

Solution 1. Rearranging the expression:

$$K_w = [H^+(aq)][OH^-(aq)]$$

gives $$[OH^-(aq)] = \frac{K_w}{[H^+(aq)]}$$

2. Substitute the given values of K_w and $[H^+(aq)]$:

$$[OH^-(aq)] = \frac{1.0 \times 10^{-14}\ \text{mol}^2\ \text{dm}^{-6}}{0.10\ \text{mol dm}^{-3}} = \mathbf{1.0 \times 10^{-13}\ mol\ dm^{-3}}$$

Notice that even in strongly acidic solutions, there are hydroxide ions present.

EXERCISE 7.2
Answers on page 261

Given that $K_w = 1.0 \times 10^{-14}$ mol^2 dm^{-6}, calculate the hydroxide ion concentrations of these acid solutions at 25 °C. Assume that both acids are fully dissociated into ions.
a 0.010 M HCl
b 0.10 M H_2SO_4

By rearranging the expression for the ionic product of water to solve for $[H^+(aq)]$, you can calculate the hydrogen ion concentration given the hydroxide ion concentration.

EXERCISE 7.3
Answers on page 262

Given that $K_w = 1.0 \times 10^{-14}$ mol^2 dm^{-6} at 25 °C, calculate the hydrogen ion concentrations of the following solutions. Assume that both alkalis are fully dissociated in solution.

a 0.010 M KOH
b 0.050 M $Ba(OH)_2$

In the next section, we show you how to express the concentrations of hydrogen ions and hydroxide ions more conveniently as pH.

pH, THE HYDROGEN ION EXPONENT

In your pre-A-level course you used the pH scale rather informally as a measure of acidity without having been given a precise definition. You know that any solution having a pH above 7 is alkaline, while a solution having a pH below 7 is acidic. In this section you learn the definition of pH and understand more fully its significance.

When we compare the acidity of different substances, we generally test them in water. This is because so many acid–base reactions take place in aqueous solution that it is convenient to 'line up' acids according to their ability to donate a proton to water. For example, consider any acid, which we represent as HA, dissolving in water and ionising:

$$HA(aq) + H_2O \rightleftharpoons H_3O^+(aq) + A^-(aq)$$

or

$$HA(aq) \rightleftharpoons H^+(aq) + A^-(aq)$$

The greater the ability of the acid to donate protons to water, the greater the concentration of hydrogen ions. It would seem logical, therefore, to compare the strengths of acids by comparing the concentrations of hydrogen ions produced in aqueous solution, as shown in Table 7.1.

Table 7.1

Solution	$[H^+(aq)]$/mol dm^{-3}	
A	0.0010	$= 1.0 \times 10^{-3}$
B	0.000 001 0	$= 1.0 \times 10^{-6}$
C	0.000 000 001 0	$= 1.0 \times 10^{-9}$

However, a scale based on hydrogen ion concentration contains awkward numbers and is cumbersome to use.

The Danish biochemist, S P L Sørensen, realised this as early as 1909. It occurred to him that it was tedious to say 'the concentration of hydrogen ions in the solution is one-hundred-thousandth of a mole per cubic decimetre' when $[H^+(aq)] = 0.000\,010$ mol dm^{-3} (or 1.0×10^{-5} mol dm^{-3}) so he suggested referring to the solution as having 'pH 5'.

As you can see, 5 is the negative of the 'power of 10' in the expression for hydrogen ion concentration ('p' in 'pH' stands for 'power'). In general:

$$[H^+(aq)] = 1.0 \times 10^{-pH} \text{ mol dm}^{-3}$$

Rearranging the equation and taking logarithms, we can write:

$$\mathbf{pH = -\log_{10}([H^+(aq)]/mol\ dm^{-3})}$$

or

$$\mathbf{pH = -lg([H^+(aq)]/mol\ dm^{-3})}$$

Some textbooks give the definition:

$$\mathbf{pH = -\log[H^+]}$$

which, although accepted by some examination boards, is strictly speaking incorrect since we can take the logarithm only of a pure number and not that of a physical quantity. (Erwin Schrödinger is reputed to have told his students 'You can milk a cow but you can't take its logarithm!'.) Nevertheless, you will find that we have used pH = $-\log[H^+(aq)]$ in the answers to exercises for simplicity. You should check with your teacher which definition or definitions will be acceptable.

We now show you how to calculate the pH of solutions of fully ionised acids and alkalis from their concentrations.

WORKED EXAMPLE What is the pH of a 1.0 M HCl solution?

Solution Assume that this acid is fully dissociated so that the hydrogen ion concentration = 1.0 mol dm^{-3}.

Substitute this value into the expression which defines pH.

$$pH = -\log([H^+(aq)]/mol\ dm^{-3}) = -\log(1.0) = -(0) = \mathbf{0}$$

Now try the following Exercises yourself.

EXERCISE 7.4
Answer on page 262

Calculate the pH of a 10 M HCl solution, assuming it to be fully dissociated.

EXERCISE 7.5
Answers on page 262

What are the pH values for solutions A, B and C in Table 7.1 on page 000?

You have already made use of the ionic product of water to determine the concentration of hydrogen ions in alkaline solution. Now we take you through an Exercise making use of the ionic product of water to calculate the pH of an alkaline solution.

WORKED EXAMPLE What is the pH of a 1.0 M NaOH solution? ($K_w = 1.0 \times 10^{-14}\ mol^2\ dm^{-6}$.)

Solution Assume that NaOH is fully dissociated in solution so that $[OH^-(aq)] = 1.0\ mol\ dm^{-3}$.

1. Starting with the ionic product of water, solve for $[H^+(aq)]$:

$$[H^+(aq)][OH^-(aq)] = 1.0 \times 10^{-14}\ mol^2\ dm^{-6}$$

$$[H^+(aq)] = \frac{1.0 \times 10^{-14}\ mol^2\ dm^{-6}}{[OH^-(aq)]}$$

2. Calculate the hydrogen ion concentration:

$$[H^+(aq)] = \frac{1.0 \times 10^{-14}\ mol^2\ dm^{-6}}{1.0\ mol\ dm^{-3}} = 1.0 \times 10^{-14}\ mol\ dm^{-3}$$

3. Substitute this value into the definition of pH:

$$pH = -\log_{10}([H^+(aq)]/\text{mol dm}^{-3})$$
$$= -\log(1.0 \times 10^{-14}) = -(0 - 14) = \mathbf{14}$$

Now try the following Exercises yourself.

EXERCISE 7.6
Answer on page 262

What is the pH of a 10 M NaOH solution? (Assume complete dissociation.)

EXERCISE 7.7
Answers on page 262

Calculate the pH, at 25 °C, of solutions in which the hydroxide ion concentrations are as follows:
a 1.0×10^{-8} mol dm^{-3}
b 0.10 mol dm^{-3}
c 1.0×10^{-4} mol dm^{-3}

A simple diagram for summarising the relationship between hydrogen ion concentration, hydroxide ion concentration and pH (at 25 °C) was devised some years ago by Professor H N Alyea of Princeton University, USA and is shown in Fig. 7.1 in a modified form.

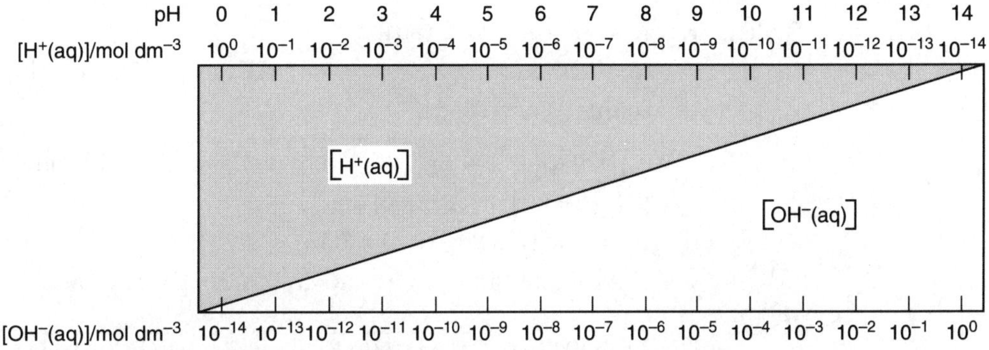

Figure 7.1
The relationship between $[H^+(aq)]$, $[OH^-(aq)]$ and pH.

In neutral solutions at 25 °C, $[H^+(aq)] = [OH^-(aq)] = 1 \times 10^{-7}$ mol dm^{-3}. However, as the concentration of one of them increases, the other must decrease accordingly. You can see this for yourself by placing a ruler vertically at any of the points indicated along the diagram; you can also deduce that $[H^+(aq)] \times [OH^-(aq)] = 1 \times 10^{-14}$ mol^2 dm^{-6}.

Calculating pH from $[H^+(aq)]$

We now take you a step further in pH calculations. For each calculation read the Worked Example, then try the Exercises which follow it.

You may find examples in textbooks that make use of logarithm tables. However, since examination boards now permit the use of electronic calculators in the examination, some Worked Examples in this chapter will show you how to solve problems with the aid of a scientific calculator. Some brands of calculator may operate in slightly different ways; if in doubt, consult the calculator's instruction manual.

WORKED EXAMPLE

Calculate the pH of a solution whose hydrogen ion concentration is 4.2×10^{-3} mol dm^{-3}.

Solution

Substitute the hydrogen ion concentration into the expression:

$$pH = -\log([H^+(aq)]/\text{mol dm}^{-3})$$
$$pH = -\log(4.2 \times 10^{-3}) = \mathbf{2.38}$$

For a calculator like a Casio Scientific press the following keys:

Keys to press	4	.	2	EXP	3	+/−	log	+/−
Effect of keys	4.2			$\times 10^{-3}$			log	−→+

Now try the following two Exercises.

EXERCISE 7.8
Answer on page 262

Calculate the pH of a solution whose hydrogen ion concentration is 7.0×10^{-7} mol dm^{-3}.

EXERCISE 7.9
Answer on page 262

Calculate the pH of a solution whose hydrogen ion concentration is 3.2×10^{-9} mol dm^{-3}.

Now we look at the effect of temperature. In working through the following Worked Example and Exercise you learn how temperature affects the pH of pure water or a neutral solution.

WORKED EXAMPLE

Calculate the pH of pure water at 10 °C, at which temperature $K_w = 0.30 \times 10^{-14}$ mol^2 dm^{-6}.

Solution

$K_w = [H^+(aq)][OH^-(aq)]$

In pure water: $[H^+(aq)] = [OH^-(aq)]$

so that $K_w = [H^+(aq)]^2$

$\therefore [H^+(aq)] = \sqrt{0.30 \times 10^{-14} \text{ mol}^2 \text{ dm}^{-6}} = 5.48 \times 10^{-8} \text{ mol dm}^{-3}$

$pH = -\log([H^+(aq)]/\text{mol dm}^{-3})$

$= -\log(5.48 \times 10^{-8}) = \mathbf{7.3}$

Again we show you how to do this using a calculator.

Keys to press	.	3	EXP	1	4	+/−	√	log	+/−
Effect of keys	0.30		$\times 10^{-14}$				√	log	−→+

Warning note!
It is important to note that when answering a question involving a calculation, particularly in an examination, the method of working, i.e. the steps in the calculation, **must** be shown as in the above example, even though the answer can be obtained with a calculator by a simple series of keystrokes.

Now try the following Exercise.

EXERCISE 7.10
Answer on page 262

Calculate the pH of pure water at 50 °C. ($K_w = 5.47 \times 10^{-14}$ mol^2 dm^{-6} at 50 °C.)

Compare your answer for this Exercise with that of the previous Worked Example and note how the pH of pure water (or a neutral solution) is affected by temperature. It has a value of 7.0 only at 25 °C.

Now we take you through the calculation of hydrogen ion concentration, given the pH of the solution.

Calculating $[H^+(aq)]$ from pH

WORKED EXAMPLE

What is the hydrogen ion concentration in a solution with a pH of 4.0?

Solution

1. Substitute the pH value in the expression:

$$pH = -\log([H^+(aq)]/mol\ dm^{-3})$$
$$4.0 = -\log([H^+(aq)]/mol\ dm^{-3})$$

2. Rearrange this expression:

$$\log([H^+(aq)]/mol\ dm^{-3}) = -4.0$$

3. Take the antilog:

$$[H^+(aq)] = 10^{-4.0}\ mol\ dm^{-3}$$

or in the standard form:

$$[H^+(aq)] = \mathbf{1.0 \times 10^{-4}\ mol\ dm^{-3}}$$

Now try the following Exercise.

EXERCISE 7.11

Answers on page 263

Calculate the hydrogen ion concentration in solutions with:
a pH = 9.0
b pH = 2.0

We now take you through a Worked Example where you are required to determine the hydrogen ion concentration from a pH value which is not an integer.

WORKED EXAMPLE

What is the hydrogen ion concentration of a solution whose pH is 3.70?

Solution

1. Substitute the pH value in the expression:

$$pH = -\log([H^+(aq)]/mol\ dm^{-3})$$
$$3.70 = -\log([H^+(aq)]/mol\ dm^{-3})$$

2. Rearrange this expression:

$$\log([H^+(aq)]/mol\ dm^{-3}) = -3.70$$

3. Take the antilog:

$$[H^+(aq)] = 10^{-3.70}\ mol\ dm^{-3}$$

4. Use the calculator to convert this to the standard form:

$$[H^+(aq)] = \mathbf{2.0 \times 10^{-4}\ mol\ dm^{-3}}$$

Step 4 is done by pressing the following keys:

Keys to press	[3] [.] [7]	[+/−]	[INV] [log]
Effect of keys	3.7	+→−	antilog

Note that on some calculators the 'INV' key is marked 'SHIFT'.
Now try the following Exercise yourself.

EXERCISE 7.12
Answers on page 262

Calculate the hydrogen ion concentration in solutions with:

a pH = 9.20
b pH = 2.63

So far, we have used pH as an indication of the strength of an acid. However, pH varies with concentration so that, to compare the strengths of different acids, we need another measure. This is the subject of the following section.

RELATIVE STRENGTHS OF ACIDS

Consider again the equation for the dissociation in water of a general acid which we represent as HA.

$$HA(aq) \rightleftharpoons H^+(aq) + A^-(aq)$$

The further the position of equilibrium lies to the right, the more hydrogen ions are present and the stronger the acid is considered to be.

Acids such as HCl, H_2SO_4 and HNO_3 are classified as **strong acids** because they can dissociate almost completely. Thus, we could write the equation for the dissociation of HCl in water as:

$$HCl(aq) \rightarrow H^+(aq) + Cl^-(aq)$$

Notice that we have removed the reverse arrow, indicating that the reaction 'goes to completion'.

Acids which are not fully dissociated in water are classified as **weak acids**. Note that the terms **weak** and **dilute** have different meanings in chemistry; the same applies to the terms **strong** and **concentrated**. Make sure you understand the difference.

In the next section you apply to acids what you learned about equilibrium constants in Chapter 6.

Dissociation constants of acids

The pH of a solution is not a very reliable measure of the strength of an acid because pH varies with concentration. There is another physical quantity that we could use which is constant at all dilutions – the equilibrium constant for the dissociation of an acid in water.

Consider the dissociation of ethanoic acid in water.

$$CH_3CO_2H(aq) + H_2O(l) \rightleftharpoons CH_3CO_2^-(aq) + H_3O^+(aq)$$

The expression for the equilibrium constant for this reaction is:

$$K_c = \frac{[CH_3CO_2^-(aq)][H_3O^+(aq)]}{[H_2O(l)][CH_3CO_2H(aq)]}$$

In practice we always omit the $[H_2O(l)]$ term in all expressions which refer to the dissociation of an acid in water. The water is present in such great excess that its concentration remains virtually constant and can be combined in the expression with the constant, K_c.

If we put the $[H_2O(l)]$ term on the other side of the expression above, it becomes:

$$K_c[H_2O(l)] = \frac{[CH_3CO_2^-(aq)][H_3O^+(aq)]}{[CH_3CO_2H(aq)]}$$

Since constant × constant = another constant, we let $K_c \times [H_2O(l)] = K_a$.

We call K_a the **acid dissociation constant** or **ionisation constant**, which we can now express as:

$$K_a = \frac{[CH_3CO_2^-(aq)][H_3O^+(aq)]}{[CH_3CO_2H(aq)]}$$

This can also be written in the form:

$$K_a = \frac{[CH_3CO_2^-(aq)][H^+(aq)]}{[CH_3CO_2H(aq)]}$$

This last expression is based on the equation:

$$CH_3CO_2H(aq) \rightleftharpoons CH_3CO_2^-(aq) + H^+(aq)$$

in which the water is omitted. This form of writing the expression for the dissociation constant, K_a, of an acid is the one most commonly used in textbooks and is the form we have chosen to adopt in this book. So, for a general acid HA dissociating in water, we will normally write:

$$HA(aq) \rightleftharpoons H^+(aq) + A^-(aq)$$

Hence, we will usually write the equilibrium law expression for the dissociation of an acid in the form:

$$\boldsymbol{K_a = \frac{[H^+(aq)][A^-(aq)]}{[HA(aq)]}}$$

Warning note!

We must warn you, as we did in Chapter 6, that some examination boards require you to use the subscript 'eqm' after each concentration term in dissociation constant expressions.

K_a values, unlike pH values, are unaffected by concentration changes and are influenced only by changes in temperature. So the value of K_a provides an accurate measure of the extent to which an acid dissociates, i.e. the strength of that acid.

It is important to note that K_a is a constant for **weak acids only**; for strong acids it is large and variable. You must remember that K_a is an equilibrium constant expressed in terms of concentration and it was pointed out in Chapter 6, that calculations involving such equilibrium constants may be reasonably accurate, provided that ionic concentration is low. This will always be true for a solution of a weak acid at **any** concentration because an increase in the concentration of the acid will result in a decrease in the degree of ionisation.

Where to draw the borderline between weak and strong acids is a matter of judgement; it is simply that for higher values of K_a the calculation results become less accurate.

In the next two sections we show you how to calculate K_a from pH for a weak acid solution, and the reverse process. For each type of calculation, read the Worked Example and then try the Exercises which follow it.

Calculating K_a from $[H^+(aq)]$ (or from pH)

We can use the expression for the dissociation constant of an acid, HA:

$$K_a = \frac{[H^+(aq)][A^-(aq)]}{[HA(aq)]}$$

to calculate a value of K_a for an acid solution if we know the initial concentration of the acid and its hydrogen ion concentration (or pH). You can see how this is done by working through the following Worked Example.

WORKED EXAMPLE A 0.100 M solution of an acid, HA, has a hydrogen ion concentration of 7.94×10^{-6} mol dm^{-3} at 25 °C. Calculate the dissociation constant of this acid at this temperature.

Solution 1. Write the equation and above it insert the initial concentrations.

$$\begin{array}{llll} \text{Initial concn./mol dm}^{-3} & 0.100 & 0 & 0 \\ & HA(aq) \rightleftharpoons & H^+(aq) + & A^-(aq) \end{array}$$

2. Write the equilibrium concentrations under the equation. There must be the same concentration of A^- ions produced as hydrogen ions, i.e. 7.94×10^{-6} mol dm^{-3}.

The equilibrium concentration of HA = initial concentration − concentration reacted
$= (0.100 - 7.94 \times 10^{-6})$ mol dm^{-3}

$$\begin{array}{lcccccc} \text{Initial concn./mol dm}^{-3} & 0.100 & & 0 & & 0 \\ & HA(aq) & \rightleftharpoons & H^+(aq) & + & A^-(aq) \\ \text{Equilibrium concn./mol dm}^{-3} & (0.100 - 7.94 \times 10^{-6}) & & 7.94 \times 10^{-6} & & 7.94 \times 10^{-6} \end{array}$$

Note that we have ignored the hydrogen ions which arise from the ionisation of water. We can do this because, even in pure water, $[H^+(aq)]$ is as small as 1.0×10^{-7} mol dm^{-3}, and the addition of an acid suppresses the ionisation still further.

3. Write the expression for K_a for this acid and substitute the equilibrium concentrations:

$$K_a = \frac{[H^+(aq)][A^-(aq)]}{[HA(aq)]}$$

$$= \frac{(7.94 \times 10^{-6}\ \text{mol dm}^{-3})^2}{(0.100 - 7.94 \times 10^{-6})\ \text{mol dm}^{-3}}$$

But $7.94 \times 10^{-6} \ll 0.100$

$\therefore (0.100 - 7.94 \times 10^{-6}) \simeq 0.100$

i.e. the ionisation makes virtually no difference to the concentration of HA.

$$\therefore K_a = \frac{(7.94 \times 10^{-6}\ \text{mol dm}^{-3})^2}{(0.100\ \text{mol dm}^{-3})}$$

$$= \mathbf{6.30 \times 10^{-10}\ mol\ dm^{-3}}$$

The following two Exercises are similar: you must decide for yourself whether you can make the approximation which we made in the Worked Example.

EXERCISE 7.13
Answer on page 263

The pH of 0.010 M carbonic acid solution, H_2CO_3, is 4.17 at 25 °C. Calculate the value of the dissociation constant, K_a, at this temperature.

EXERCISE 7.14
Answer on page 263

The pH of 5.0×10^{-3} M benzoic acid solution, $C_6H_5CO_2H$, is 3.28 at 25 °C. Calculate the value of the dissociation constant at this temperature.

Calculating pH from K_a

Now we take you through a calculation where you are given a value for the dissociation constant of an acid and you have to determine its pH at a particular concentration, i.e. the reverse of our previous Exercise.

WORKED EXAMPLE

What is the pH of a 0.100 M solution of ethanoic acid at 25 °C? ($K_a = 1.7 \times 10^{-5}$ mol dm^{-3}.)

Solution

1. Write the equation for the dissociation of ethanoic acid, leaving space above and below for initial and equilibrium concentrations:

Initial
concn./mol dm^{-3} $CH_3CO_2H(aq) \rightleftharpoons CH_3CO_2^-(aq) + H^+(aq)$
Equilibrium
concn./mol dm^{-3}

2. Indicate the initial concentrations above the equation:

Initial concn./mol dm^{-3}	0.100	0	0
	$CH_3CO_2H(aq) \rightleftharpoons$	$CH_3CO_2^-(aq)$ +	$H^+(aq)$

3. Let $[H^+(aq)] = x$ mol dm^{-3}. There must also be the same concentration of ethanoate ions produced.
If x mol dm^{-3} of ions are produced, this must leave $(0.100 - x)$ mol dm^{-3} of ethanoic acid undissociated.

4. Now write the equilibrium concentrations under the equation:

Initial concn./mol dm^{-3}	0.100	0	0
	$CH_3CO_2H(aq) \rightleftharpoons$	$CH_3CO_2^-(aq)$ +	$H^+(aq)$
Equilibrium concn. /mol dm^{-3}	$0.100 - x$	x	x

5. Write the expression for K_a for this reaction and substitute the equilibrium concentrations in the expression:

$$K_a = \frac{[CH_3CO_2^-(aq)][H^+(aq)]}{[CH_3CO_2H(aq)]}$$

$$1.7 \times 10^{-5} \text{ mol dm}^{-3} = \frac{(x \text{ mol dm}^{-3})^2}{(0.100 - x) \text{ mol dm}^{-3}}$$

6. Decide whether x is small enough to put $(0.100 - x) \simeq 0.100$. (If this is the case, your calculation is simplified; if not, you must solve a quadratic equation, as you did in Chapter 6.)

Clearly, x is small because K_a is small, but is it small enough to be ignored relative to 0.100? A simple 'rule-of-thumb' is that if the initial acid concentration, c_o, divided by K_a, is greater than 1000, then $c_o - x \simeq c_o$.

$$\text{i.e. if } \frac{c_o}{K_a} > 1000 \text{ then } c_o - x \simeq c_o$$

7. Here, $\dfrac{c_o}{K_a} = \dfrac{0.10}{1.7 \times 10^{-5}} = 5880$

   $\therefore c_o - x \simeq c_o$

   and you can write:

$$1.7 \times 10^{-5} = \frac{x^2}{0.100}$$

$$x^2 = 1.7 \times 10^{-5} \times 0.100$$

$$x = \sqrt{(1.7 \times 10^{-6})} = 1.3 \times 10^{-3}$$

$$\therefore [H^+(aq)] = 1.3 \times 10^{-3}\ \text{mol dm}^{-3}$$

8. Substitute this value of $[H^+(aq)]$ into the expression which defines pH:

$$\text{pH} = -\log([H^+(aq)]/\text{mol dm}^{-3}) = -\log(1.3 \times 10^{-3}) = \mathbf{2.9}$$

   The validity of the simplifying assumption made in step 7 can be shown by solving the quadratic equation at the end of step 5 – this gives the same answer, but by a tedious route.

   A simpler check is to compare the calculated value of x with the initial concentration:

$$c_o - x = (0.100 - 0.0013)\ \text{mol dm}^{-3} = 0.099\ \text{mol dm}^{-3}$$

   i.e. $c_o - x \simeq c_o$ and the assumption was justified.

To sum up then, we can say that, for a solution of a weak acid, if $c_o - x \simeq c_o$, the dissociation constant expression simplifies to $K_a = [H^+(aq)]^2/c_o$ and so $[H^+(aq)] = \sqrt{K_a \times c_o}$.

If, however, x is too large to make the simplifying assumption, we can still avoid the chore of solving a quadratic equation by using the method of successive approximations. The following Worked Example should make this clear.

WORKED EXAMPLE What is the pH of a 0.100 M solution of nitric(III) acid, HNO_2, at 25 °C?

$$HNO_2(aq) \rightleftharpoons H^+(aq) + NO_2^-(aq)$$
$$(K_a = 4.7 \times 10^{-4}\ \text{mol dm}^{-3})$$

Solution

1. Proceed as in the previous Worked Example, steps 1 to 4, i.e. write the equation and the initial and equilibrium concentrations.

Initial concn./mol dm^{-3}	0.100	0	0
	$HNO_2(aq) \rightleftharpoons$	$H^+(aq)$ +	$NO_2^-(aq)$
Equilibrium concn./mol dm^{-3}	$0.100 - x$	x	x

2. Write the expression for K_a for the reaction and substitute the equilibrium concentrations in the expression:

$$K_a = \frac{[H^+(aq)][NO_2^-(aq)]}{[HNO_2(aq)]}$$

$$4.7 \times 10^{-4}\ \text{mol dm}^{-3} = \frac{(x\ \text{mol dm}^{-3})^2}{(0.100 - x)\ \text{mol dm}^{-3}}$$

3. Put $(0.100 - x) = 0.100$ and solve the equation for x:

$$4.7 \times 10^{-4} = \frac{x^2}{0.100}$$

$$x^2 = 4.7 \times 10^{-4} \times 0.100$$

$$x = \sqrt{4.7 \times 10^{-5}} = 6.9 \times 10^{-3}$$

4. Substitute this value for x in the denominator of the expression written in step 2 and again solve the equation for x:

$$x = \sqrt{4.7 \times 10^{-4}(0.1 - 6.9 \times 10^{-3})} = 6.6 \times 10^{-3}$$

If this new value for x is substituted again in the denominator of the expression and x is recalculated, then almost the same value for x is obtained. There is rarely any need to recalculate x more than once.

$$\therefore [H^+(aq)] = 6.6 \times 10^{-3}\ \text{mol dm}^{-3}$$

5. Calculate the pH of the solution.

$$\text{pH} = -\log([H^+(aq)]/\text{mol dm}^{-3}) = -\log(6.6 \times 10^{-3}) = \mathbf{2.2}$$

In the following two Exercises, you must judge whether or not to make the simplifying assumption made in the Worked Example on page 145.

EXERCISE 7.15
Answers on page 263

a Given that the ionic product of water, $[H^+][OH^-]$, is 1.00×10^{-14} mol^2 dm^{-6} at 298 K, calculate to three significant figures the pH at this temperature of a 0.0500 molar aqueous solution of sodium hydroxide.
b The dissociation constant of ethanoic acid at 5 °C is 1.69×10^{-5} mol dm^{-3}. Calculate to two significant figures the pH of a 0.100 molar solution of this acid at 5 °C.

EXERCISE 7.16
Answer on page 263

Calculate the pH of a 0.100 M chloric(III) acid solution, $HClO_2$, at 25 °C.

$$HClO_2(aq) \rightleftharpoons H^+(aq) + ClO_2^-(aq)$$
$$(K_a = 0.010\ \text{mol dm}^{-3})$$

In the next section, you will apply what you have learned about pH calculations to the calculation of pH titration curves.

pH CHANGES DURING TITRATIONS

If an alkali is run into an acid in a titration, the pH of the mixture starts at a low value, depending on the strength and concentration of the acid and, when excess alkali has been added, the pH reaches a high value, depending on the strength and concentration of the alkali.

The four types of pH curve correspond to four types of titration using a strong or weak acid and a strong or weak base.

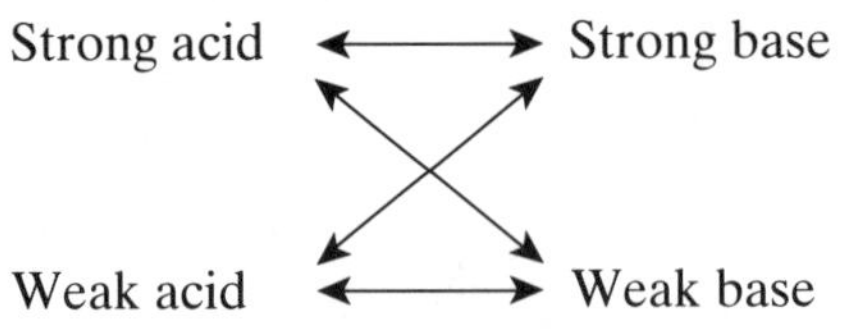

Specimen curves are shown below. Note in particular how the pH changes near the equivalence point.

Figure 7.2

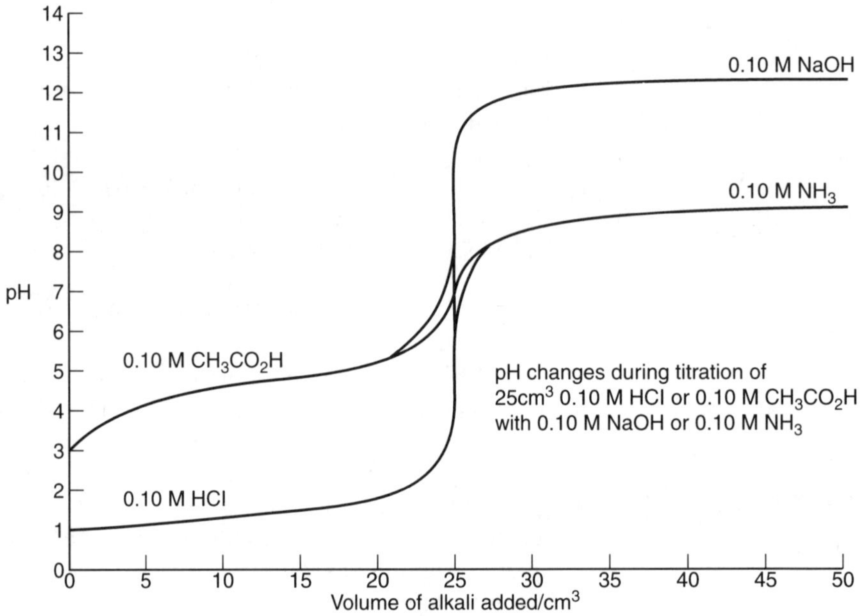

pH curves can be drawn from calculated values of pH, as we show in the next section.

pH curves by calculation

It is possible to calculate accurately the pH at any point during a titration from a knowledge of the dissociation constants for the acid and alkali, and their concentrations. We show you only the simplest type of calculation here because it is frequently included in examination papers; this is the calculation of pH for strong acid–strong base titrations.

You can calculate the pH at any stage during the titration by calculating the amount of excess acid before the equivalence point, or the amount of excess alkali after the equivalence point. From this amount you can calculate the pH as previously described.

WORKED EXAMPLE Calculate the pH of the solution resulting from adding 7.00 cm^3 of 0.100 M NaOH to 25.0 cm^3 of 0.100 M HCl.

Solution Amount of acid initially = cV = 0.100 mol dm^{-3} × 0.0250 dm^3 = 2.50 × 10^{-3} mol
Amount of alkali added = cV = 0.100 mol dm^{-3} × 0.007 00 dm^3 = 7.00 × 10^{-4} mol
According to the equation:

$$NaOH(aq) + HCl(aq) \rightarrow NaCl(aq) + H_2O(l)$$

the amount of acid reacting = the amount of alkali added.

The amount of acid left = initial amount − amount reacted

$$= (2.50 \times 10^{-3} - 7.00 \times 10^{-4})\ \text{mol} = 1.80 \times 10^{-3}\ \text{mol}$$

Total volume of solution = $25.0\ \text{cm}^3 + 7.0\ \text{cm}^3 = 32.0\ \text{cm}^3$

$$\therefore [\text{H}^+(\text{aq})] = [\text{HCl}(\text{aq})] = n/V = \frac{0.001\,80\ \text{mol}}{0.0320\ \text{dm}^3} = 0.0563\ \text{mol dm}^{-3}$$

$$\therefore \text{pH} = -\log([\text{H}^+(\text{aq})]/\text{mol dm}^{-3}) = -\log(0.0563) = \mathbf{1.25}$$

WORKED EXAMPLE Calculate the pH of the solution resulting from adding 25.2 cm^3 of 0.100 M NaOH to 25.0 cm^3 of 0.0500 M H_2SO_4.

Solution Amount of acid initially = $cV = 0.0500\ \text{mol dm}^{-3} \times 0.0250\ \text{dm}^3 = 1.25 \times 10^{-3}$ mol
Amount of alkali added = $cV = 0.100\ \text{mol dm}^{-3} \times 0.0252\ \text{dm}^3 = 2.52 \times 10^{-3}$ mol
According to the balanced equation:

$$\text{H}_2\text{SO}_4(\text{aq}) + 2\text{NaOH}(\text{aq}) \rightarrow \text{Na}_2\text{SO}_4(\text{aq}) + 2\text{H}_2\text{O}(\text{l})$$

the amount of alkali reacting = twice the initial amount of acid.
The amount of excess alkali = amount added − amount reacted

$$= 2.52 \times 10^{-3}\ \text{mol} - (2 \times 1.25 \times 10^{-3})\ \text{mol}$$
$$= 2.0 \times 10^{-5}\ \text{mol}$$

Total volume of solution = $25.0\ \text{cm}^3 + 25.2\ \text{cm}^3 = 50.2\ \text{cm}^3$

$$\therefore [\text{OH}^-(\text{aq})] = [\text{NaOH}(\text{aq})] = n/V = \frac{2.00 \times 10^{-5}\ \text{mol}}{0.0502\ \text{dm}^3} = 3.98 \times 10^{-4}\ \text{mol dm}^{-3}$$

But $K_w = 1.00 \times 10^{-14}\ \text{mol}^2\ \text{dm}^{-6} = [\text{H}^+(\text{aq})][\text{OH}^-(\text{aq})]$

$$\therefore [\text{H}^+(\text{aq})] = \frac{K_w}{[\text{OH}^-(\text{aq})]} = \frac{1.00 \times 10^{-14}\ \text{mol}^2\ \text{dm}^{-6}}{3.98 \times 10^{-4}\ \text{mol dm}^{-3}} = 2.51 \times 10^{-11}\ \text{mol dm}^{-3}$$

$$\therefore \text{pH} = -\log([\text{H}^+(\text{aq})]/\text{mol dm}^{-3}) = -\log(2.51 \times 10^{-11}) = \mathbf{10.6}$$

Note that a difference of only 0.2 cm^3 of alkali between this point and the equivalence point at 25.0 cm^3 gives a large difference in pH, 7.0 → 10.6.

Use a method similar to that in the worked examples to do the next Exercise.

EXERCISE 7.17
Answers on page 264

25.0 cm^3 of 0.100 M HCl were titrated with 0.100 M NaOH.
Calculate the pH of the solution for each of the following volumes of NaOH:
0.0, 5.0, 10.0, 20.0, 24.0, 25.0, 26.0 and 30.0 cm^3.

BUFFER SOLUTIONS

Solutions containing substances that control the pH are known as buffers. The composition of a buffer is such that the addition of a small amount of a strong acid or a strong base, or even a moderate amount of water, has very little effect on the pH.

Buffer solutions usually contain a mixture of an acid and its conjugate base, one of which is weak.

We begin this section by looking again at a pH titration curve.

Buffering regions of titration curves

Consider the titration of ethanoic acid with sodium hydroxide. The pH curve for exactly 0.1 M solutions has this shape:

Figure 7.3

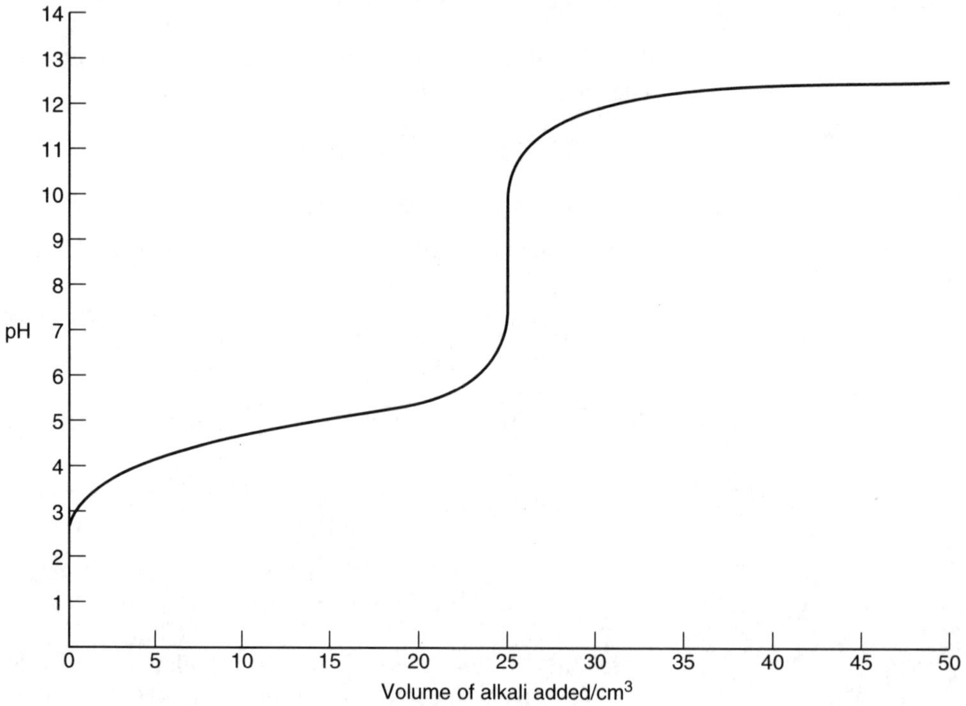

Notice that the curve flattens out in the region before neutralisation, i.e. there is a very small change in pH for the addition of a large amount of alkali. At this point in the titration, some ethanoic acid molecules have been neutralised by hydroxide ions from the alkali:

$$CH_3CO_2H(aq) + OH^-(aq) \rightleftharpoons CH_3CO_2^-(aq) + H_2O(l)$$

Although hydroxide ions are continually being added, the change in pH is slowed down by buffering action. As the above equation shows, the amount of acid anions increases as hydroxide ions are added. When the concentration of anions reaches about the same level as the concentration of un-ionised acid, and both are much greater than the concentration of free hydrogen ions, the solution is a buffer.

The acid dissociation constant is given by the expression:

$$K_a = \frac{[H^+(aq)][CH_3CO_2^-(aq)]}{[CH_3CO_2H(aq)]}$$

$$\therefore [H^+(aq)] = K_a \frac{[CH_3CO_2H(aq)]}{[CH_3CO_2^-(aq)]}$$

Small changes in the concentration of either un-ionised acid or its conjugate base have very little effect on the concentration of hydrogen ions so that the pH remains relatively constant.

In the next Exercise, you calculate the pH at the middle of the buffering region when ethanoic acid is titrated with sodium hydroxide.

EXERCISE 7.18
Answers on page 264

25.0 cm^3 of a solution of 0.10 M ethanoic acid are titrated with 0.10 M sodium hydroxide. When sufficient sodium hydroxide has been added to neutralise half the ethanoic acid (the 'half neutralisation point'), calculate:

a the concentrations of ethanoic acid and ethanoate ions,
b the pH of the solution.
$K_a(CH_3CO_2H) = 1.7 \times 10^{-5}$ mol dm^{-3}.

The relationship between pH, K_a and composition of a buffer

Consider an 'acid' buffer solution containing a weak acid HA and a salt of that acid Na^+A^-. For calculations we use the now familiar dissociation constant expression,

$$K_a = \frac{[H^+(aq)][A^-(aq)]}{[HA(aq)]}$$

In such a buffer solution we assume that the anion concentration $[A^-(aq)]$ is the same as that of the salt and that the concentration of the undissociated form of the acid [HA(aq)] is the same as the total acid concentration. Although not strictly true, these assumptions are accurate enough so long as [HA(aq)] and $[A^-(aq)]$ do not become too small, i.e. the solution is not too dilute and the ratio $[HA]/[A^-]$ is not too large or too small. Of course, K_a must also be small.

It is sometimes convenient to use a logarithmic form of the above equation where $-\log[H^+(aq)]$ can be replaced by pH and $-\log K_a$ by pK_a. In the next Exercise you derive the expression for pH of a buffer containing the weak acid HA.

EXERCISE 7.19
Answer on page 264

For a buffer solution made from the weak acid, HA, and a salt, Na^+A^-, derive the relationship:

$$pH = pK_a - \log \frac{[HA(aq)]}{[A^-(aq)]}$$

It is worth noting that when $[HA(aq)] = [A^-(aq)]$, i.e., when the ratio $[HA(aq)]/[A^-(aq)] = 1$ (when the solution exhibits its greatest buffering capacity):

$$[H^+(aq)] = K_a \quad \text{or} \quad pH = pK_a$$

The logarithmic form of the equation should also help you to realise that, provided that [HA(aq)] and $[A^-(aq)]$ are sufficiently large, the addition of relatively small amounts of acid or alkali will not significantly alter the ratio $[HA(aq)]/[A^-(aq)]$, and therefore the pH will change very little.

In the next two sections we show you how to use the dissociation constant expression (either in its original or in its logarithmic form) to calculate the pH or the composition of a buffer solution.

pH of a buffer solution

As you have just found out in Exercise 7.18, it is fairly simple to calculate the pH of a buffer solution, knowing K_a for the weak acid and the concentrations of acid and conjugate base in solution. In that Exercise it was particularly easy because the acid and base concentrations were the same. When these concentrations are different, a little more arithmetic is needed.

In the days when examination candidates had to use logarithm tables, there was sometimes an advantage in using the logarithmic form of the dissociation constant expression. However, now that you are allowed to use electronic calculators, it makes little difference which form of the expression that you use. (Remember that if you use the logarithmic form of the expression you may be expected to derive it first!)

For completeness, we now show you how to calculate the pH of a buffer solution, using both expressions.

WORKED EXAMPLE Calculate the pH of a buffer solution containing 0.50 mol dm^{-3} of methanoic acid, HCO_2H, and 2.5 mol dm^{-3} of potassium methanoate, HCO_2K. K_a for methanoic acid = 1.6×10^{-4} mol dm^{-3}.

Solution **a** Using the dissociation constant expression:

$$K_a = \frac{[H^+(aq)][HCO_2^-(aq)]}{[HCO_2H(aq)]}$$

$$\therefore [H^+(aq)] = K_a \frac{[HCO_2H(aq)]}{[HCO_2^-(aq)]}$$

$$= 1.6 \times 10^{-4} \text{ mol dm}^{-3} \times \frac{0.5 \text{ mol dm}^{-3}}{2.5 \text{ mol dm}^{-3}}$$

$$= 3.2 \times 10^{-5} \text{ mol dm}^{-3}$$

$$\therefore \text{pH} = -\log[3.2 \times 10^{-5}) = \mathbf{4.5}$$

Alternative Solution **b** Using the logarithmic form of the expression:

$$\text{pH} = \text{p}K_a - \log \frac{[HCO_2H(aq)]}{[HCO_2^-(aq)]}$$

$$= -\log(1.6 \times 10^{-4}) - \log \frac{0.5}{2.5}$$

$$= -(-3.8) - (-0.7) = \mathbf{4.5}$$

To make sure that you can do this type of calculation, try the next two Exercises, using either of the above methods. (In the specimen answers we use method **a**.)

EXERCISE 7.20
Answer on page 264

Calculate the pH of a solution which contains 0.10 mol dm^{-3} of propanoic acid and 0.050 mol dm^{-3} of sodium propanoate. $K_a(CH_3CH_2CO_2H) = 1.3 \times 10^{-5}$ mol dm^{-3}.

EXERCISE 7.21
Answers on page 265

Calculate the pH of the following buffer solutions containing butanoic acid and potassium butanoate. $K_a(CH_3CH_2CH_2CO_2H) = 1.5 \times 10^{-5}$ mol dm^{-3}.

Table 7.2

$[CH_3CH_2CH_2CO_2H]$/mol dm^{-3}	$[CH_3CH_2CH_2CO_2^-K^+]$/mol dm^{-3}
a 1.0	1.0
b 2.0	1.0
c 2.0	0.5
d 0.5	2.0

Composition of a buffer solution

Again we make use of the dissociation constant expression, either in its original or its logarithmic form. We give a Worked Example for each form. Study these, then try the Exercises which follow, using either method.

WORKED EXAMPLE

A buffer of pH 3.50 is to be made by adding sodium iodoethanoate, $ICH_2CO_2^-Na^+$, to 0.500 M iodoethanoic acid, ICH_2CO_2H. What concentration of sodium iodoethanoate is needed? $K_a(ICH_2CO_2H) = 6.8 \times 10^{-4}$ mol dm^{-3}.

Solution

Using the original form of the dissociation constant expression:

1. Write the dissociation constant expression for iodoethanoic acid:

$$K_a = \frac{[H^+(aq)][ICH_2CO_2^-(aq)]}{[ICH_2CO_2H(aq)]}$$

2. Rearrange to give an expression for the concentration of the salt:

$$[ICH_2CO_2^-(aq)] = \frac{K_a[ICH_2CO_2H(aq)]}{[H^+(aq)]}$$

3. Substitute the given values (remembering that $[H^+] = 10^{-pH}$) and complete the arithmetic:

$$[ICH_2CO_2^-(aq)] = \frac{6.8 \times 10^{-4}\ \text{mol dm}^{-3} \times 0.5\ \text{mol dm}^{-3}}{10^{-3.5}\ \text{mol dm}^{-3}}$$

$$= \frac{6.8 \times 10^{-4} \times 0.5\ \text{mol dm}^{-3}}{3.16 \times 10^{-4}}$$

$$= 1.07\ \text{mol dm}^{-3}$$

∴ the concentration of sodium iodoethanoate needed is **1.07 mol dm^{-3}**

WORKED EXAMPLE

In what proportions must 0.020 M solutions of benzoic acid, $C_6H_5CO_2H$, and sodium benzoate, $C_6H_5CO_2Na$, be mixed to give a buffer solution of pH 4.5? K_a for benzoic acid is 6.4×10^{-5} mol dm^{-3}.

Solution

Using the logarithmic form of the dissociation constant expression:

1. Write the logarithmic form of the dissociation constant expression for benzoic acid (or preferably derive it):

$$pH = pK_a - \log \frac{[C_6H_5CO_2H(aq)]}{[C_6H_5CO_2^-(aq)]}$$

2. Rearrange the expression:

$$\log \frac{[C_6H_5CO_2H(aq)]}{[C_6H_5CO_2^-(aq)]} = pK_a - pH$$

3. Substitute the known values of K_a and pH and complete the calculation:

$$\log \frac{[C_6H_5CO_2H(aq)]}{[C_6H_5CO_2^-(aq)]} = -\log(6.4 \times 10^{-5}) - 4.5$$

$$= -0.3$$

$$\therefore \frac{[C_6H_5CO_2H(aq)]}{[C_6H_5CO_2^-(aq)]} = 10^{-0.3}$$

$$= 0.5$$

4. Since K_a is small, assume that the acid remains undissociated and that the salt is completely ionised.
   Then the mixture should be made in the proportions of one volume of 0.02 M benzoic acid to two volumes of 0.02 M sodium benzoate.

EXERCISE 7.22
Answers on page 265

In what ratio must 0.50 M solutions of ethanoic acid and sodium ethanoate be mixed to produce buffer solutions of:
a pH 4.70,
b pH 4.40?
$pK_a(CH_3CO_2H) = 4.74$.

EXERCISE 7.23
Answers on page 265

What amount of sodium methanoate, $HCO_2^-Na^+$, must be added per dm^3 of 0.50 M solutions of methanoic acid, HCO_2H, to produce buffer solutions of:
a pH = 3.8,
b pH = 4.1?
$K_a(HCO_2H) = 1.58 \times 10^{-4}$ mol dm^{-3}.

Basic buffers

A weak base and one of its salts in solution can be used as a buffer system. In solution, a weak base such as phenylamine, $C_6H_5NH_2$, tends to accept a proton to become its conjugate acid:

$$C_6H_5NH_2(aq) + H_2O(l) \rightleftharpoons C_6H_5NH_3^+(aq) + OH^-(aq)$$

For this reaction:

$$K_c = \frac{[C_6H_5NH_3^+(aq)][OH^-(aq)]}{[C_6H_5NH_2(aq)][H_2O(l)]}$$

Since water is present in excess, its concentration can be regarded as constant and we can combine it with K_c to give $K_c \times [H_2O(l)] = K_b$, the base dissociation constant. The expression then becomes:

$$K_b = \frac{[C_6H_5NH_3^+(aq)][OH^-(aq)]}{[C_6H_5NH_2(aq)]}$$

However, since any calculation of pH involves hydrogen ion concentration at some point, it is more convenient to treat the system instead as a weak acid dissociating and use K_a in the usual way. The above system becomes:

$$C_6H_5NH_3^+(aq) \rightleftharpoons C_6H_5NH_2(aq) + H^+(aq)$$

and

$$K_a = \frac{[C_6H_5NH_2(aq)][H^+(aq)]}{[C_6H_5NH_3^+(aq)]}$$

Remembering that $K_w = [H^+(aq)][OH^-(aq)]$, you can see from the expressions for K_a and K_b that:

$$K_b = \frac{K_w}{K_a}$$

To make sure that you are confident with this way of treating basic buffers, read through the next Worked Example and try the Exercise following it.

WORKED EXAMPLE A buffer solution is prepared by mixing 750 cm^3 of 0.500 M phenylmethylamine, $C_6H_5CH_2NH_2$, and 250 cm^3 of 1.00 M phenylmethylamine hydrochloride, $C_6H_5CH_2NH_3^+Cl^-$. Assuming that K_a for phenylmethylamine hydrochloride is 4.30×10^{-10} mol dm^{-3}, calculate the pH of the buffer solution.

Solution

1. Write an equation representing the equilibrium as the dissociation of an acid:

$$C_6H_5CH_2NH_3^+(aq) \rightleftharpoons C_6H_5CH_2NH_2(aq) + H^+(aq)$$

2. Write an expression for K_a and rearrange it to give an expression for $[H^+(aq)]$:

$$K_a = \frac{[C_6H_5CH_2NH_2(aq)][H^+(aq)]}{[C_6H_5CH_2NH_3^+(aq)]}$$

$$\therefore [H^+(aq)] = K_a \frac{[C_6H_5CH_2NH_3^+(aq)]}{[C_6H_5CH_2NH_2(aq)]}$$

Note that this is simply another example of the general equation:

$$[H^+(aq)] = K_a \times \frac{[\text{acid}]}{[\text{conjugate base}]}$$

3. Calculate the amount of each substance added:
$n(C_6H_5CH_2NH_2) = cV = 0.500 \text{ mol dm}^{-3} \times 0.750 \text{ dm}^3 = 0.375$ mol
$n(C_6H_5CH_2NH_3^+Cl^-) = cV = 1.00 \text{ mol dm}^{-3} \times 0.250 \text{ dm}^3 = 0.250$ mol
4. Since K_a is so small, assume that these amounts are also the equilibrium amounts, and substitute them in the expression for $[H^+(aq)]$:

$$[H^+(aq)] = K_a \times \frac{[C_6H_5CH_2NH_3^+(aq)]}{[C_6H_5CH_2NH_2(aq)]}$$

$$= 4.30 \times 10^{-10} \text{ mol dm}^{-3} \times \frac{0.250 \text{ mol}/V}{0.375 \text{ mol}/V}$$

$$= 2.87 \times 10^{-10} \text{ mol dm}^{-3}$$

5. Calculate the pH:

$$\text{pH} = -\log([\text{H}^+(\text{aq})]/\text{mol dm}^{-3})$$
$$= -\log(2.87 \times 10^{-10}) = \mathbf{9.54}$$

Now try a similar problem.

EXERCISE 7.24
Answer on page 265

Calculate the pH of a buffer solution made by mixing 750 cm^3 of 0.20 M ammonium chloride and 750 cm^3 of 0.10 M ammonia solution. Assume that K_a for the ammonium ion is 6.00×10^{-10} mol dm^{-3}.

In the next section, you learn how the colour changes of indicators can be explained by considering them as weak acids.

ACID–BASE INDICATORS

All acid–base indicators can be regarded as weak acids and, in general discussion, a simplified formula, HIn, is often used to represent an indicator. In aqueous solution, the indicator dissociates according to the equation:

$$\text{HIn(aq)} \rightleftharpoons \text{H}^+\text{(aq)} + \text{In}^-\text{(aq)}$$

Generally, the two species, HIn and In$^-$, have quite distinct colours, so that colour changes are observed if the equilibrium is shifted to the left, by adding hydrogen ions, or to the right, by removing hydrogen ions.

$$\underset{\text{Colour A}}{\text{HIn(aq)}} \rightleftharpoons \text{H}^+\text{(aq)} + \underset{\text{Colour B}}{\text{In}^-\text{(aq)}}$$

The **end-point** during an acid–base titration is the stage where the indicator used has a colour half-way between its extreme colours; i.e. when:

$$\underset{\text{Colour A}}{[\text{HIn(aq)}]} = \underset{\text{Colour B}}{[\text{In}^-\text{(aq)}]}$$

One of the secrets of successful titration is to choose the indicator so that the end-point coincides with the equivalence point. The correct choice of indicators depends on a simple mathematical relationship which you derive in the following Revealing Exercise.

REVEALING EXERCISE

Q1 Write a general equation for the dissociation of an indicator, HIn.

A1 $\text{HIn(aq)} \rightleftharpoons \text{H}^+\text{(aq)} + \text{In}^-\text{(aq)}$.

Q2 Write an expression for the equilibrium constant, K_{In}.

A2 $$K_{\text{In}} = \frac{[\text{H}^+\text{(aq)}][\text{In}^-\text{(aq)}]}{[\text{HIn(aq)}]}$$

Q3 Rearrange the expression to solve for $[H^+(aq)]$.

A3 $$[H^+(aq)] = K_{In} \times \frac{[HIn(aq)]}{[In^-(aq)]}$$

Q4 Take logarithms of the above expression.
Hint: $\log(A \times B) = \log A + \log B$.
Use this to separate $\log K_{In}$ and $\log \dfrac{[HIn(aq)]}{[In^-(aq)]}$.

A4 $$\log[H^+(aq)] = \log K_{In} + \log \frac{[HIn(aq)]}{[In^-(aq)]}$$

Q5 Multiply the above expression by -1.

A5 $$-\log[H^+(aq)] = -\log K_{In} - \log \frac{[HIn(aq)]}{[In^-(aq)]}$$

Q6 Replace $-\log[H^+(aq)]$ by pH and $-\log K_{In}$ by pK_{In}. (It is not only in pH that the prefix p is used to mean '$-$log'.)

A6 $$pH = pK_{In} - \log \frac{[HIn(aq)]}{[In^-(aq)]}$$

Q7 Simplify **A6** to obtain an expression for the pH at the end-point for the indicator.

A7 The end-point is when $[HIn(aq)] = [In^-(aq)]$.

Then, $\log \dfrac{[HIn(aq)]}{[In^-(aq)]} = \log 1 = 0$

$$\therefore \textbf{pH at the end-point} = \mathbf{p}K_{In}$$

The equation you derived in **A6** is, in fact, a particular case of a more general equation:

$$\mathbf{pH = p}K_a - \mathbf{log} \frac{\textbf{[acid]}}{\textbf{[conjugate base]}}$$

which you will find useful in other contexts too, as you have seen in calculations involving buffers. It is not easy to learn the equation, but you should certainly learn to do the derivation yourself.

Now you can use the equation to determine the pH range of an indicator.

■ Calculating the pH range of an indicator

If we assume that the 'acid' colour is completely obscured by the 'base' colour when the concentration of the base form is ten times the concentration of the acid form; and if we assume that the 'base' colour is obscured by the 'acid' colour when the concentration of the acid form is ten times that of the base form, we can establish the pH range of an indicator.

Colour change starts when $\frac{[HIn(aq)]}{[In^-(aq)]} = 10$ and $pH = pK_{In} - \log 10$
$= pK_{In} - 1$

Colour change ends when $\frac{[HIn(aq)]}{[In^-(aq)]} = \frac{1}{10}$ and $pH = pK_{In} - \log \frac{1}{10}$

$= pK_{In} + 1$

∴ the pH range = $(pK_{In} - 1)$ to $(pK_{In} + 1)$

Thus, the pH range of an indicator is usually about two pH units, one each side of pK_{In}, but the assumption of a ratio of ten to one in concentrations of the two forms for complete colour masking is not always precisely true. If you look at your data book, you will see that pH ranges vary a little from an average of two pH units.

Now try the last two Exercises.

EXERCISE 7.25
Answers on page 266

The equivalence point in a titration between 0.10 M HCl and 0.10 M NH_3 occurs at pH = 5. Consult your data book and:

a give one reason why phenolphthalein is not a suitable indicator,
b select an indicator you would expect to be suitable.

EXERCISE 7.26
Answers on page 266

The indicator methyl red has a pK_{In} value of 5.1. Its colours are red in acid solution and yellow in alkaline solution. The ionisation of this indicator is represented by:

$$HIn(aq) \rightleftharpoons H^+(aq) + In^-(aq)$$

a What is the colour of the In^- ion in solution?
b On the assumption that when the concentration of one coloured species exceeds the other by a factor of ten the only detectable colour is due to the species in excess, calculate the pH range over which methyl red changes colour.
c Calculate the pH of the resulting solution when 1.0 cm^3 of 0.010 M HCl is added to 499 cm^3 of water.
d A few drops of methyl red are added to the solution in **c**. Calculate the ratio of the concentration of the red form to that of the yellow form in solution.

END-OF-CHAPTER QUESTIONS

Answers on page 266

Questions 7.1 to 7.4 each contain five suggested answers. Select the best answer in each case. They are based on the following information.

A glass electrode connected to a pH meter was placed in 20 cm^3 of a solution containing 0.1 mol dm^{-3} of a substance X and a reading of the pH was taken. A solution of another substance Y was added to the first solution a little at a time and the pH noted after each addition. The resulting graph of pH against volume of solution Y added was as follows:

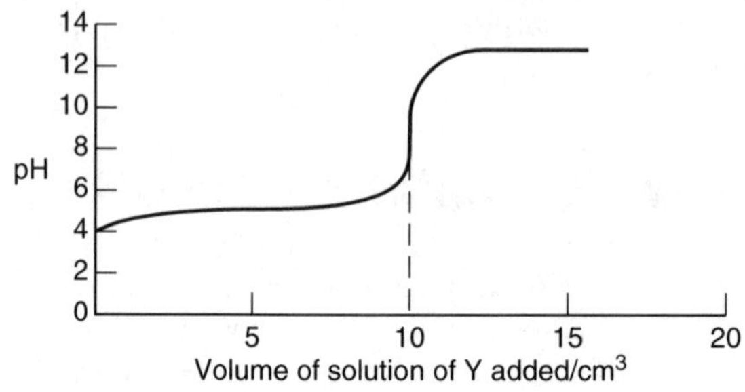

7.1 The graph shows that the most effective buffer solution, i.e. the solution which has the smallest change in pH on addition of small quantities of acid or alkali, would be:
A Solution X alone,
B A mixture of 20 cm^3 of solution X + 5 cm^3 of solution Y,
C A mixture of 20 cm^3 of solution X + 10 cm^3 of solution Y,
D A mixture of 20 cm^3 of solution X + 15 cm^3 of solution Y,
E Solution Y alone.

7.2 Which would be the best indicator to show the end-point of the reaction between X and Y?
A Congo red, $K_a = 10^{-4}$
B Methyl red, $K_a = 10^{-5}$
C Bromothymol blue, $K_a = 10^{-7}$
D Thymol blue, $K_a = 10^{-9}$
E Alizarin yellow, $K_a = 10^{-12.5}$

7.3 The hydrogen ion concentration in mol dm^{-3} in the solution of X was:
A 10^{-10}
B 10^{-4}
C 4
D 10^4
E 10^{10}

7.4 If 1 mol of X reacts with 1 mol of Y, the concentration in mol dm^{-3} of Y in its original solution was:
A 0.025
B 0.05
C 0.1
D 0.2
E 0.4

7.5 Define the term **pH of a solution** and calculate the pH values of:
a 5.00×10^{-4} M hydrochloric acid,
b 5.00×10^{-4} M sodium hydroxide solution.
$K_w = [H^+][OH^-] = 1.00 \times 10^{-14}$ mol^2 dm^{-6}.

7.6 **a** Calculate the pH of hydrochloric acid of concentration 1.00×10^{-2} mol dm^{-3} and ethanoic acid of concentration 1.00 mol dm^{-3} and use these results to distinguish between a dilute solution of a strong acid and a concentrated solution of a weak acid.
K_a for ethanoic acid is 1.74×10^{-5} mol dm^{-3}.
b 'The pH of pure water is 7.00 only at 298 K when the ionic product for water is 1.00×10^{-14} mol^2 dm^{-6}.' Explain what this statement means and calculate the pH of water at 363 K when the ionic product is 5.48×10^{-14} mol^2 dm^{-6}.

7.7 The pH of aqueous 0.01 M ethanoic acid is 3.4.
a Calculate the concentration of hydrogen ions in 0.01 M ethanoic acid.
b State the concentration of ethanoate ions in equilibrium with the hydrogen ions in 0.01 M ethanoic acid.
c Calculate K_a, the dissociation constant for ethanoic acid.

7.8 Write down an expression for the acid dissociation constant (K_a) for the equilibrium:

$$CH_3COOH(aq) \rightleftharpoons H^+(aq) + CH_3COO^-(aq)$$

and calculate pK_a, given that K_a is 1.78×10^{-5} mol l^{-1} at 25 °C.

7.9 The following data show how the pH changes during a titration when aqueous 0.100 M NaOH is added to 10 cm^3 of aqueous ethanoic (*acetic*) acid.

Table 7.3

Volume 0.100 M NaOH/cm^3	0.0	1.0	2.0	4.0	6.0	7.0	8.0	8.5	10.0	14.0
pH	2.9	4.0	4.3	4.7	5.2	5.5	6.4	11.2	12.0	12.4

a Use the data to plot a titration curve on graph paper.
b Calculate the initial concentration of the ethanoic acid.
c Using any suitable pH value, calculate K_a for ethanoic acid. Show your working.

7.10 The colour of blackberries is due to a compound known as cyanidin. At low pH, cyanidin (*Cy*) exists as *Cy*H^+, which is red, and at high pH as *Cy*, which is purple:

$$\underset{\text{red}}{CyH^+} \rightleftharpoons \underset{\text{purple}}{Cy} + H^+$$

a Write an expression for the acid dissociation constant, K_a, of *Cy*H^+.
b In a buffer of pH = 5.00, the ratio of the red to the purple form is 1:5. Calculate a value for K_a.
c Calculate the ratio of the red to the purple form in a fruit juice buffered at pH = 3.00, and hence predict its colour.

7.11 In a buffered solution:

$$pH = -\log K_a + \log \frac{[\text{base}]_{\text{eqm}}}{[\text{acid}]_{\text{eqm}}}$$

Human plasma is buffered mainly by dissolved carbon dioxide which has reacted to form carbonic(IV) acid:

$$H_2CO_3(aq) \rightleftharpoons H^+(aq) + HCO_3^-(aq)$$

a When the concentrations of carbonic(IV) acid and hydrogencarbonate ion are equal, the concentration of hydrogen ions is 7.9×10^{-7} mol dm^{-3}. Calculate the value of log K_a for carbonic(IV) acid.
b Usually the pH of human plasma is about 7.4. Calculate the **ratio** of the concentrations of hydrogencarbonate ion and carbonic(IV) acid in plasma.
c If the total concentration of hydrogencarbonate ion and carbonic(IV) acid was equivalent to 2.52×10^{-2} mol dm^{-3} of carbon dioxide, calculate the separate **concentrations** of the hydrogencarbonate ion and the carbonic(IV) acid in plasma.

Notice that the expression given for the pH is not quite the same as the one that you have used so far. However, the two forms are equivalent because:

$$\log \frac{[\text{base}]}{[\text{acid}]} = -\log \frac{[\text{acid}]}{[\text{base}]}$$

CHAPTER 8

EQUILIBRIUM III – REDOX REACTIONS

INTRODUCTION AND PRE-KNOWLEDGE

In this chapter we assume that you are familiar with the definitions of oxidation and reduction in terms of electron transfer (see also Chapter 2). We extend these ideas to account for the potential differences generated between two different metals when placed in an electrolyte and introduce the concept of standard electrode potentials, which are very useful in considering whether a particular chemical reaction is likely to occur in aqueous solution.

STANDARD ELECTRODE POTENTIALS

When a metal electrode is dipped into a solution of its ions, a potential difference is set up between the two. This can occur in two ways; either the metal atoms tend to form positive ions leaving the metal with a surplus of electrons, or the aqueous ions tend to gain electrons from the metal leaving the electrode with a net positive charge, as shown in Fig. 8.1.

Figure 8.1
Absolute electrode potentials.

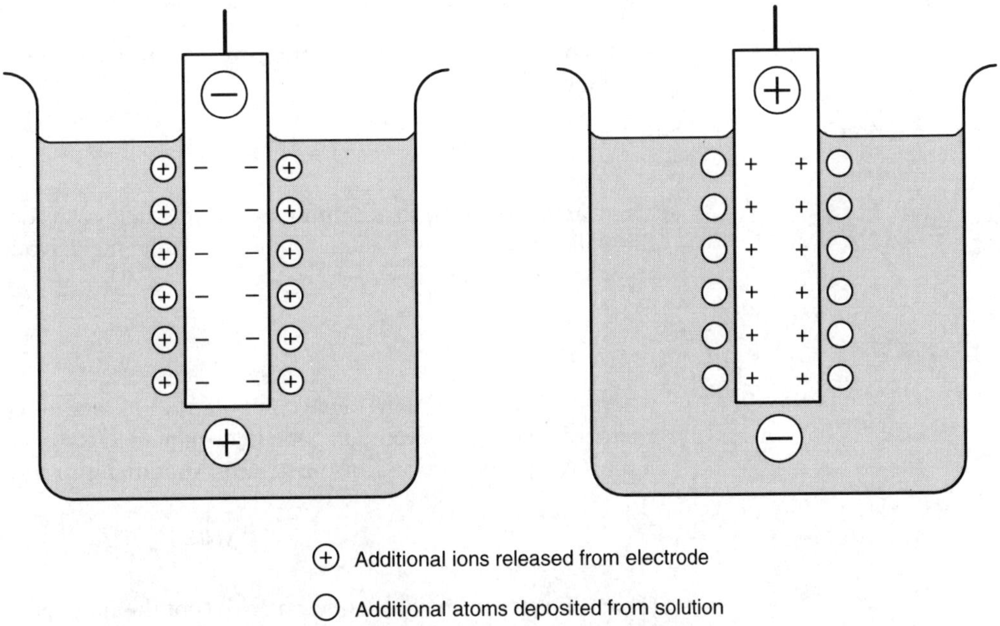

However, we cannot be quite certain which of these processes occurs in a particular case because we cannot measure the potential difference between electrode and solution (known as the absolute electrode potential). Measurement would require an electrical connection to be made between a voltmeter and the solution by inserting a metal wire or strip, which would inevitably act as another electrode with its own electrode potential.

That is to say, we can only measure the **difference** between two electrode potentials, never a single electrode potential. To overcome this difficulty, we arbitrarily assign an electrode potential of zero to one particular half-cell and compare all other half-cells with this standard. By international agreement, the standard hydrogen electrode has been chosen as the reference electrode.

The standard hydrogen electrode

Fig. 8.2 represents a standard hydrogen electrode connected by a salt bridge to another electrode, or half-cell, to make a cell. This cell generates a potential difference, known as the e.m.f., which can be measured by a high resistance voltmeter (i.e. one that allows no current to pass).

Figure 8.2

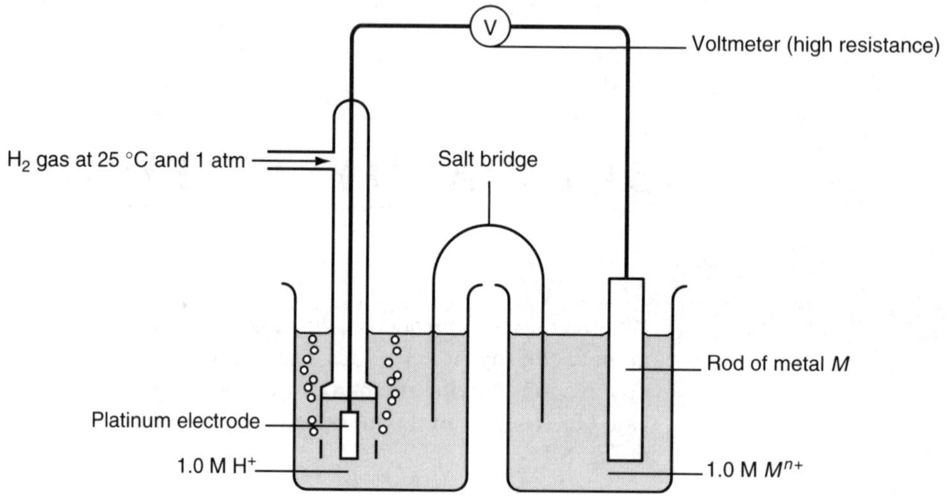

In this book we use the conventions by which such a cell is represented in a cell diagram, as follows:

$$\text{Pt(s), H}_2\text{(g)} \mid \text{H}^+\text{(aq, 1.0 M)} \;\vdots\vdots\; M^{n+}\text{(aq, 1.0 M)} \mid M\text{(s)}$$

The standard electrode potential, $E^\ominus$, of the metal M is defined as equal to the e.m.f. of this cell, $\Delta E^\ominus$, since we assume that the hydrogen electrode has zero potential under standard conditions, i.e.

$$\Delta E^\ominus = E_R{}^\ominus - E_H{}^\ominus$$

where $E_R{}^\ominus$ is the standard electrode potential of the right-hand electrode, and $E_H{}^\ominus$ is the standard electrode potential of the hydrogen electrode = 0.00 V.

This is a particular case of the general statement for any cell under any conditions:

$$\boldsymbol{\Delta E = E_R - E_L}$$

(R and L refer to the electrode potentials of the half-cells written on the right and left, respectively, of the cell diagrams.)

Note that, by convention, the **sign** of the e.m.f. is the polarity of the right-hand electrode in the cell diagram, and the hydrogen electrode is always written on the left. Because of this convention, standard electrode potentials of half-cells are tabulated as reduction reactions since they always refer to the half-cell diagram which represents the half-equations written as reduction:

$$M^{n+}\text{(aq)} \mid M\text{(s)} \quad \text{refers to} \quad M^{n+}\text{(aq)} + ne^- \rightarrow M\text{(s)}$$

This is why standard electrode potentials are sometimes called **standard reduction potentials**, or **standard redox potentials**.

To check that you understand the conventions, try the next Exercise.

EXERCISE 8.1
Answers on page 266

Three cells were set up similar to the one shown in Fig. 8.2 using three different metals. The potential difference between the two electrodes, and the polarity of the metal, were as shown below.

Zinc negative, hydrogen positive	0.76 V
Copper positive, hydrogen negative	0.35 V
Silver positive, hydrogen negative	0.81 V

a For each cell, write a cell diagram and state the e.m.f.

b Write an equation for each metal half-cell together with the appropriate value for the standard electrode potential.

The electrode processes you have studied so far in this chapter have consisted of a metal in contact with a solution of its ions. We now extend this to cover other electrode systems involving ion/ion equilibria such as:

$$Fe^{3+}(aq) + e^- \rightleftharpoons Fe^{2+}(aq)$$

and non-metal/non-metal-ion equilibria such as:

$$I_2(aq) + 2e^- \rightleftharpoons 2I^-(aq)$$

Such standard half-cells can be set up by dipping a platinum electrode into a mixture in which each reacting species has a concentration of 1.0 mol dm^{-3}. For instance, the standard $Fe^{3+}(aq)$, $Fe^{2+}(aq)$ half-cell would look like this:

Figure 8.3

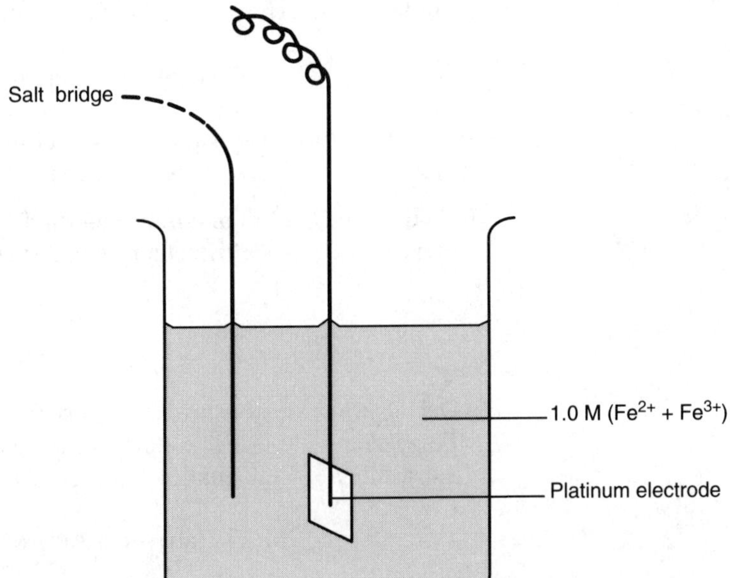

The platinum electrode is inert and serves only to transfer electrons. It does not participate in the cell reaction except, perhaps, as a catalyst.

The half-cell diagram for this electrode is:

$$Fe^{3+}(aq),Fe^{2+}(aq) \mid Pt(s)$$

When a solution contains more than one type of ion, these are shown separated by commas, with the more reduced species nearer the electrode.

Sometimes the electrode system contains more than two chemical species which take part in the cell reaction. For instance, the equilibrium between aqueous manganate(VII) and manganese(II) ions is established only in the presence of hydrogen ions, as shown below:

$$MnO_4^-(aq) + 8H^+(aq) + 5e^- \rightleftharpoons Mn^{2+}(aq) + 4H_2O(l)$$

The half-cell is shown in tables of electrode potentials as follows:

$$[MnO_4^-(aq) + 8H^+(aq)],[Mn^{2+}(aq) + 4H_2O(l)] \mid Pt$$

The square brackets are used to separate the oxidised and reduced forms of the equilibrium mixture. However, when writing a complete cell diagram, the half-cell may be abbreviated to:

$$MnO_4^-(aq),H^+(aq),\ Mn^{2+}(aq) \mid Pt$$

In the following Worked Example, we show you how to relate the information in a half-cell diagram to the equation for the half-reaction.

WORKED EXAMPLE For the half-cell shown below, write down the half-equation corresponding to the standard electrode potential:

$$[MnO_4^-(aq) + 8H^+(aq)],[Mn^{2+}(aq) + 4H_2O] \mid Pt \qquad E^\ominus = +1.51\text{ V}$$

Solution The half-equation corresponding to a standard electrode potential is always written as reduction:

$$\text{oxidised form} + ne^- \rightarrow \text{reduced form}$$

The problem, therefore, reduces to calculating n in the equation:

$$MnO_4^-(aq) + 8H^+(aq) + ne^- \rightarrow Mn^{2+}(aq) + 4H_2O(l) \qquad (1)$$

You can either use the oxidation number method (method 1) or the balancing of charge method (method 2). We consider each method in turn.

Method 1 – the oxidation number method

1. Identify the atom which changes its oxidation number in the half-cell:

$$\overset{+7}{Mn}O_4^- \rightarrow \overset{+2}{Mn}{}^{2+}$$

2. Decide how many electrons are necessary to effect the change in oxidation number. To go from +7 to +2, five electrons are gained per Mn atom.
3. Substitute $n = 5$ in equation (1):

$$MnO_4^-(aq) + 8H^+(aq) + 5e^- \rightarrow Mn^{2+}(aq) + 4H_2O(l)$$

Method 2 – balancing charge method

1. Count the charges on the left of equation (1) (excluding the electrons):

$$1(-1) + 8(+1) = +7$$

2. Count the charges on the right of equation (1):

$$1(+2) = +2$$

3. Work out the number of electrons which must be added so that the charge on the left equals the charge on the right:

$$+7 + 5e^- = +2$$

4. Write the complete half-equation including the number of electrons:

$$MnO_4^-(aq) + 8H^+(aq) + 5e^- \rightarrow Mn^{2+}(aq) + 4H_2O(l)$$

Decide which method you prefer and then apply it in the next Exercise.

EXERCISE 8.2
Answers on page 266

Write down the half-equations for the reactions which correspond to the standard electrode potentials for the following half-cells:

a $[2NO_3^-(aq) + 4H^+(aq)],[N_2O_4(g) + 2H_2O(l)] \mid Pt$ $E^\ominus = +0.80$ V
b $[2H_2SO_3(aq) + 2H^+(aq)],[S_2O_3^{2-}(aq) + 3H_2O(l)] \mid Pt$ $E^\ominus = +0.40$ V
c $[IO^-(aq) + H_2O(l)],[I^-(aq) + 2OH^-(aq)] \mid Pt$ $E^\ominus = +0.49$ V

In the following sections we show you how useful standard electrode potentials are in making predictions about likely reactions.

Calculating cell e.m.f. from tabulated $E^\ominus$ values

Instead of measuring cell e.m.f.s experimentally, we can use tabulated $E^\ominus$ values to calculate them. We show you how to do this in a Worked Example.

WORKED EXAMPLE

Calculate the magnitude and the sign of the e.m.f. of the cell represented below, and write an equation for the cell reaction:

$$Mn(s) \mid Mn^{2+}(aq) \;\vdots\vdots\; Co^{2+}(aq) \mid Co(s)$$

Solution

1. Look up the standard electrode potential for each half-cell reaction:

$$Co^{2+}(aq) \mid Co(s) \quad E^\ominus = -0.28 \text{ V}$$
$$Mn^{2+}(aq) \mid Mn(s) \quad E^\ominus = -1.19 \text{ V}$$

2. Write equations for the half-cell reactions.
   The right-hand side of the cell diagram corresponds to the electrode potential as tabulated in a data book, i.e. as a reduction:

$$Co^{2+}(aq) + 2e^- \rightleftharpoons Co(s) \quad E^\ominus = -0.28 \text{ V}$$

   The left-hand side of the cell diagram corresponds to the **reverse** of the electrode potential as tabulated, so the sign must be reversed:

$$Mn(s) \rightleftharpoons Mn^{2+}(aq) + 2e^- \quad E^\ominus = +1.19 \text{ V}$$

3. Add the two half-cell reactions together:

$$Co^{2+}(aq) + 2e^- \rightleftharpoons Co(s) \quad E^\ominus = -0.28 \text{ V}$$
$$Mn(s) \rightleftharpoons Mn^{2+}(aq) + 2e^- \quad E^\ominus = +1.19 \text{ V}$$

$$Mn(s) + Co^{2+}(aq) \rightarrow Co(s) + Mn^{2+}(aq) \quad \Delta E^\ominus = +0.91 \text{ V}$$

The overall reaction occurs from left to right as shown, because $\Delta E^\ominus$ is positive. The right-hand electrode (cobalt) is positive, i.e. electron deficient because it is supplying electrons to the cobalt ions in the solution. A negative value would show that the reaction proceeds in the opposite direction.

4. Complete the cell diagram by inserting $\Delta E^\ominus$:

$$Mn(s) \mid Mn^{2+}(aq) \;\vdots\vdots\; Co^{2+}(aq) \mid Co(s) \qquad \Delta E^\ominus = +0.91\ V$$

A short cut

The steps followed above can be summarised in the simple rule you have already met (page 162):

$$\Delta E^\ominus = E_R^\ominus - E_L^\ominus$$

($E_R^\ominus$ refers to the right-hand electrode in the cell diagram and $E_L^\ominus$ refers to the left-hand electrode in the cell diagram.)

So, a short cut through the calculation is achieved by substituting into this expression:

$$\Delta E^\ominus = E_R^\ominus - E_L^\ominus = -0.28\ V - (-1.19\ V) = +0.91\ V$$

Since $\Delta E^\ominus$ is positive, the reaction proceeds in the direction indicated by the cell diagram:

$$Mn(s) \mid Mn^{2+}(aq) \;\vdots\vdots\; Co^{2+}(aq) \mid Co(s) \qquad \Delta E^\ominus = +0.91\ V$$
$$Mn(s) + Co^{2+}(aq) \rightarrow Mn^{2+}(aq) + Co(s)$$

You can apply this rule in the next Exercise. Use the table of standard electrode potentials in your data book.

EXERCISE 8.3
Answers on page 267

For each of the following cells calculate the cell e.m.f. and write the equation to represent the reaction which would occur if the cell were short-circuited. (Assume the temperature is 25 °C and all solutions are of concentration 1.0 mol dm^{-3}.)

a $Pt(s),H_2(g) \mid H^+(aq) \;\vdots\vdots\; Ag^+(aq) \mid Ag(s)$
b $Ni(s) \mid Ni^{2+}(aq) \;\vdots\vdots\; Zn^{2+}(aq) \mid Zn(s)$
c $Ni(s) \mid Ni^{2+}(aq) \;\vdots\vdots\; [NO_3^-(aq) + 3H^+(aq)],[HNO_2(aq) + H_2O(l)] \mid Pt(s)$
d $Pt(s) \mid 2H_2SO_3(aq),[4H^+(aq) + S_2O_6^{2-}(aq)] \;\vdots\vdots\; Cr^{3+}(aq) \mid Cr(s)$

In the next section we take these ideas one small step further.

Using $E^\ominus$ values to predict spontaneous reactions

We have shown that, given a cell diagram and the sign on the cell e.m.f., we are able to predict the cell reaction. We now show you a similar method for predicting whether the reaction described by a given equation could occur or not.

WORKED EXAMPLE

Use values of standard electrode potentials to decide whether or not the following reaction could occur:

$$Pb(s) + Mn^{2+}(aq) \rightarrow Pb^{2+}(aq) + Mn(s)$$

Solution

1. Identify the two half-equations and write them down:

$$Pb(s) \rightleftharpoons Pb^{2+}(aq) + 2e^-$$
$$Mn^{2+}(aq) + 2e^- \rightleftharpoons Mn(s)$$

2. Combine these in a cell diagram:

$$Pb(s) \mid Pb^{2+}(aq) \;\vdots\vdots\; Mn^{2+}(aq) \mid Mn(s)$$

3. Look up the standard electrode potentials and write them down:

$$E^{\ominus}(Pb^{2+} \mid Pb) = -0.13\ V \qquad E^{\ominus}(Mn^{2+} \mid Mn) = -1.18\ V$$

4. Determine the e.m.f. of this cell by substituting into the expression:

$$\Delta E^{\ominus} = E_R^{\ominus} - E_L^{\ominus}$$
$$\therefore \Delta E^{\ominus} = -1.18\ V - (-0.13\ V) = -1.05\ V$$

The negative sign shows that the given reaction will **not** proceed spontaneously:

$$Pb(s) + Mn^{2+}(aq) \not\rightarrow Pb^{2+}(aq) + Mn(s)$$

(However, we should expect the reverse reaction to occur, because it has a positive value, +1.05 V, of $\Delta E^{\ominus}$.)

In the next Exercise we ask you to predict whether or not certain reactions are likely to be spontaneous.

EXERCISE 8.4
Answers on page 267

Use $E^{\ominus}$ values to predict whether the following reactions are spontaneous under standard conditions.

a $Br_2(aq) + 2I^-(aq) \rightarrow 2Br^-(aq) + I_2(aq)$
b $Br_2(aq) + 2Cl^-(aq) \rightarrow 2Br^-(aq) + Cl_2(aq)$
c $Zn(s) + 2Fe^{3+}(aq) \rightarrow Zn^{2+}(aq) + 2Fe^{2+}(aq)$
d $2MnO_4^-(aq) + 16H^+(aq) + 10Br^-(aq) \rightarrow 8H_2O(l) + 2Mn^{2+}(aq) + 5Br_2(aq)$
e $2MnO_4^-(aq) + 16H^+(aq) + 5Cu(s) \rightarrow 8H_2O(l) + 2Mn^{2+}(aq) + 5Cu^{2+}(aq)$
f $3S_2O_8^{2-}(aq) + 2Cr^{3+}(aq) + 7H_2O(l) \rightarrow 6SO_4^{2-}(aq) + Cr_2O_7^{2-}(aq) + 14H^+(aq)$

Limitations of predictions made using $E^{\ominus}$ values

Consideration of $E^{\ominus}$ values in Exercise 8.4 **f** led you to expect the following reaction to occur:

$$3S_2O_8^{2-}(aq) + 2Cr^{3+}(aq) + 7H_2O(l) \rightarrow Cr_2O_7^{2-}(aq) + 6SO_4^{2-}(aq) + 14H^+(aq)$$

However, the prediction is **not** borne out by experiment. This illustrates one of the limitations of using $E^{\ominus}$ values; they cannot predict reaction rates. Reactions like the one above are considered energetically favourable yet kinetically unfavourable. So we can only say that a reaction **could** occur, not that it definitely will.

A second possible reason for the non-occurrence of a reaction predicted from $E^{\ominus}$ values is that some alternative reaction may be more favoured.

Thirdly, you must remember that $E^{\ominus}$ values refer to standard conditions only; you will see later that changing the conditions can change the value of ΔE for a reaction, though not usually by very much.

We now describe another method for predicting spontaneous reactions. It simply applies in a different way the principle you have already learned.

The 'anticlockwise rule'

We illustrate this rule by a Worked Example.

WORKED EXAMPLE

Predict which reaction could occur on combining the following half-cells:

$I_2(aq) + 2e^- \rightleftharpoons 2I^-(aq) \qquad E^{\ominus} = +0.53\ V$
$Br_2(aq) + 2e^- \rightleftharpoons 2Br^-(aq) \qquad E^{\ominus} = +1.06\ V$

Solution

1. Write the half-cells with the more negative electrode potential written above the other. This is the order in which they are usually tabulated.
2. Draw in anticlockwise arrows as shown below.

$$\begin{aligned} I_2(aq) + 2e^- &\rightleftharpoons 2I^-(aq) \\ Br_2(aq) + 2e^- &\rightleftharpoons 2Br^-(aq) \end{aligned}$$

(This can easily be justified using the principles you have already learned. The direction of the arrows merely follows what you would expect from the values of $E^\ominus$.)

3. Re-write the half-equations following these arrows:

$$\begin{aligned} Br_2(aq) + 2e^- &\rightarrow 2Br^-(aq) \\ 2I^-(aq) &\rightarrow I_2(aq) + 2e^- \end{aligned}$$

Add them to obtain the predicted reaction:

$$Br_2(aq) + 2I^-(aq) \rightarrow 2Br^-(aq) + I_2(aq)$$

Now attempt the following two Exercises.

EXERCISE 8.5
Answers on page 268

Predict the reactions which might occur on combining the following half-cells:

a $Mn^{2+}(aq) + 2e^- \rightleftharpoons Mn(s)$ $E^\ominus = -1.19$ V
$Pb^{2+}(aq) + 2e^- \rightleftharpoons Pb(s)$ $E^\ominus = -0.13$ V
b $I_2(aq) + 2e^- \rightleftharpoons 2I^-(aq)$ $E^\ominus = +0.54$ V
$S(s) + 2e^- \rightleftharpoons S^{2-}(aq)$ $E^\ominus = -0.48$ V
c $2H^+(aq) + 2e^- \rightleftharpoons H_2(g)$ $E^\ominus = 0.00$ V
$Ag^+(aq) + e^- \rightleftharpoons Ag(s)$ $E^\ominus = +0.80$ V

EXERCISE 8.6
Answers on page 268

Below is a list of standard electrode potentials for a number of half-reactions.

Half-reaction	$E^\ominus$/V
$Al^{3+}(aq) + 3e^- \rightleftharpoons Al(s)$	−1.66
$I_2(aq) + 2e^- \rightleftharpoons 2I^-(aq)$	+0.54
$Fe^{3+}(aq) + e^- \rightleftharpoons Fe^{2+}(aq)$	+0.77
$H_2O_2(l) + 2H^+(aq) + 2e^- \rightleftharpoons 2H_2O(l)$	+1.77
$Co^{3+}(aq) + e^- \rightleftharpoons Co^{2+}(aq)$	+1.88

a State the substances among those appearing in this table that can be used to convert Fe^{2+} into Fe^{3+}.
b Write balanced equations for these conversions.

We have shown you several different methods for tackling problems involving cells and the use of $E^\ominus$ values. The next section helps you to choose which is most appropriate in a particular case.

Summary

Which method you use in predicting spontaneous reactions largely depends on the way the question is presented.

If you are given the cell diagram, then the direction of the spontaneous reaction can be predicted from the sign of the cell e.m.f. If you have to calculate the e.m.f., use $\Delta E^\ominus = E_R^\ominus - E_L^\ominus$.

If you are given a list of standard electrode potentials, then the direction of the spontaneous reaction can be predicted by writing your own cell diagram or using the anticlockwise rule.

Whichever method you choose it is important that you understand that both use the same basic principles – oxidation occurs at the more negative electrode (i.e. the electrode with the more negative electrode potential), and reduction occurs at the more positive electrode (i.e. the electrode with the more positive electrode potential) in an electrochemical cell. All you need to know to be able to predict a cell reaction is the polarity of the electrodes.

The cells studied so far have been made up of standard half-cells, i.e. electrodes dipping into solutions where each ionic species has a concentration of 1.0 mol dm^{-3}. We now consider non-standard half-cells and investigate the effect on the electrode potential and on the cell e.m.f. of changing the concentration.

The effect of concentration on electrode potential and cell e.m.f.

Le Chatelier's principle allows us to predict the effect of diluting the solution in a metal/metal-ion half-cell, for example:

$$Ag^{+}(aq) + e^{-} \rightleftharpoons Ag(s)$$

If $[Ag^{+}(aq)]$ is decreased, the equilibrium position will shift to the left. This produces additional electrons which make the silver electrode more negative, or less positive, and so the electrode potential changes accordingly.

EXERCISE 8.7

Answers on page 268

Apply Le Chatelier's principle to the cell:

$$Zn(s) \mid Zn^{2+}(aq) \;\vdots\vdots\; Cu^{2+}(aq) \mid Cu(s) \qquad \Delta E^{\ominus}(298\ K) = +1.10\ V$$

and consider how the e.m.f. of the cell would change if:

a the solution of copper ions were diluted,
b the solution of zinc ions were diluted.

Hint: Consider first the effect on the position of equilibrium in the half-cell.

In the next Exercise you will use experimental results to test the predictions made in the previous Exercise.

EXERCISE 8.8

Answers on page 268

The cell shown in Exercise 8.7 was set up and the e.m.f. measured at different concentrations of zinc and copper ions. The results are shown below.

Table 8.1

$[Zn^{2+}(aq)]$/mol dm^{-3}	$[Cu^{2+}(aq)]$/mol dm^{-3}	ΔE/V
1.0	1.0	+1.06
1.0	0.10	+1.03
0.10	1.0	+1.09

a How does the e.m.f. of the cell change as the copper ion solution becomes more dilute?
b How does the e.m.f. of the cell change as the zinc ion solution becomes more dilute?
c Were your predictions made in Exercise 8.7 verified by this experiment?

We now consider how the **magnitude** of $\Delta E^{\ominus}$ gives an indication as to the **extent** of a reaction, and the relationships between $\Delta E^{\ominus}$, $\Delta G^{\ominus}$ and K_c.

RELATIONSHIPS BETWEEN $\Delta E^\ominus$, $\Delta G^\ominus$ AND K_c

As you learned in Chapter 6, when studying equilibrium constants, K_c, it is useful to think of **all** reactions as proceeding to an equilibrium position. Reactions which appear to 'go to completion' have very large values of K_c: reactions which 'do not go at all' have very small values of K_c.

For instance, when we say that the following reaction proceeds spontaneously:

$$Zn(s) + Cu^{2+}(aq) \rightarrow Zn^{2+}(aq) + Cu(s) \qquad \Delta E^\ominus = +1.1\ V$$

we mean that, under standard conditions, it proceeds to an equilibrium mixture in which the concentration of $Zn^{2+}(aq)$ is many times greater than the concentration of $Cu^{2+}(aq)$.

So far, we have simplified the situation by assuming that if $\Delta E^\ominus$ is positive the reaction 'goes' and if $\Delta E^\ominus$ is negative the reaction does not 'go'. It is better to say that the more positive $\Delta E^\ominus$, the larger the value of K_c, and the more negative $\Delta E^\ominus$, the smaller the value of K_c.

For values of $\Delta E^\ominus$ between about +0.3 V and −0.3 V, the values of K_c are such that the equilibrium mixtures contain significant amounts of products and reactants. In these situations, it may be possible to reverse predictions of the direction of reaction by using non-standard concentrations, since this alters electrode potentials somewhat and causes small changes in the e.m.f.s of cells.

You already know from Chapter 4 that the standard free energy change, $\Delta G^\ominus$, can also be used to decide whether a particular reaction is likely to occur. You also know of the relationship between $\Delta G^\ominus$ and K_c from Chapter 6. As you can see, we now have three parameters ($\Delta G^\ominus$, $\Delta E^\ominus$ and K_c) which all indicate the extent of a reaction. You won't be surprised, therefore, to find that they are all mathematically related. First we look at the relationship between $\Delta E^\ominus$ and $\Delta G^\ominus$.

Calculating the energy generated by a redox reaction

Consider the cell:

$$Cu(s)\ |\ Cu^{2+}(aq)\ \vdots\vdots\ Ag^{+}(aq)\ |\ Ag(s) \qquad \Delta E^\ominus = +0.46\ V$$

The positive value of $\Delta E^\ominus$ shows that the reaction below should proceed spontaneously in the direction shown:

$$Cu(s) + 2Ag^{+}(aq) \rightarrow Cu^{2+}(aq) + 2Ag(s)$$

Now imagine the reaction proceeding to a very small extent, corresponding to the transfer of just a few electrons at a constant potential difference, $\Delta E^\ominus$ (if the reaction goes any further, ΔE falls). The energy change associated with this transfer can be calculated from the relationship:

$$\underset{\text{(joules)}}{\text{electrical energy}} = \underset{\text{(volts)}}{\text{potential difference}} \times \underset{\text{(coulombs)}}{\text{charge transfer}}$$

Scaling up this energy change to the amounts shown by the equation gives us the theoretical maximum energy available from the reaction to do useful work, i.e. the standard free energy change, $\Delta G^{\ominus}$. Since the charge transfer for 'one mole of equation' is two faradays ($2 \times 9.65 \times 10^4$ C mol^{-1}), we have:

$$\text{electrical energy} = 0.46\text{ V} \times 2 \times 9.65 \times 10^4\text{ C mol}^{-1}$$
$$= 89 \times 10^3\text{ C V mol}^{-1} = 89\text{ kJ mol}^{-1}$$

Since this energy is potentially released, we give it a negative sign, so that:

$$\Delta G^{\ominus} = -89\text{ kJ mol}^{-1}$$

This information may be expressed in the form of an energy diagram.

Figure 8.4

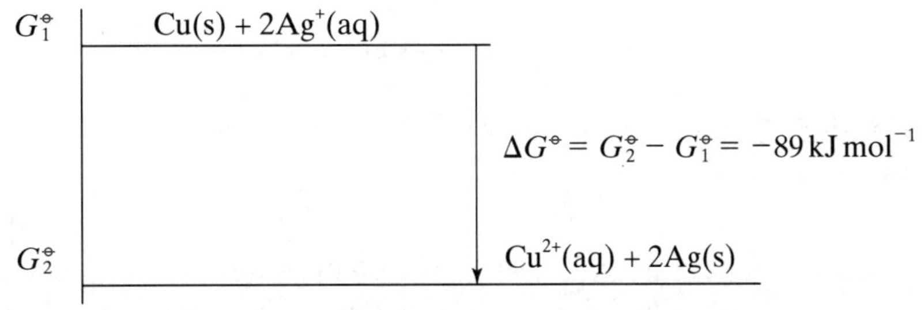

The general equation which relates $\Delta G^{\ominus}$ to the standard e.m.f. of the cell, $\Delta E^{\ominus}$, is as follows:

$$\Delta G^{\ominus} = -zF\,\Delta E^{\ominus}$$

z = number of electrons transferred (some textbooks use n)
F = Faraday constant (9.65×10^4 C mol^{-1} = 9.65×10^4 J V^{-1} mol^{-1})
$\Delta E^{\ominus}$ = standard cell e.m.f.

This equation shows why you can use values of $\Delta E^{\ominus}$ just as well as values of $\Delta G^{\ominus}$ to determine the energetic feasibility of a reaction. Remember, however, that whenever $\Delta G^{\ominus}$ is negative, $\Delta E^{\ominus}$ is positive, and vice versa.

Use the relationship to calculate some values of $\Delta G^{\ominus}$ in the next Exercise.

EXERCISE 8.9
Answers on page 269

For each of the following standard cells, calculate the e.m.f. from standard electrode potentials, and write an equation for the spontaneous reaction which occurs. Calculate $\Delta G^{\ominus}$ for each reaction.

a $Zn(s)\ |\ Zn^{2+}(aq)\ \vdots\vdots\ Cd^{2+}(aq)\ |\ Cd(s)$
b $Cu(s)\ |\ Cu^{2+}(aq)\ \vdots\vdots\ Cd^{2+}(aq)\ |\ Cd(s)$
c $Ag(s)\ |\ Ag^{+}(aq)\ \vdots\vdots\ Cd^{2+}(aq)\ |\ Cd(s)$

In the next section we remind you of a relationship we presented in Chapter 6.

■ The relationship between $\Delta E^{\ominus}$ and K_c

In Chapter 4 you learned that values of the equilibrium constant, K, show us the extent to which a given reaction will 'go to the right'. In Chapter 6 you also encountered an expression which relates the equilibrium constant to the standard free energy change:

$$\Delta G^{\ominus} = -RT \ln K$$

For reactions involving gases, K is K_p: otherwise we can use K_c.

(Note: many textbooks give the equation $\Delta G^{\ominus} = -2.3\,RT \log K_c$ because calculations used to be done with the aid of logarithm tables (to the base 10). With electronic calculators it is simpler to use ln terms, thus avoiding the need for the factor 2.3.)

We can use this expression to determine K_c for a redox reaction by substituting a value for $\Delta G^\ominus$ which has been obtained from e.m.f. measurements. We can summarise this two-step calculation as follows:

$\Delta E^\ominus$	$\rightarrow$	$\Delta G^\ominus$	$\rightarrow$	K_c
(from experiment or tabulated $E^\ominus$ values)		(from the expression $\Delta G^\ominus = -zF\Delta E^\ominus$)		(from the expression $\Delta G^\ominus = -RT \ln K_c$)

Of course, K_c can be calculated directly from:

$$zF\Delta E^\ominus = RT \ln K_c$$

You use this expression in the next three Exercises.

EXERCISE 8.10
Answers on page 269

The following cell is set up and short-circuited:

$$Cu(s) \mid Cu^{2+}(aq) \;\vdots\vdots\; Br_2(aq), 2Br^-(aq) \mid Pt$$

a Write an equation for the resulting cell equilibrium.
b Calculate K_c for this equilibrium and comment on the result.

EXERCISE 8.11
Answers on page 269

The oxidation of iron(II) ions by aqueous acidified dichromate(VI) ions proceeds according to the equation:

$$6Fe^{2+}(aq) + Cr_2O_7^{2-}(aq) + 14H^+(aq) \rightleftharpoons 6Fe^{3+}(aq) + 2Cr^{3+}(aq) + 7H_2O(l)$$

a Write half-equations for the two component half-reactions.
b Write a cell diagram for an electrochemical cell in which this reaction occurs.
c Calculate the standard e.m.f. of the cell at 298 K.
d Calculate the value of the equilibrium constant, K_c.

EXERCISE 8.12
Answers on page 269

For the following reactions at 25 °C, calculate the equilibrium constant, K_c, from values of $E^\ominus$.
a $Ag^+(aq) + Fe^{2+}(aq) \rightleftharpoons Ag(s) + Fe^{3+}(aq)$
b $3Cl_2(aq) + 2Cr^{3+}(aq) + 7H_2O(l) \rightleftharpoons 6Cl^-(aq) + Cr_2O_7^{2-}(aq) + 14H^+(aq)$

END-OF-CHAPTER QUESTIONS

Answers on page 269

A

8.1 You are provided with the following data which may be helpful.

$H_2(g) + 2e^- \rightarrow 2H^-(aq)$	$E^\ominus = -2.25$ V
$2H^+(aq) + 2e^- \rightarrow H_2(g)$	$E^\ominus = 0.00$ V
$Fe^{3+}(aq) + e^- \rightarrow Fe^{2+}(aq)$	$E^\ominus = +0.77$ V
$O_2(g) + 4H^+(aq) + 4e^- \rightarrow 2H_2O(l)$	$E^\ominus = +1.23$ V
$Cl_2(g) + 2e^- \rightarrow 2Cl^-(aq)$	$E^\ominus = +1.36$ V

For each of parts **a** and **b**, choose from the list of equations, A to E, the reaction described.

A $\quad 2Fe^{2+}(aq) + Cl_2(g) \rightarrow 2Fe^{3+}(aq) + 2Cl^-(aq)$
B $\quad 2H_2(g) + O_2(g) \rightarrow 2H_2O(l)$

C $4Fe^{3+}(aq) + 2H_2O(l) \rightarrow 4Fe^{2+}(aq) + O_2(g) + 4H^+(aq)$
D $H^-(aq) + H_2O(l) \rightarrow OH^-(aq) + H_2(g)$
E $Cl_2(g) + H_2O(l) \rightarrow HOCl(aq) + H^+(aq) + Cl^-(aq)$

a A reaction which represents the equation for a cell with a standard e.m.f. of +0.59 V.
b A reaction which cannot occur spontaneously because it represents a cell with a negative e.m.f.

8.2 Parts **a**, **b** and **c** of this question concern the apparatus shown below.

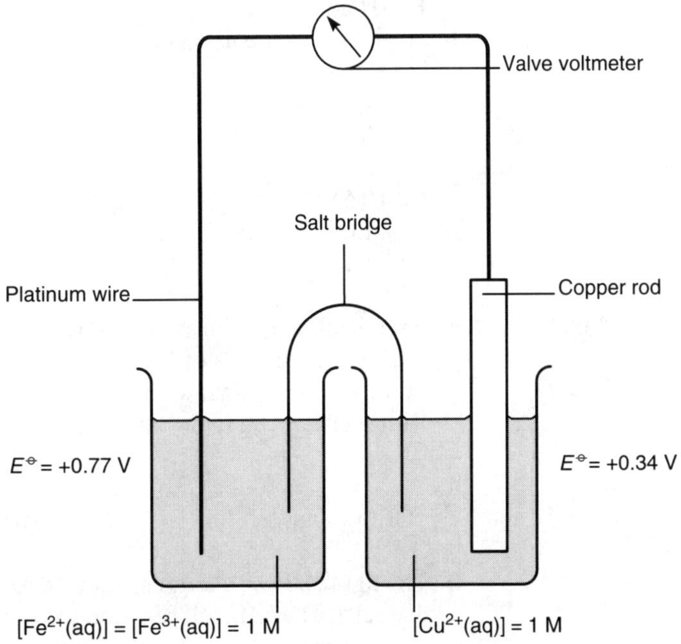

a Which of the following is a correct formulation of the cell shown?
A $Fe^{3+}(aq) + Fe^{2+}(aq) \mid Pt(s) \mathrel{\vdots\vdots} Cu(s) \mid Cu^{2+}(aq)$
B $Cu^{2+}(aq) \mid Cu(s) \mathrel{\vdots\vdots} Pt(s) \mid Fe^{2+}(aq), Fe^{3+}(aq)$
C $Pt(s) \mid Fe^{3+}(aq), Fe^{2+}(aq) \mathrel{\vdots\vdots} Cu^{2+}(aq) \mid Cu(s)$
D $Cu(s) \mid Cu^{2+}(aq) \mathrel{\vdots\vdots} Fe^{3+}(aq), Fe^{2+}(aq) \mid Pt(s)$
E $Fe^{3+}(aq), Fe^{2+}(aq) \mid Pt(s) \mathrel{\vdots\vdots} Cu^{2+}(aq) + Cu(s)$

b A large amount of water is added **first** to the $Fe^{2+}(aq)/Fe^{3+}(aq)$ [thereby diluting, but maintaining the ratio of concentrations of $Fe^{2+}(aq)$ to $Fe^{3+}(aq)$] **and then** to the $Cu^{2+}(aq)$. Which of the following changes in the meter reading would be observed?
A A rise followed by a fall.
B A rise followed by another rise.
C A fall followed by another fall.
D No change followed by a fall.
E No change followed by a rise.

c Which of the following equations represents the cell reaction?
A $Cu^{2+}(aq) + 2Fe^{2+}(aq) \rightarrow Cu(s) + 2Fe^{3+}(aq)$
B $Cu(s) + Fe^{3+}(aq) \rightarrow Cu^+(aq) + Fe^{2+}(aq)$
C $Cu^{2+}(aq) + Fe^{2+}(aq) \rightarrow Cu^+(aq) + Fe^{3+}(aq)$
D $Cu(s) + 2Fe^{3+}(aq) \rightarrow Cu^{2+}(aq) + 2Fe^{2+}(aq)$
E $Cu(s) + Fe^{2+}(aq) \rightarrow Cu^{2+}(aq) + Fe(s)$

8.3 Parts **a** and **b** of this question refer to the standard electrode potentials of some electrodes, which are as follows:

I	$Zn^{2+}(aq) \mid Zn(s)$	−0.76 V
II	$Fe^{2+}(aq) \mid Fe(s)$	−0.44 V
III	$Sn^{2+}(aq) \mid Sn(s)$	−0.14 V
IV	$Cu^{2+}(aq) \mid Cu(s)$	+0.34 V
V	$Ag^{+}(aq) \mid Ag(s)$	+0.80 V

a The cell with the largest e.m.f. is made up of electrodes:
A I and II
B I and V
C II and III
D II and IV
E III and V

b A cell was made using electrodes IV and V. A 1 M solution of Cu^{2+} ions was used in IV but a much more dilute solution of Ag^{+} ions was used in V. The e.m.f. of the cell could be:
A +0.42 V
B +0.46 V
C +0.54 V
D +1.06 V
E +1.14 V

8.4 **a** Excess dilute sulphuric acid is added to a freshly prepared solution containing equimolar amounts of $Fe^{2+}(aq)$, $Fe^{3+}(aq)$, $Cr^{3+}(aq)$ and $Cr_2O_7^{2-}(aq)$. By reference to the following data, deduce what happens and explain the role of each reactant. Write a balanced ionic equation for the reaction.

$$Fe^{3+}(aq) + e^- \rightarrow Fe^{2+}(aq) \qquad E^\ominus = +0.77\ V$$
$$Cr_2O_7^{2-}(aq) + 14H^+(aq) + 6e^- \rightarrow 2Cr^{3+}(aq) + 7H_2O(l) \qquad E^\ominus = +1.33\ V$$

b The figure shows a cross-section of a common type of torch battery (a Leclanché dry cell). The half-cell reaction for the negative electrode is shown in equation (1) and one possibility for the other half-cell reaction is shown in equation (2).

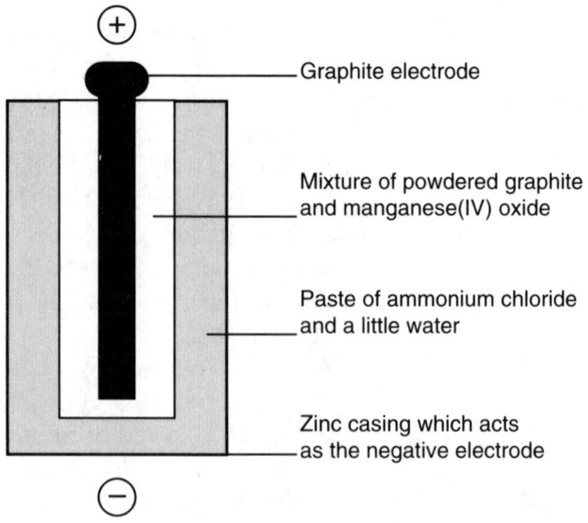

$$[Zn(NH_3)_4]^{2+}(aq) + 2e^- \rightarrow Zn(s) + 4NH_3(aq) \qquad E^\ominus = -1.03\ V\ (1)$$
$$NH_4^+(aq) + MnO_2(s) + H_2O(l) + e^- \rightarrow Mn(OH)_3(s) + NH_3(aq) \qquad E^\ominus = +1.00\ V\ (2)$$

i) Calculate what the e.m.f. of this cell would be under standard conditions.
ii) Suggest a reason why such cells tend to leak when nearing exhaustion.

iii) The actual e.m.f. of this cell is +1.50 V. Suggest why this value is different from the one calculated in i).

iv) Calculate the value of the free energy change, ΔG, for the actual cell. (Faraday constant = 9.65×10^4 C mol^{-1}.)

8.5 The table below gives the standard electrode potentials of various processes.

Table 8.2

Process	$E^{\ominus}/V$
$2HCO_2^- + 2CO_2 + 6H^+(aq) + 6e^- \rightleftharpoons (C_4H_4O_6)^{2-} + 2H_2O$	+0.20
$O_2 + 2H^+(aq) + 2e^- \rightleftharpoons H_2O_2$	+0.68
$H_2O_2 + 2H^+(aq) + 2e^- \rightleftharpoons 2H_2O$	+1.77

a Hydrogen peroxide can act as an oxidant or a reductant. Hydrogen peroxide solution is added to an acid solution containing HCO_2^- and $(C_4H_4O_6)^{2-}$ ions, under standard conditions. Predict and explain the reaction which is likely to occur between hydrogen peroxide and these species.

b Write the equation for this reaction.

c State what is meant by **disproportionation**.

d By considering the standard electrode potentials given, explain the disproportionation of hydrogen peroxide when treated with a manganese(IV) oxide catalyst.

8.6 The standard electrode potentials of four half-reactions are given below:

$$Sn^{2+}(aq) + 2e^- \rightleftharpoons Sn(s); \quad E^{\ominus} = -0.14\ V$$
$$Fe^{3+}(aq) + e^- \rightleftharpoons Fe^{2+}(aq); \quad E^{\ominus} = +0.77\ V$$
$$2Hg^{2+}(aq) + 2e^- \rightleftharpoons Hg_2^{2+}(aq); \quad E^{\ominus} = +0.92\ V$$
$$\tfrac{1}{2}Br_2(aq) + e^- \rightleftharpoons Br^-(aq); \quad E^{\ominus} = +1.07\ V$$

Based on this information, which of the following reactions will probably take place?

A $Sn(s) + 2Hg^{2+}(aq) \rightarrow Sn^{2+}(aq) + Hg_2^{2+}(aq)$

B $2Fe^{2+}(aq) + 2Hg^{2+}(aq) \rightarrow 2Fe^{3+}(aq) + Hg_2^{2+}(aq)$

C $2Br^-(aq) + Sn^{2+}(aq) \rightarrow Br_2(aq) + Sn(s)$

D $2Fe^{2+}(aq) + Br_2(aq) \rightarrow 2Fe^{3+}(aq) + 2Br^-(aq)$

8.7 **a** Write down the equations for the half-cell reactions corresponding to these three standard electrode potentials:

$Cr^{3+}(aq),Cr^{2+}(aq) \mid Pt$; $E^{\ominus} = -0.41$ V

$Ce^{4+}(aq),Ce^{3+}(aq) \mid Pt$; $E^{\ominus} = +1.70$ V

$[Cr_2O_7^{2-}(aq) + H^+(aq)],[Cr^{3+}(aq) + H_2O(l)] \mid Pt$; $E^{\ominus} = +1.33$ V

b Find the e.m.f.s of the following cells (all concentrations are 1.0 mol dm^{-3}):

i) $Pt \mid Cr^{2+}(aq),Cr^{3+}(aq) \,\vdots\vdots\, Ce^{4+}(aq),Ce^{3+}(aq) \mid Pt$

ii) $Pt \mid [Cr^{3+}(aq) + H_2O(l)],[Cr_2O_7^{2-}(aq) + H^+(aq)] \,\vdots\vdots\, Ce^{4+}(aq),Ce^{3+}(aq) \mid Pt$

c Write down the cell reactions which would occur in cells i) and ii) in part **b** above if the cells were short-circuited.

CHAPTER 9

VAPOUR PRESSURE AND RAOULT'S LAW

INTRODUCTION AND PRE-KNOWLEDGE

If a liquid is introduced into a closed container, some of it evaporates until equilibrium is reached. Taking water as an example:

$$H_2O(l) \rightleftharpoons H_2O(g)$$

The equilibrium constant, K_c, for this system is equal to the partial pressure of water, pH_2O, which is known as the **saturated vapour pressure** of water. (The word 'saturated' is often omitted.)

Vapour pressure varies from substance to substance and also with temperature, as shown in Fig. 9.1. The diagram also shows that the boiling-point of a liquid is the temperature at which the vapour pressure is equal to atmospheric pressure (760 mmHg). Liquids with lower vapour pressures have higher boiling points, and vice versa.

Figure 9.1

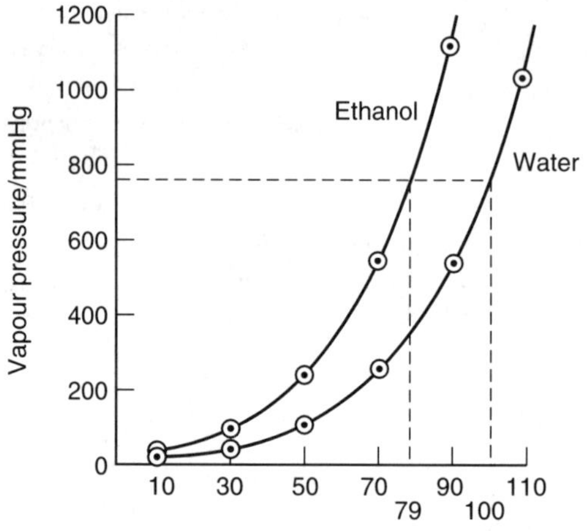

In the following sections, we assume that you are familiar with Dalton's law of partial pressures and the concept of mole fraction (see Chapter 5).

THE VAPOUR PRESSURE OF TWO-COMPONENT SYSTEMS

The vapour pressure of a mixture of two liquids is the sum of two partial vapour pressures, one for each liquid. If the liquids are similar to one another, for instance, hexane and heptane, the variation of total vapour pressure with mole fraction of each component is linear, as shown in Fig. 9.2.

Such a system is said to be an 'ideal mixture' or an 'ideal solution'. When the vapour pressure curve is *not* linear, which is more often the case, the mixture is 'non-ideal'. You will probably study non-ideal solutions from a qualitative point of view, but we shall deal with calculations only for ideal solutions.

The linear relationship between total vapour pressure and composition is often expressed as Raoult's Law, which we now consider.

Figure 9.2

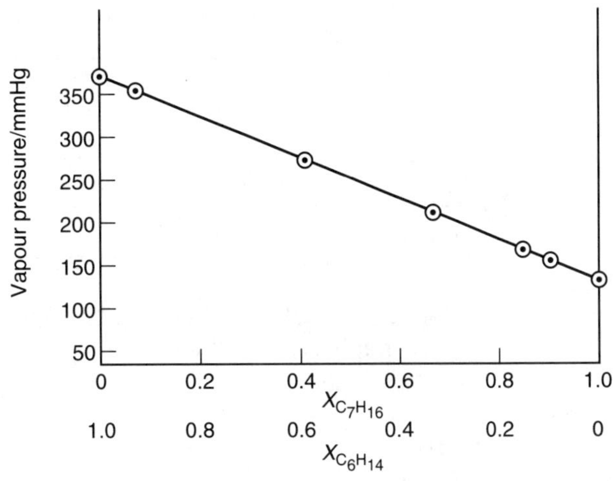

Raoult's Law

Raoult's Law states that the partial vapour pressure of any volatile component, A, of an ideal solution is equal to the vapour pressure of pure A multiplied by the mole fraction of A in the solution.

Raoult's Law can be expressed mathematically as follows:

$$p_A = p^\circ_A X_A$$

where p_A = partial vapour pressure of A in the solution
p°_A = vapour pressure of pure A
X_A = mole fraction of A in the solution.
Similarly, for liquid B we have:

$$p_B = p^\circ_B X_B$$

A formula very similar to this is covered in Chapter 5 where methods of calculating partial pressures are considered. Since our ideal solution (A, B) consists of two volatile components, they both must contribute to the total vapour pressure.

Thus:

$$p_T = p_A + p_B \quad \text{(Dalton's law of partial pressures)}$$

where p_T is the total vapour pressure of the solution at a fixed temperature, or:

$$p_T = p^\circ_A X_A + p^\circ_B X_B$$

In the next Exercise you will use Raoult's Law to construct a vapour pressure/composition curve for the hexane/heptane mixture.

EXERCISE 9.1
Answers on page 270

a Use Raoult's Law to determine the partial pressures of hexane, C_6H_{14}, and heptane, C_7H_{16}, at the mole fractions shown in Table 9.1, given that, at 47 °C,

$$p^\circ_{\text{hexane}} = 372 \text{ mmHg} \quad \text{and} \quad p^\circ_{\text{heptane}} = 127 \text{ mmHg}$$

Enter your answers in a copy of Table 9.1.

Table 9.1

X_{hexane}	0.00	0.200	0.400	0.600	0.800	1.00
p_{hexane}						
$X_{heptane}$	0.00	0.200	0.400	0.600	0.800	1.00
$p_{heptane}$						

b Plot a graph of partial pressure of i) hexane, and ii) heptane against mole fraction. The graph axes should be labelled as shown in Fig. 9.3.

Figure 9.3

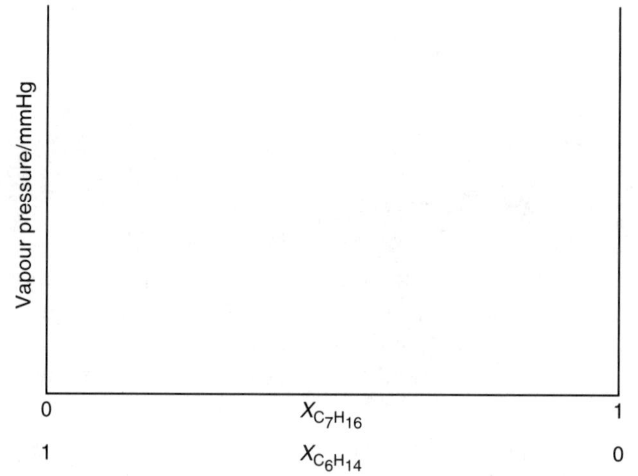

c Now plot total vapour pressure, p_T, against mole fraction on the same graph. Use the equation:

$$p_T = p_{hexane} + p_{heptane}$$

d What is the total vapour pressure of an equimolar mixture?
e What is the composition of a mixture which has a vapour pressure of 200 mmHg?

You can see that for two-liquid mixtures which obey Raoult's law, the vapour pressure curve can be drawn simply by joining the vapour pressures of pure components with a straight line.

Now attempt the next two Exercises which are taken from A-level questions. The calculations are similar, but you need not plot a graph.

EXERCISE 9.2

Answers on page 270

Two liquids, A and B, have vapour pressures of 75 mmHg and 130 mmHg respectively, at 25 °C. What is the total vapour pressure of the following ideal mixtures?
a 1 mol of A and 1 mol of B.
b 3 mol of A and 1 mol of B.
c 1 mol of A and 4 mol of B.

EXERCISE 9.3

Answers on page 270

Hexane and heptane are totally miscible and form an ideal two-component system. If the vapour pressures of the pure liquids are 56 000 and 24 000 N m^{-2} at 51 °C calculate **a** the total vapour pressure, and **b** the mole fraction of heptane in the vapour above an equimolar mixture of hexane and heptane.

The vapour pressure/composition curves you have considered up to now have shown the composition of the liquid only. The composition of the **vapour** in equilibrium with each liquid is different, as you see in the next Exercise, which refers to Fig. 9.4.

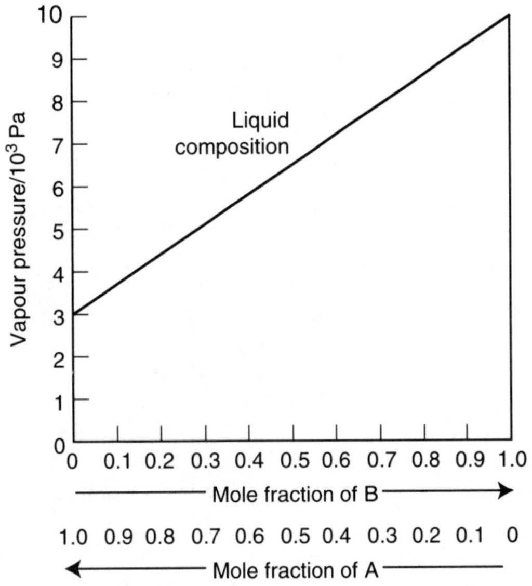

Figure 9.4
Vapour pressures of an ideal solution showing liquid composition.

EXERCISE 9.4
Answers on page 270

Consider an ideal solution of components A and B. The vapour pressure/composition curve for this solution is shown in Fig. 9.4.

a By reference to Fig. 9.4 state the vapour pressures of pure A and pure B.

b Consider a solution in which the mole fractions of A and B (X_A, X_B) are equal, i.e. $X_A = X_B = 0.50$.

i) Use Raoult's law to calculate the partial pressures of A and B (p_A and p_B) at this composition.

ii) Calculate the mole fraction of A in the vapour (X'_A) given that:

$$X'_A = \frac{p_A}{p_A + p_B}$$

iii) Similarly, calculate the mole fraction of B in the vapour (X'_B).

c Is the mole fraction of B greater in the liquid or the vapour?

d Repeat the calculation for $X_B = 0.1, 0.2, 0.4, 0.6$ and 0.8. Plot a vapour composition line on a copy of Figure 9.4.

You have already learnt that fractional distillation can be used to separate some mixtures of miscible liquids. In the next section we deal with the separation of immiscible liquids by a technique known as **steam distillation**, in which one of the immiscible liquids is water.

Steam distillation

Consider a mixture of immiscible liquids such as nitrobenzene and water. Figure 9.5 shows the vapour pressure curves for this mixture. Note that since the liquids do not mix, each exerts the same vapour pressure as it does in the pure state. Study the diagram and answer the Exercise which follows.

EXERCISE 9.5
Answers on page 271

a In Fig. 9.5, can you see a relationship between the total vapour pressure of the mixture and the separate vapour pressures of water and nitrobenzene?

b What is the boiling point of the mixture? Explain how you arrive at your answer.

c Is the boiling point of the mixture higher or lower than the boiling point of the individual pure liquids? Explain.

d At what temperature will the mixture steam distil?

Figure 9.5

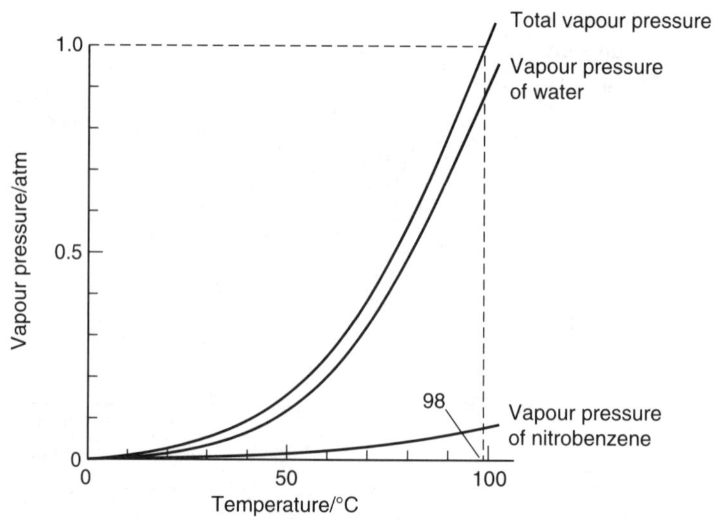

Water and another liquid, Y, undergoing steam distillation, each exert a vapour pressure proportional to the mole fraction present in the vapour. The distillate has the same composition as the vapour. From this it follows that:

$$\frac{p_{\text{water}}}{p_{\text{Y}}} = \frac{\text{amount of water}}{\text{amount of Y}} = \frac{\text{mass of water}}{18} \times \frac{\text{relative molecular mass of Y}}{\text{mass of Y}}$$

This rearranges to:

$$\frac{\text{mass of Y}}{\text{mass of water}} = \frac{\text{r.m.m. of Y}}{\text{r.m.m. of water}} \times \frac{p_{\text{Y}}}{p_{\text{water}}}$$

Use this equation in the following Exercises.

EXERCISE 9.6

Answer on page 271

In the extraction of bromobenzene by steam distillation at 101 kPa (1 atm) pressure, the mixture boils at 95.5 °C. At that temperature the vapour pressure of pure water is 85.8 kPa. Calculate the percentage by mass of bromobenzene in the distillate.

EXERCISE 9.7

Answers on page 271

The vapour pressures of pure water and pure phenylamine (aniline) at the stated temperatures are:

Table 9.2

Temperature/°C		**85**	**90**	**95**	**100**	**105**
Vapour pressure of water	/kPa	57.9	70.1	84.5	101.3	120.5
	/mmHg	434	526	634	760	906
Vapour pressure of phenylamine	/kPa	3.0	3.9	4.9	6.1	7.3
	/mmHg	22.9	29.2	36.5	45.7	55.0

With the aid of suitable graphs

a find to the nearest °C the temperature at which a mixture of phenylamine (aniline) and water will steam distil under standard atmospheric pressure,

b calculate the percentage by mass of phenylamine (aniline) which would be expected to be present in the distillate.

END-OF-CHAPTER QUESTIONS

Answers on page 271

9.1 Mixtures of water and methanol obey Raoult's vapour pressure law.
If the vapour pressure of pure water at 298 K is 24 mmHg, calculate the partial vapour pressure of water in a mixture of 36 g water and 32 g methanol at this temperature.

9.2 Hexane and heptane have the following properties.

Table 9.3

	Relative molar mass	Boiling point /°C	Vapour pressure at 25 °C/Pa
Hexane	86.2	68.7	20180
Heptane	100.2	98.1	6018

For a liquid mixture containing 23.0 g of hexane and 42.0 g of heptane at 25 °C calculate:

a the mole fraction of hexane in the liquid,
b the partial vapour pressure of each component in the vapour,
c the total vapour pressure above the mixture,
d the mole fraction of hexane in the vapour,
e the difference between your answers in **a** and **b** and state what technique depends on this difference.

CHAPTER 10

REACTION KINETICS

INTRODUCTION AND PRE-KNOWLEDGE

The rate at which a reaction proceeds may be defined in a number of ways and it is important to be clear which definition you use in a particular case. The most useful definition is the **rate of increase in concentration of one specified product of the reaction**. However, others are sometimes used because they are related to the means of measurement, e.g. the rate at which the **mass** of a **reactant decreases**, or the rate at which the **volume** of a gaseous **product increases**.

We assume that you have already investigated **qualitatively** the effect on the rate of reaction of such factors as concentration of reactant, surface area of solid reactant and the presence of a catalyst. In this chapter we consider some of these factors **quantitatively**. In particular, we introduce an expression known as the rate equation, which relates the rate of reaction at a constant temperature to the concentrations of the reactants.

We also assume that you can plot graphs accurately and obtain from them values for the **slope** at any point and the **intercept**.

THE RATE OF A REACTION

The study of reaction kinetics relies on experimental work. We begin by presenting you with a set of experimental results.

We have chosen a reaction where a single substance (the reactant) breaks down. We will take you through a series of Exercises on the data, to show you how to find the **rate equation** for the reaction. Our purpose is to illustrate the key ideas of reaction rate, order of reaction and rate constant.

2,4,6-trinitrobenzoic acid in solution loses carbon dioxide when heated, as shown by the following equation:

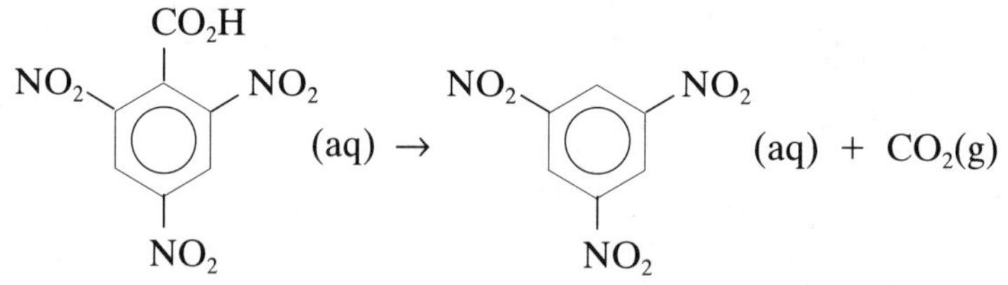

Since one of the products is gaseous, the rate of reaction could be studied by measuring the volume of carbon dioxide produced. Alternatively, you could follow the decrease in the amount of 2,4,6-trinitrobenzoic acid itself.

The scientists who did the experiment in 1931 (E A Moelwyn-Hughes and O N Hinshelwood) chose the second method. They set up several mixtures at 90 °C. After various reaction times they withdrew a sample and added a large volume of iced water to quench (i.e. stop) the reaction. They then titrated each mixture with 5.0×10^{-3} M barium hydroxide solution using bromothymol blue as indicator.

We have taken the following figures from their results:

Table 10.1

Time/min	**Concentration of 2,4,6-trinitrobenzoic acid/mol dm^{-3}**
0	2.77×10^{-4}
18	2.32×10^{-4}
31	2.05×10^{-4}
55	1.59×10^{-4}
79	1.26×10^{-4}
157	0.58×10^{-4}
Infinity	0.00

The first step in analysing this sort of data is to plot a graph of concentration of reactant against time. You do this in Exercise 10.1. Square brackets will be used to denote concentration throughout this chapter, i.e.

$$[X(aq)] = \text{concentration of X in mol dm}^{-3} \text{ of aqueous solution}$$

EXERCISE 10.1

Answers on page 272

a Plot a graph of concentration of 2,4,6-trinitrobenzoic acid (vertical axis) against time (horizontal axis). Use a large piece of graph paper, as you will have to draw tangents to the curve and you will need plenty of working space for this.

b Draw tangents to the curve at 10, 50, 100 and 150 minutes. Calculate their slopes and complete a copy of Table 10.2. Bear in mind that where the slope is negative, the value of the slope, but not the sign, gives the rate.

Table 10.2

Time /min	**Concentration /mol dm^{-3}**	**Slope /mol dm^{-3} min^{-1}**	**Rate /mol dm^{-3} min^{-1}**
10			
50			
100			
150			

c Plot a graph of rate of reaction (y-axis) against concentration of 2,4,6-trinitrobenzoic acid (x-axis) using your values in Table 10.2. Now answer the following questions:
   i) Does your graph go through the origin? Explain why it should.
   ii) Use your graph to state the relationship between rate of reaction and concentration of reactant both in words and mathematically.

The expression you have just worked out in Exercise 10.1 is called the **rate equation** for the reaction. We now consider rate equations in more detail.

RATE EQUATIONS, RATE CONSTANTS AND ORDERS OF REACTION

A rate equation (rate expression) relates the rate of a reaction at a fixed temperature to the concentrations of the reacting species by means of constants called **a** the **rate constant** (velocity constant), and **b** the **order(s) of reaction**.

The general form of a rate equation is:

$$\text{rate} = k\,[\text{A}]^a\,[\text{B}]^b\,[\text{C}]^c$$

k = rate constant (velocity constant) at the particular temperature
a = order of reaction with respect to reactant A
b = order of reaction with respect to reactant B
c = order of reaction with respect to reactant C
$a + b + c$ = overall order of reaction

EXERCISE 10.2
Answers on page 272

Use the rate equation which you worked out in Exercise 10.1 to state:
a the order of reaction with respect to 2,4,6-trinitrobenzoic acid,
b the overall order of reaction.

You should be able to work out a value for the rate constant for a first order reaction like this one. Try the next Exercise.

EXERCISE 10.3
Answers on page 272

Using both your graph and your rate equation from Exercise 10.1, work out a value for the rate constant (k) of the decarboxylation of 2,4,6-trinitrobenzoic acid at 90 °C.
What is the unit of the rate constant for this reaction?

Now try the next Exercise, where you are given rate/concentration data for a different reaction and asked to deduce the rate expression, the order of reaction and the rate constant.

EXERCISE 10.4
Answers on page 273

Table 10.3 contains some data for the decomposition of dinitrogen pentoxide in tetrachloromethane (tcm) solution:

$$2N_2O_5(\text{tcm}) \rightarrow 4NO_2(\text{tcm}) + O_2(\text{g})$$

Table 10.3

Concentration of N_2O_5/mol dm^{-3}	Rate of reaction as decrease in concentration of N_2O_5 per second /10^{-5} mol dm^{-3} s^{-1}
2.21	2.26
2.00	2.10
1.79	1.93
1.51	1.57
1.23	1.20
0.92	0.95

Plot a graph of the rate of the reaction against the concentration of N_2O_5, and try to answer these questions:

a What is the rate expression for the reaction?

b What order is this reaction with respect to N_2O_5?

c What is the value of the constant in the rate expression? Include the correct unit.

In the next section we aim to extend your knowledge of rate equations and orders of reaction to more complicated reactions.

Rate equations involving higher orders of reaction

So far you have studied reactions in which one compound decomposes. We now take a brief look at reactions involving more than one reactant.

To see that you have understood the difference between overall order and order of reaction with respect to a single reactant, try the following three short Exercises.

EXERCISE 10.5
Answers on page 273

The reaction between mercury(II) chloride and ethanedioate ions takes place according to the equation:

$$2HgCl_2(aq) + C_2O_4^{2-}(aq) \rightarrow 2Cl^-(aq) + 2CO_2(g) + Hg_2Cl_2(s)$$

The rate equation for the reaction is:

$$\text{rate} = k\,[HgCl_2(aq)][C_2O_4^{2-}(aq)]^2$$

a What is the order of reaction with respect to each reactant?

b What is the overall order of reaction?

In the next Exercise you may be surprised to find a fractional order in the rate equation. You will not come across fractional orders very often at A-level, but you should be aware that they exist and are a sign of a more complex reaction mechanism.

EXERCISE 10.6
Answers on page 273

For the following reaction:

$$C_2H_4(g) + I_2(g) \rightarrow C_2H_4I_2(g)$$

the rate equation is:

$$\text{rate} = k[C_2H_4(g)][I_2(g)]^{3/2}$$

a What is the order of reaction with respect to each reactant?

b What is the overall order of reaction?

The next Exercise deals with the overall order of a hypothetical reaction.

EXERCISE 10.7
Answer on page 273

Substances P and Q react together to form products and the overall order of reaction is three. Which of the following rate equations could not be correct?

A rate = $k[P]^2[Q]$

B rate = $k[P]^0[Q]^3$

C rate = $k[P][Q]^2$

D rate = $k[P][Q]^3$

E rate = $k[P][Q]^2[H^+]^0$

Now do the next Exercise, to check that you can work out units for the rate constant, *k*, for a reaction which is not first order.

EXERCISE 10.8
Answer on page 273

The equation for the reaction between peroxodisulphate ions and iodide ions is:

$$S_2O_8^{2-}(aq) + 2I^-(aq) \rightarrow 2SO_4^{2-}(aq) + I_2(aq)$$

The rate equation is:

$$\text{rate} = k[S_2O_8^{2-}(aq)][I^-(aq)]$$

If concentrations are measured in mol dm^{-3} and rate in mol dm^{-3} s^{-1}, determine the unit of *k*. Show your working clearly.

In the next section we take a look at another aspect of first order reactions.

Half-life of first order reactions

If you are asked to show that a reaction is of the first order you can often do this by calculating its half-life. As you will see, there is only one graph to plot!

You have probably come across the idea of half-life of a reaction in connection with the decay of radioactive isotopes. All radioactive isotopes decay by first order kinetics. The rate equation is:

$$\text{rate} = k\,[\text{reactant}]$$

Each radioactive isotope has its own half-life. Since the intensity of the radioactivity of a sample is proportional to the amount of the isotope it contains, **the half-life is the time taken for the radioactivity to fall to half its original value**.

To revise the idea of radioactive decay and half-life, try the next Exercise.

EXERCISE 10.9
Answers on page 273

a State what is meant by the term 'half-life'.

b The following is a decay curve for a radioactive element X.

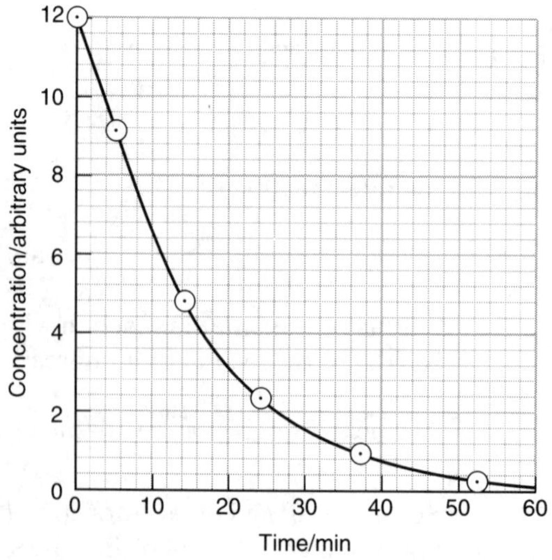

Figure 10.1

i) Determine the half-life of X.

ii) What is the order of the decay reaction?

c On a copy of the axes below, sketch the approximate relationship between rate of decay and concentration of X.

Figure 10.2

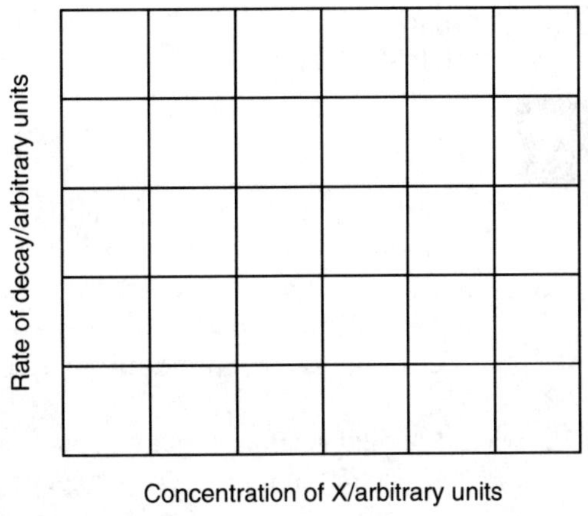

Other first order reactions, besides radioactive decay processes, have constant half-lives. You use this fact in the next Exercise.

EXERCISE 10.10

Answers on page 274

The following chemical reaction, under certain conditions, proceeds with the order shown.

$$2N_2O_5 \rightarrow 4NO_2 + O_2 \qquad \text{first order}$$

a In one experiment on the reaction, the initial concentration of N_2O_5 was 32 units and the half-life was five minutes. On a copy of the graph below, plot the relation between the concentration of N_2O_5 and time for this experiment.

Figure 10.3

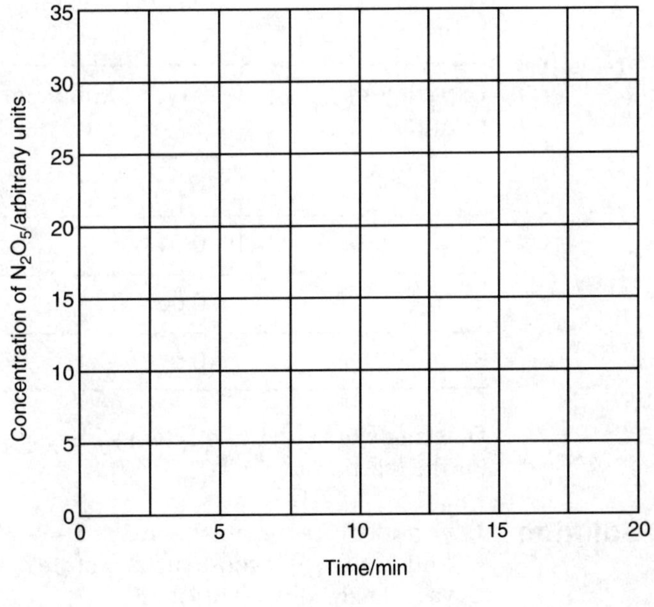

b When the reaction was carried out in tetrachloromethane, it was found that an initial concentration of N_2O_5 of 2.00 mol dm^{-3} gave an initial rate of reaction of 2.10×10^{-5} mol dm^{-3} s^{-1}. Calculate the rate constant for the reaction and state the units.

In the next Exercise you apply the half-life concept to a reaction you may have considered experimentally.

EXERCISE 10.11
Answer on page 274

The decomposition of benzenediazonium chloride in aqueous solution is a reaction of the first order which proceeds according to the equation:

$$C_6H_5N_2Cl(aq) \rightarrow C_6H_5Cl(l) + N_2(g)$$

A certain solution of benzenediazonium chloride contains initially an amount of this compound which gives 80 cm^3 of nitrogen on complete decomposition. It is found that, at 30 °C, 40 cm^3 of nitrogen are evolved in 40 minutes. How long after the start of the decomposition will 70 cm^3 of nitrogen have been evolved? (All volumes of nitrogen refer to the same temperature and pressure.)

In the next section we show you how to determine the order of reaction with respect to one reactant at a time, from experimental data, in a reaction which involves two or more reactants.

The order of reaction with respect to individual reactants

We show you how to determine the order of reaction with respect to individual reactants in the following Worked Example.

WORKED EXAMPLE

The data in Table 10.4 were obtained for the reaction between nitrogen monoxide and hydrogen at 700 °C. The stoichiometric equation for the reaction is:

$$2H_2(g) + 2NO(g) \rightarrow 2H_2O(g) + N_2(g)$$

Table 10.4

Experiment number	Initial concentration /mol dm^{-3}		Initial rate /mol dm^{-3} s^{-1}
	H_2	NO	
1	0.01	0.025	2.4×10^{-6}
2	0.005	0.025	1.2×10^{-6}
3	0.01	0.0125	0.6×10^{-6}

Determine the order of reaction with respect to NO and to H_2. Write the rate equation for the reaction.

Solution

1. Consider experiments 1 and 2 in Table 10.4.
   Initial [NO(g)] is constant. Therefore, any variation in initial rate is due to the variation in initial [H_2(g)].
2. Deduce the order of reaction with respect to hydrogen.
   Halving initial [H_2(g)] halves the rate of reaction.

$$\therefore \text{rate} \propto [H_2]^1 \quad \text{i.e. the order with respect to } H_2 = 1$$

3. Consider experiments 1 and 3 in Table 10.4.
   Initial [H_2(g)] is constant. Therefore, any variation in initial rate is due to the variation in initial [NO(g)].

4. Deduce the order of reaction with respect to nitrogen monoxide.
   Halving initial [NO(g)] gives one quarter the rate, i.e. $(\frac{1}{2})^2$.

$$\therefore \text{rate} \propto [\text{NO(g)}]^2 \quad \text{i.e. the order with respect to NO} = 2$$

5. Write the rate equation:

$$\text{rate} = k[\text{H}_2\text{(g)}][\text{NO(g)}]^2$$

Use the same method for the following Exercises.

EXERCISE 10.12

Answers on page 274

Two gases, A and B, react according to the stoichiometric equation:

$$\text{A(g)} + 3\text{B(g)} \rightarrow \text{AB}_3\text{(g)}$$

A series of experiments carried out at 298 K in order to determine the order of this reaction gave the following results.

Table 10.5

Experiment	**Initial concentration of A** c**/mol dm**$^{-3}$	**Initial concentration of B** c**/mol dm**$^{-3}$	**Initial rate of formation of AB**$_3$ **/mol dm**$^{-3}$ **min**$^{-1}$
1	0.100	0.100	0.00200
2	0.100	0.200	0.00798
3	0.100	0.300	0.01805
4	0.200	0.100	0.00399
5	0.300	0.100	0.00601

a What is the order of the reaction between A and B with respect to:
  i) substance A,
  ii) substance B?

b Write down a **rate equation** for the reaction between A and B.

c Using the experimental data given for Experiment 1, in Table 10.5, calculate the **rate constant**, k, for the reaction. Give the appropriate units of k.

EXERCISE 10.13

Answers on page 274

A series of experiments was carried out on the reaction:

$$2H_2(g) + 2NO(g) \rightarrow 2H_2O(g) + N_2(g)$$

The initial rate of reaction at 750 °C was determined by noting the rate of formation of nitrogen, and the following data recorded:

Table 10.6

Experiment	**Initial concentration of nitrogen monoxide/mol dm^{-3}**	**Initial concentration of hydrogen /mol dm^{-3}**	**Rate of formation of nitrogen /mol dm^{-3} s^{-1}**
1	6.0×10^{-3}	1.0×10^{-3}	2.88×10^{-3}
2	6.0×10^{-3}	2.0×10^{-3}	5.77×10^{-3}
3	6.0×10^{-3}	3.0×10^{-3}	8.62×10^{-3}
4	1.0×10^{-3}	6.0×10^{-3}	0.48×10^{-3}
5	2.0×10^{-3}	6.0×10^{-3}	1.92×10^{-3}
6	3.0×10^{-3}	6.0×10^{-3}	4.30×10^{-3}

The rate equation for the reaction is:

$$\text{rate} = k[H_2]^m[NO]^n$$

Deduce the order of the reaction with respect to:
a hydrogen,
b nitrogen monoxide.

EXERCISE 10.14

Answers on page 274

Samples of bromomethane and 2-bromo-2-methylpropane were dissolved in dilute aqueous ethanol and reacted with sodium hydroxide solution. Several experiments were carried out, at constant temperature, using different initial concentrations of each bromoalkane and hydroxide ions. The initial rate of reaction was determined in each case.

Table 10.7
Data for the hydrolysis of bromomethane, CH_3Br

Experiment	**[CH_3Br] /mol dm^{-3}**	**[OH^-] /mol dm^{-3}**	**Rate /mol dm^{-3} s^{-1}**
A	0.010	0.0050	0.107
B	0.010	0.010	0.216
C	0.010	0.020	0.425
D	0.020	0.020	0.856
E	0.040	0.020	1.72
F	0.080	0.020	3.42

Table 10.8 Data for the hydrolysis of 2-bromo-2-methylpropane, $(CH_3)_3CBr$

Experiment	$[(CH_3)_3CBr]$ /mol dm^{-3}	$[OH^-]$ /mol dm^{-3}	Rate /mol dm^{-3} s^{-1}
A	0.020	0.010	20.2
B	0.020	0.020	20.1
C	0.020	0.030	20.2
D	0.040	0.030	40.1
E	0.060	0.030	60.2
F	0.080	0.030	80.0

Use the data to establish a rate equation for each reaction.

EXERCISE 10.15

Answer on page 275

A^P

In the *Journal of the Chemical Society* for 1950, Hughes, Ingold and Reed report some kinetic studies on aromatic nitration. In one experiment ethanoic acid containing 0.2% water was used as a solvent and pure nitric acid was added to make a 7 M solution.

The kinetics of the nitration of ethylbenzene, $C_6H_5C_2H_5$, were studied with the following results at 20 °C.

Table 10.9

Time /min	Concentration of ethylbenzene/mol dm^{-3}
0	0.090
8.0	0.063
11.0	0.053
13.0	0.049
16.0	0.037
21.0	0.024
25.0	0.009

Determine an order of reaction from these results.

The kinetics of the iodination of propanone can be investigated by a sampling method such as that described in the next Exercise.

EXERCISE 10.16

Answers on page 275

A

Under conditions of acid catalysis, propanone reacts with iodine as follows:

$$(CH_3)_2CO(aq) + I_2(aq) \rightarrow CH_2ICOCH_3(aq) + HI(aq)$$

50 cm^3 of 0.02 M I_2(aq) and 50 cm^3 of acidified 0.25 M propanone were mixed together. 10 cm^3 portions of the reaction mixture were removed at five-minute intervals and rapidly added to an excess of 0.5 M $NaHCO_3$(aq). The iodine remaining was titrated against aqueous sodium thiosulphate.

Figure 10.4

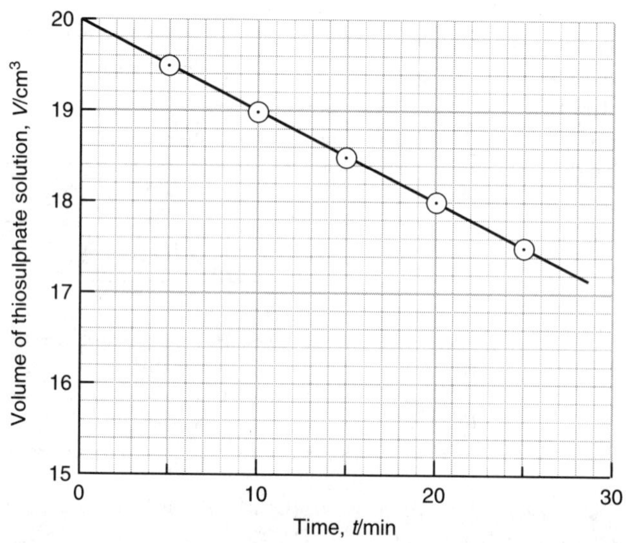

The graph (Fig. 10.4) records the volume of aqueous sodium thiosulphate required to react with the iodine remaining at different times after mixing the reactants.

a Why are the 10 cm^3 portions of the reaction mixture added to aqueous sodium hydrogencarbonate before titration with sodium thiosulphate?
b How does the rate of change of iodine concentration vary during the experiment?
c What is the rate of reaction in terms of cm^3 of sodium thiosulphate per minute?
d What is the order of the reaction with respect to iodine?
e Suppose the reaction is first order with respect to propanone. What would be the rate of reaction (in cm^3 of sodium thiosulphate min^{-1}) if 0.50 M propanone were used instead of 0.25 M propanone?
f Indicate by means of a sketch how the volume of aqueous sodium thiosulphate used would vary with time if no catalytic acid was present in the reacting mixture of propanone and iodine.
g Explain your answer to **f**.

Finally, in this section, we present an Exercise where you are given the rate equation and the rate constant for a reaction, at a particular temperature, and asked to calculate the rates at different initial concentrations of reactants.

EXERCISE 10.17

Answers on page 275

Hydrogen and iodine react together to produce hydrogen iodide:

$$H_2(g) + I_2(g) \rightleftharpoons 2HI(g)$$

The rate equation for the reaction is:

$$\text{rate} = k[H_2(g)][I_2(g)]$$

and the rate constant at 374 °C is $8.58 \times 10^{-5}\ mol^{-1}\ dm^3\ s^{-1}$.

a Work out the rate of reaction at each of the following initial concentrations of hydrogen and iodine, in experiments A, B and C.

Table 10.10

Experiment	Initial $[H_2]$ /mol dm^{-3}	Initial $[I_2]$ /mol dm^{-3}
A	0.010	0.050
B	0.020	0.050
C	0.020	0.10

b From your answers to **a**, what can you say about the effect of concentration of reactants on reaction rate?

We now go on to consider the effect of temperature on the rate of reaction, and introduce the important concept of **activation energy**.

ACTIVATION ENERGY AND THE EFFECT OF TEMPERATURE ON REACTION RATE

The rate of many gaseous and aqueous reactions often doubles for a temperature rise of only 10 K. Why is it that such a small rise in temperature can cause a large percentage increase in the reaction rate? We consider the question in this section, which also helps to explain why some reactions, which we might expect to proceed on the basis of values of ΔH° or ΔG°, do not appear to occur.

Activation energy

Consider the following reaction:

$$C_8H_{18}(l) + 12\tfrac{1}{2}O_2(g) \rightarrow 8CO_2(g) + 9H_2O(g) \quad \Delta H^\circ = -5498 \text{ kJ mol}^{-1}$$

This reaction is highly exothermic and can occur with explosive force under the right conditions. However, it does not take place at room temperature!

There are many other examples like this, where there is an 'energy barrier' to be overcome before reaction can take place. This barrier is called the activation energy for the reaction. Since collisions between reactant particles are always occurring it seems reasonable to assume that particles do not always react when they collide. A reaction occurs as a result of collisions between particles which possess more than a certain minimum amount of energy – the activation energy, E_a (sometimes E_A).

An energy profile diagram is a way of showing how the potential energy of reacting molecules changes. To ensure that you can distinguish between the activation energy and the enthalpy change for a reaction, try the next Exercise.

EXERCISE 10.18
Answers on page 275

a Draw an energy profile to represent the reaction:

$$2N_2O(g) \rightleftharpoons 2N_2(g) + O_2(g) \quad \Delta H^\circ = -164.0 \text{ kJ mol}^{-1}$$

The activation energy, E_a, for the forward reaction is 250 kJ mol^{-1}. Use graph paper and let 1 cm represent 50 kJ mol^{-1}.

b Use your energy profile diagram to calculate the activation energy for the back reaction.

Next we go on to consider what proportion of particles in a gaseous reaction mixture reach the activation energy and how the proportion is affected by temperature.

Fraction of particles with energy greater than E_a

In your studies you may have learnt about the Maxwell distribution curves for a sample of gas at different temperatures. You can use these curves to calculate the fraction of molecules with kinetic energies in a certain range at a particular temperature. For our purposes, in this section, it would be useful to know the fraction of molecules with energy greater than the energy of activation, E_a.

Consider Fig. 10.5 below, which shows the distribution of kinetic energies of particles in the gas phase and the activation energy, E_a.

Figure 10.5

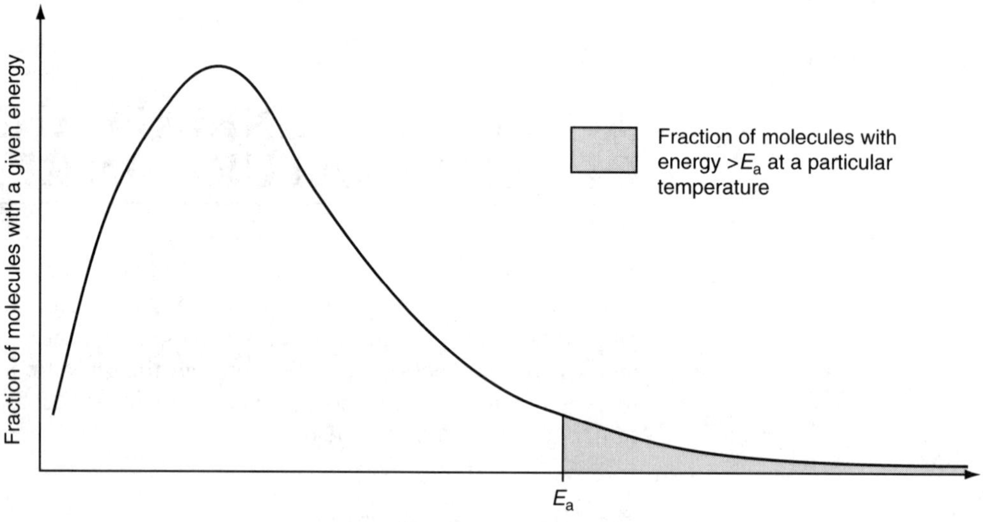

The area beneath the curve is proportional to the total number of particles involved and the shaded area under the curve is proportional to the number of particles with energy greater than E_a. Hence the fraction of particles with energy greater than E_a is given by the ratio:

$$\frac{\text{shaded area under curve}}{\text{total area under curve}} = \frac{\text{number of particles with } E > E_a}{\text{total number of particles}}$$

Maxwell and Boltzmann derived a useful mathematical expression for this fraction in terms of the activation energy, E_a, the gas constant, R, and the absolute temperature, T:

$$\textbf{fraction of particles with energy} > E_a = e^{-E_a/RT}$$

The following Worked Example shows you how to use this expression to calculate the number of molecules with energy greater than a given energy.

WORKED EXAMPLE

Calculate the number of molecules in 1.0 mol of gas at 25 °C with energy greater than 55.0 kJ mol^{-1}.

Solution

1. Calculate the fraction of molecules with energy greater than 55.0 kJ mol^{-1}.

   Fraction of molecules with $E > 55.0$ kJ $mol^{-1} = e^{-E/RT}$ where $R = 8.31$ J K^{-1} mol^{-1}, $T = 298$ K and $E = 55\,000$ J mol^{-1}.

   $$\therefore \text{fraction of molecules} = e^{-(55\,000 \text{ J mol}^{-1})/(8.31 \text{ J K}^{-1} \text{ mol}^{-1})(298 \text{ K})}$$
   $$= e^{-22.2} = 2.28 \times 10^{-10}$$

2. Calculate the total number of molecules with energy greater than 55.0 kJ mol^{-1}.

   The total number of molecules present in 1.0 mol of gas is given by:
   $L = 6.01 \times 10^{23}$ mol^{-1} where L is the Avogadro constant.

   Since fraction of molecules
   $$\text{with } E > 55.0 \text{ kJ mol}^{-1} = \frac{\text{number of molecules with } E > 55.0 \text{ kJ mol}^{-1}}{\text{total number of molecules}}$$

Then number with $E > 55.0\ \text{kJ mol}^{-1} = Le^{-E/RT}$
$= 6.02 \times 10^{23}\ \text{mol}^{-1} \times 2.28 \times 10^{-10}$
$= \mathbf{1.37 \times 10^{14}\ mol^{-1}}$

Now you do a similar calculation for a different temperature.

EXERCISE 10.19
Answers on page 276

a Calculate the number of molecules in 1.0 mol of gas at 35 °C with energy greater than $55.0\ \text{kJ mol}^{-1}$.
b Compare this value to that worked out at 298 K.

You have just calculated that for a rise in temperature of only 10 K, twice as many molecules exceed the energy barrier of $55.0\ \text{kJ mol}^{-1}$. Thus, on the Maxwell distribution curves the shaded area under the $T + 10$ K curve will be twice that on the T curve (where T is the absolute temperature) as shown in Fig. 10.6.

Figure 10.6

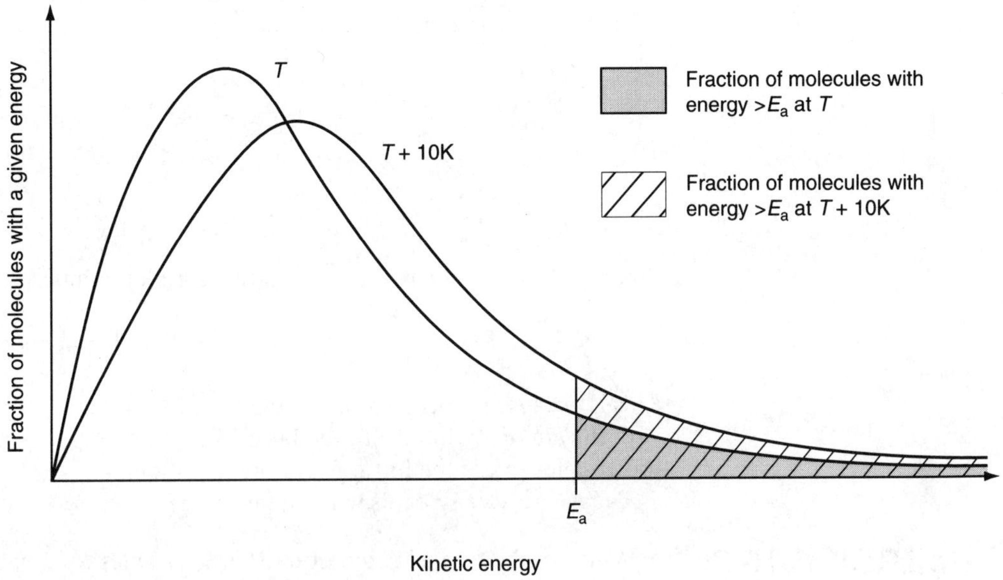

We now go on to consider the relationship between the rate constant, k, for a reaction and the fraction of particles with energy greater than E_a.

■ The Arrhenius equation

For particles to react, their energy must exceed E_a for that reaction. This suggests that at a given temperature:

$$\textbf{rate} \propto e^{-E_a/RT} \qquad (1)$$

Since the rate changes during the progress of a reaction, it is more useful to use k (rate constant).

Consider the general reaction:

$$A(g) + B(g) \rightarrow \text{products}$$
$$\text{rate} = k[A]^x[B]^y \text{ at a room temperature } T_1$$

If the experiment is repeated using the same concentrations of A and B at a higher temperature, T_2, the rate increases. Since $[A]^x$ and $[B]^x$ have the same initial value, then

k must increase. Therefore, we can say that the rate constant is a general measure of the reaction rate at a particular temperature. Equation (1) becomes:

$$k \propto e^{-E_a/RT}$$

$$\therefore \boldsymbol{k = A e^{-E_a/RT}} \qquad (2)$$

A is a constant, sometimes called the Arrhenius constant or pre-exponential factor. Equation (2) is often called the Arrhenius equation, named after the Swedish chemist, Svante Arrhenius, who formulated it in 1880.

Using the Arrhenius equation

The equation provides an extremely useful means of getting values for the activation energy and the pre-exponential factor for a reaction. It is usually changed to a logarithmic form to make it more manageable:

$$k = A e^{-E_a/RT}$$

$$\therefore \ln k = \ln(A e^{-E_a/RT}) \quad (\ln = \log_e)$$

$$= \ln A + \ln e^{-E_a/RT}$$

$$\boldsymbol{\ln k = \ln A - \frac{E_a}{RT}} \qquad (3)$$

Some textbooks may give an alternative form of equation (3):

$$\log_{10} k = \log_{10} A - \frac{E_a}{2.3RT}$$ (The factor 2.3 appears because $\ln x = 2.3 \log x$.)

This is because in former years students were expected to perform their calculations with the aid of log tables (to the base 10). However, using modern electronic calculators it is simpler to use the $\log_e$ form of the equation.

The next Exercise deals with equation (3).

EXERCISE 10.20

Answers on page 276

a Which of the terms in equation (3) are variables for a particular reaction?
b Compare equation (3) with the equation for a straight line, i.e. $y = mx + c$. Which of the terms in equation (3) are analogous to y, m, x and c?
c Sketch the shape of the graph you would expect to obtain by plotting $\ln k$ against $1/T$.
d Show how you could calculate A and E_a from your graph.

Use the method you have just devised in the following Exercises.

EXERCISE 10.21

Answers on page 276

A

When gaseous hydrogen iodide decomposes in accordance with the equation:

$$2HI(g) \rightleftharpoons H_2(g) + I_2(g)$$

the reaction is found to be second order with respect to hydrogen iodide. In a series of experiments, the rate constant for this reaction was determined at several temperatures. The results obtained are shown in the table below.

Table 10.11

Temperature (T) /K	$1/T$ K^{-1}	Rate constant (k)/dm^3 mol^{-1} s^{-1}	ln k
633		1.78×10^{-5}	
666		1.07×10^{-4}	
697		5.01×10^{-4}	
715		1.05×10^{-3}	
781		1.51×10^{-2}	

Arrhenius deduced that for a reaction:

$$\ln k = -\frac{E}{RT} + \ln A$$

where R is the gas constant and A is a constant.

a Use the data given to determine the activation energy E for the reaction.

b Determine by what factor the rate increases when the temperature rises from 300 K to 310 K.

You may wish to use the alternative form of the Arrhenius equation:

$$k = Ae^{-E/RT}$$

c The kinetic energy of a fixed mass of gas is directly proportional to its temperature measured on the Kelvin scale. Calculate the ratio of the kinetic energies of a fixed mass of gas at 310 K and at 300 K.

d Compare this ratio with the increase in the rate of reaction determined from experimental data. Discuss the significance of this result.

EXERCISE 10.22

Answers on page 277

In this Exercise you determine the activation energy for the oxidation of iodide ions by peroxodisulphate(VI) (persulphate) ions. The experiment is an example of a 'clock' reaction, in which you measure the time taken for a reaction to reach a certain stage.

The table below shows the results of a clock reaction experiment. t is the time taken at different temperatures for sufficient iodine to be formed to react with a fixed amount of sodium thiosulphate. Further production of iodine turns the solution blue because starch is also present. $1/t$ is then proportional to the rate of reaction.

Plot a graph of $\ln(1/t)$ (y-axis) against $1/T$ (x-axis) and use it to calculate a value for the activation energy.

Table 10.12

Temperature/°C	30	36	39	45	51
Temperature, T/K	303	309	312	318	324
Time t/s	204	138	115	75	55
$\ln\frac{1}{t}$					
$\frac{1}{T}/10^{-3}\,K^{-1}$ (or $10^3\,K/T$)					

EXERCISE 10.23

Answers on page 277

This Exercise refers to the same reaction as the last one but a catalyst was present.

Plot a graph of $\ln(1/t)$ (y-axis) against $1/T$ (x-axis) and use it to calculate a value for the activation energy.

Table 10.13

Temperature/°C	15	19.5	26	35	42
Temperature, T/K	288	292.5	299	308	315
Time, t/s	10.0	7.0	5.0	3.5	2.5
$\ln(1/t)$					
$\frac{1}{T}/10^{-3}\,K^{-1}$					

In the next Exercise we present data for a reaction between peroxodisulphate(VI) ions and cobalt metal. In this case, you take the units of k as loss in mass of metal per minute and work out a value for the activation energy for the reaction.

EXERCISE 10.24

Answers on page 278

When strips of cobalt foil are rotated rapidly at a constant rate in sodium peroxodisulphate(VI) solution, there is a slow reaction and the cobalt dissolves. The reaction can be followed by removing and weighing the foil at intervals.

Table 10.14

Experiment 1 at 0.5 °C			Experiment 2 at 13.5 °C			Experiment 3 at 25 °C		
Time /min	**Mass /mg**	**Loss /mg**	**Time /min**	**Mass /mg**	**Loss /mg**	**Time /min**	**Mass /mg**	**Loss /mg**
0	130		0	130		0	130	
20	120		10	115		4	116	
60	98		15	106		8	103	
80	86		30	86		12	87	

Determine a rate constant k at each temperature as a loss in mass per minute (mg min^{-1}), and then determine the activation energy E of the reaction using the relationship:

$$k \propto e^{-E/RT}$$

What difference in the results would you predict if the cobalt foil were **not** rotated?

Sometimes data may be provided for only two temperatures. In this case, instead of a graphical method, the activation energy may be calculated by using simultaneous equations. You can try this in the next Exercise.

EXERCISE 10.25

Answer on page 278

Data are given below for the dissociation of hydrogen iodide:

$$2HI(g) \rightleftharpoons H_2(g) + I_2(g)$$

Table 10.15

Temperature T/K	Rate constant $k/10^{-5}$ dm^3 mol^{-1} s^{-1}
647	8.58
700	116

Calculate the activation energy for the reaction.

You have seen in Exercises 10.22 and 10.23 that a catalyst reduces the activation energy of a given reaction. You explore this graphically in the next Exercise.

EXERCISE 10.26
Answers on page 278

The decomposition of hydrogen iodide is catalysed by platinum metal. The following diagram, Fig. 10.7, represents the energy changes that take place during the course of the uncatalysed reaction, with the changes during the catalysed reaction superimposed.

Figure 10.7

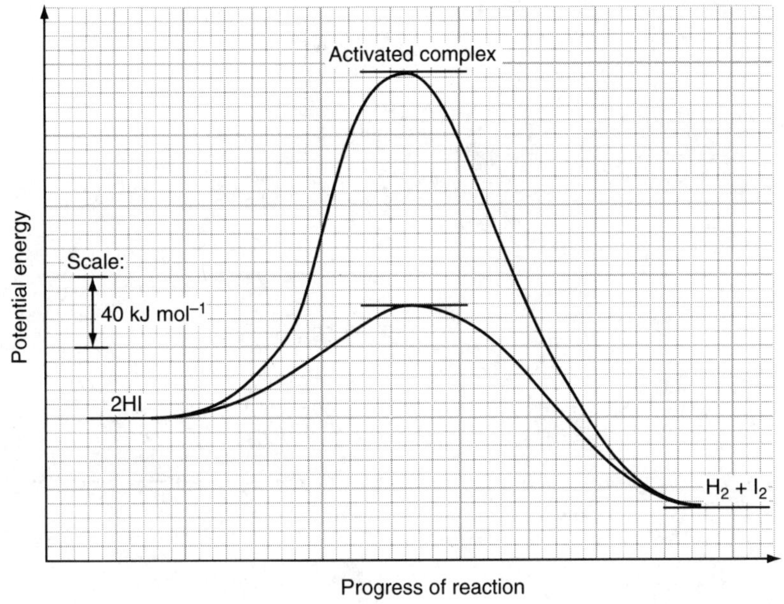

Use the scale in Fig. 10.7 to work out the following for both the forward and back reactions:

a the activation energies (uncatalysed),
b the activation energies (catalysed),
c the enthalpy changes.

Autocatalysis

Reactions in which a product acts as a catalyst are called **autocatalytic**. Once the reaction starts, more catalyst is produced, so that the rate of reaction **increases** with time for at least part of its duration. Although we did not discuss autocatalysis then, you have already come across one example of a reaction like this in Exercise 10.16 – the acid-catalysed iodination of propanone:

$$CH_3COCH_3(aq) + I_2(aq) \rightarrow CH_2ICOCH_3(aq) + H^+(aq) + I^-(aq)$$

The next Exercise is about the autocatalytic oxidation of ethanedioate ions, $C_2O_4^{2-}$, by manganate(VII) ions, MnO_4^-, catalysed by manganese(II) ions, Mn^{2+}:

$$2MnO_4^-(aq) + 16H^+(aq) + 5C_2O_4^{2-}(aq) \rightarrow 10CO_2(g) + 2Mn^{2+}(aq) + 8H_2O(l)$$

EXERCISE 10.27

Answers on page 279

The graph in Fig. 10.8 refers to the reaction between manganate(VII) ions and ethanedioate ions.

Figure 10.8

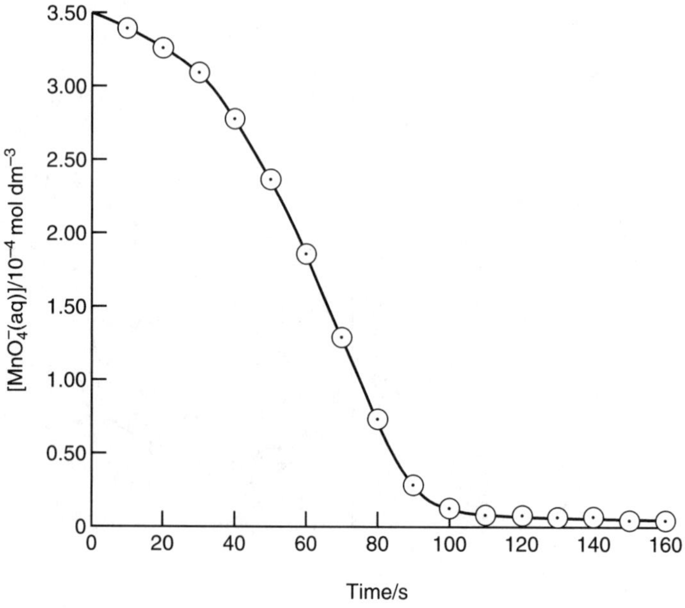

a Estimate from the graph the rates of reaction at the beginning of the process and after 50 seconds. How do you explain the difference in these values?

b The rate of reaction suddenly drops again after about 90 seconds. Give a reason for this.

END-OF-CHAPTER QUESTIONS

Answers on page 279

10.1 The half-life of radioactive $^{233}_{91}P$ is 28 days. How many days will it take for the radioactivity to fall to one eighth of the initial value?

10.2 Chemists can determine the age of organic matter by measuring the proportion of $^{14}_{6}C$ present. Assuming that carbon-14 has a half-life of 5600 years, what is the age of a piece of wood found to contain ⅛ as much carbon-14 as living material?

10.3 In an experiment designed to investigate the effect of acid concentration on the reaction between an acid and a metal, a large excess of hydrochloric acid of concentrations varying from 0.2 M to 1.0 M reacts with a fixed length of magnesium ribbon (5 cm in each case). The reaction times required for the evolution of 15 cm^3 of hydrogen gas are determined at a constant pressure of 1 atm and at 300 K.

The following table summarises the results obtained by a group of students.

Table 10.16

Concentration of hydrochloric acid /mol dm^{-3}	Time required for the evolution of 15 cm^3 of hydrogen at 1 atm and 300 K/s
0.25	160
0.35	80
0.50	40
0.70	20
1.00	10

a Suggest a reason why the experiments were carried out with a **large excess** of hydrochloric acid.

b i) Consider the data in the table. Give a relationship between reaction time t and the concentration of hydrochloric acid.

ii) What is the order of the reaction with respect to hydrochloric acid?

10.4 The table shows the results of an investigation into the rate of reaction between hydroxide ions, OH^-, and phosphinate ions, $PH_2O_2^-$, at 80 °C. The overall equation for the reaction is:

$$PH_2O_2^-(aq) + OH^-(aq) \rightarrow PHO_3^{2-}(aq) + H_2(g)$$

Table 10.17

Experiment	Initial concentrations		Initial rate of hydrogen production /cm^3 min^{-1}
	[$PH_2O_2^-$(aq)] /mol dm^{-3}	[OH^-(aq)] /mol dm^{-3}	
1	0.6	1.0	2.4
2	0.6	2.0	9.6
3	0.6	3.0	21.5
4	0.1	6.0	14.4
5	0.2	6.0	28.8
6	0.3	6.0	43.2

a How does the initial rate of this reaction depend on the concentration of hydroxide ions?

b How does the initial rate of this reaction depend on the concentration of phosphinate ions?

c What is the rate expression for this reaction?

d In what units will the rate constant be expressed?

10.5 The hydrolysis of CH_3Br proceeds by an S_N2 mechanism. The rate of reaction at 60 °C was found to be 8.13×10^{-5} mol dm^{-3} s^{-1} when the concentrations of hydroxide ion and CH_3Br were 0.2 and 0.05 mol dm^{-3}, respectively.

a Give the rate equation for the reaction and calculate the rate constant at 60 °C.

b At 80 °C the rate constant was measured as 50.0×10^{-3} dm^3 mol^{-1} s^{-1}. Calculate the activation energy, E, for the hydrolysis using the Arrhenius equation which may be written either as $k = A\,e^{-E/RT}$ or as $\ln k = \ln A - E/RT$ ($R = 8.31$ J K^{-1} mol^{-1}).

10.6 This Exercise concerns an experiment to determine the rate equation for the reaction between bromide and bromate(V) ions in aqueous solution.
Bromide and bromate(V) ions in acid solution react according to the equation:

$$5Br^{-}(aq) + BrO_3^{-}(aq) + 6H^{+}(aq) \rightarrow 3Br_2(aq) + 3H_2O(l) \quad (1)$$

In order to follow the reaction, two other substances are added to the reaction mixture.

i) A precisely known, small amount of phenol. This reacts immediately with the bromine produced, removing it from solution:

$$3Br_2(aq) + C_6H_5OH(aq) \rightarrow C_6H_2Br_3OH(aq) + 3H^{+}(aq) + 3Br^{-}(aq) \quad (2)$$

ii) Methyl orange solution, which is bleached colourless by free bromine:

$$\underset{}{Br_2(aq)} + \underset{\text{(acid form: pink)}}{\text{methyl orange}} \rightarrow \underset{\text{(colourless)}}{\text{bleached methyl orange}} \quad (3)$$

As soon as all the phenol has reacted with bromine produced in reaction (1), free bromine will appear in solution and bleach the methyl orange. If the time taken for the methyl orange solution to be bleached is t, then the rate of reaction (1) is proportional to $1/t$.

The following tables show the effect of varying the volume of one reactant at a time, with the volumes of other reagents and the total volume kept constant, at constant temperature.

Table 10.18

Volume of Br^{-}(aq) cm^3	10.0	8.0	6.0	5.0	4.0	3.0
Time t/s	22.5	26.0	34.0	37.5	48.0	65.0
$\frac{1}{t}/10^{-2}\ s^{-1}$						

Table 10.19

Volume of BrO_3^{-}(aq) cm^3	10.0	8.0	6.0	5.0	4.0	3.0
Time t/s	18.5	24.0	31.0	35.0	45.0	69.0
$\frac{1}{t}/10^{-2}\ s^{-1}$						

Table 10.20

Volume of acid/cm^3	10.0	8.0	6.0	5.0	4.0	3.0
Time t/s	13.0	19.0	30.0	39.0	60.0	97.0
$\frac{1}{t}/10^{-2}\ s^{-1}$						

a For each part of the experiment, plot a graph of $1/t$ against volume of the reactant under consideration. $1/t$ is proportional to the rate of reaction, and the volume of reactant is proportional to the concentration, since the total volume is constant.

b Deduce from each graph whether or not the reaction is first order with respect to the reactant under consideration.

c If you think the reaction is **not** first order, plot another graph, as explained below.

Suppose that, for a reactant A:

$$\text{rate} = k_1[\text{A}]^n$$

Under the conditions of this experiment, it follows that:

$$\frac{1}{t} = k_2V^n \quad (V \text{ is the initial volume of reactant})$$

$$\therefore \log\frac{1}{t} = n\log V + \log k_2$$

Plotting $\log(1/t)$ against $\log V$ should therefore give a straight line with slope n.

d Write the rate equation for the reaction.

A

10.7 For the reaction $X + Y \rightarrow Z$, the rate expression is

$$\text{Rate} = k[X]^2[Y]^{\frac{1}{2}}$$

If the concentrations of X and Y are both increased by a factor of 4, by what factor will the rate increase?

A P

10.8 The rate of the reaction between peroxodisulphate ions and iodide ions in aqueous solution may be studied by measuring the amount of iodine formed after different reaction times. The stoichiometric equation is:

$$S_2O_8^{2-}(aq) + 2I^-(aq) \rightarrow 2SO_4^{2-}(aq) + I_2(aq)$$

The graphs show the results of a kinetic investigation of this reaction.

Figure 10.9

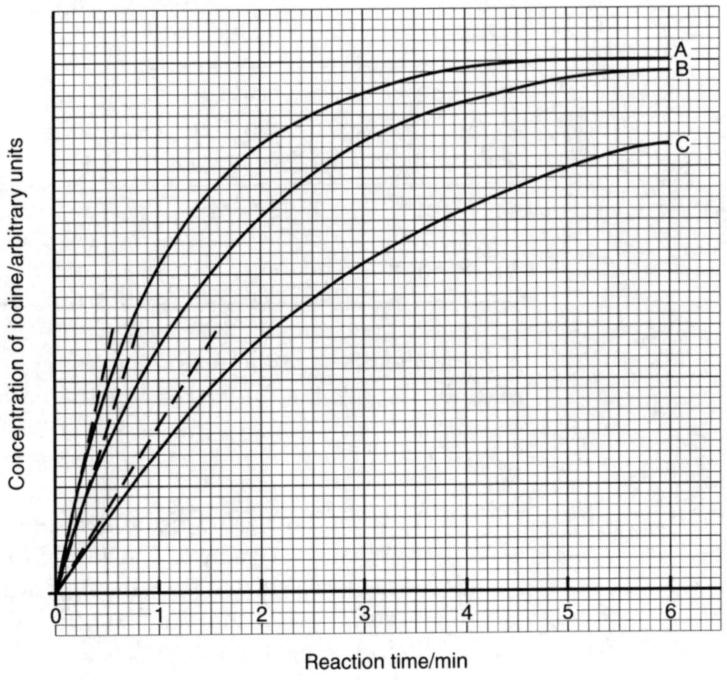

The experimental conditions were at a fixed temperature:

Table 10.21

	Initial concentration of $S_2O_8^{2-}$(aq)/mol dm^{-3}	Initial concentration of I^-(aq)/mol dm^{-3}
Curve A	0.01	0.3
Curve B	0.01	0.2
Curve C	0.01	0.1

a Evaluate the **initial reaction rates** (given as $\Delta[I_2]/\Delta t$) for the three curves.
b i) With respect to which reactant concentration do these reaction rates vary?
ii) What is the order of the reaction with respect to this reactant?
c With reference to experimental curve A, find the times required for the completion of:
i) one-half of the reaction,
ii) three-quarters of the reaction.
In the light of these results, what is the order of the reaction with respect to the second reactant?

10.9 Dinitrogen pentoxide (N_2O_5) decomposes according to the equation:

$$2N_2O_5(g) \rightarrow 2N_2O_4(g) + O_2(g)$$

The reaction proceeds at a rate which is conveniently measurable at a temperature of 45 °C; at this temperature, the following results were obtained.

Table 10.22

$[N_2O_5]$/mol dm^{-3}	Rate of disappearance of N_2O_5/mol dm^{-3} s^{-1}
22.3×10^{-3}	11.65×10^{-6}
17.4×10^{-3}	8.67×10^{-6}
13.2×10^{-3}	6.63×10^{-6}
9.5×10^{-3}	4.65×10^{-6}
4.7×10^{-3}	2.35×10^{-6}

a Plot these results on a suitable graph.
b The rate law for the reaction can be expressed in the form:

$$\text{rate} = k[N_2O_5]^x$$

i) What is the value of x as deduced from the graph?
ii) What is the value of k at 45 °C? In what units is it expressed?
c It can be calculated from the figures given that the concentration would fall from 2.0×10^{-2} mol dm^{-3} to 1.0×10^{-2} mol dm^{-3} in 1386 seconds. How much longer would it take to fall to 2.5×10^{-3} mol dm^{-3}?

10.10 The Arrhenius equation which involves the activation energy of a reaction, E, may be given in the forms:

$$\text{either} \quad k = Ae^{-E/RT} \quad \text{or} \quad \ln k = \ln A - E/RT$$

For the reaction:

$$H_2(g) + I_2(g) \rightleftharpoons 2HI(g),$$

the values of k at 590 K and 700 K are 1.4×10^{-3} $dm^3\ mol^{-1}\ s^{-1}$ and 6.4×10^{-2} $dm^3\ mol^{-1}\ s^{-1}$ respectively. ($R = 8.3\ J\ K^{-1}\ mol^{-1}$.)

Use an Arrhenius equation and the above data to calculate the activation energy for the reaction:

$$H_2(g) + I_2(g) \rightarrow 2HI(g)$$

10.11 An experiment was carried out to investigate the rate of reaction of an organic chloride of molecular formula C_4H_9Cl with hydroxide ions. The reaction was carried out in solution in a mixture of propanone and water with the following results:

Table 10.23

Time elapsed /s	**Concentration of C_4H_9Cl/mol dm^{-3}**	**Concentration of hydroxide ions/mol dm^{-3}**
0	0.0100	0.0300
294	0.0050	0.0250
595	0.0025	0.0225

a From the results given, deduce an order of reaction. Explain your answer.
b Write a rate expression for the overall reaction.

10.12 The reaction between potassium iodide and potassium peroxodisulphate(VI), $K_2S_2O_8$, in aqueous solution proceeds according to the overall equation.

$$2KI(aq) + K_2S_2O_8(aq) \rightleftharpoons 2K_2SO_4(aq) + I_2(aq)$$

The rate of this reaction is found by experiment to be directly proportional to the concentration of the potassium iodide, and directly proportional to that of the potassium peroxodisulphate(VI).

a Write the above equation in ionic form.
b What is the overall order of the reaction?
c With an initial concentration of potassium iodide of 1.0×10^{-2} mol dm^{-3} and of potassium peroxodisulphate(VI) of 5.0×10^{-4} mol dm^{-3}, it is found that at 298 K, the initial rate of disappearance of the peroxodisulphate(VI) ions is 1.02×10^{-8} mol $dm^{-3}\ s^{-1}$. What is the velocity constant of the reaction at this temperature?

APPENDIX

SIGNIFICANT FIGURES AND SCIENTIFIC MEASUREMENTS

These notes on the use of significant figures are not rigorous. We merely give some useful rules which you can apply to ordinary A-level calculations. For a detailed treatment you can read a textbook on the theory of measurements.

The numerical value of any physical measurement is an approximation which is limited by the accuracy of the measuring instrument.

Generally, the last digit in a measured quantity has an uncertainty associated with it. For example, in reading a thermometer, part of which is shown in Fig. A1, some may read the temperature as 21.1 °C and some may read it as 21.2 °C or 21.3 °C.

Figure A1

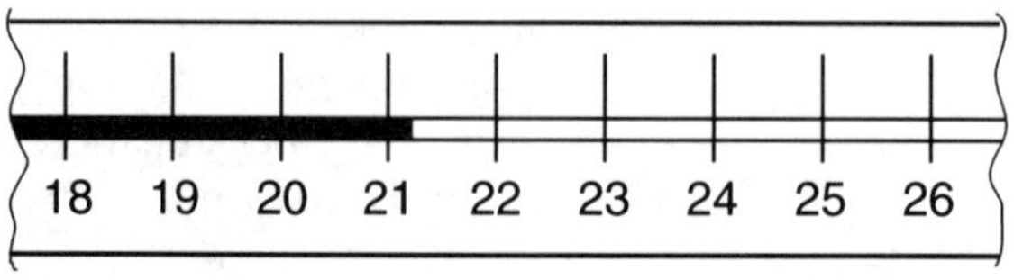

That is, there is no doubt that the temperature is between 21 and 22 degrees Celsius, but there is some uncertainty in the last place. It is for this reason that we must consider the use of significant figures. Furthermore, it is important to consider the use of significant figures when so many calculations are made using electronic calculators which give as many as ten digits on their displays. You are very rarely justified in using all of them.

Zeros

A measured mass of 23 g has two significant figures, 2 and 3. If this same mass is written as 0.023 kg, it still contains two significant figures because zeros appearing as the first figures of a number are not significant – they merely locate the decimal point. However, the mass 0.0230 kg is expressed to three significant figures (2, 3 and the last 0).

The expression 'the length is 4700 m' does not necessarily show the accuracy of the measurement. To do this, the number should be written in standard form. If the measurement is made only to the nearest 1000 m, we use only one significant figure, i.e. $l = 5 \times 10^3$ m.

A more precise measurement, to the nearest 100 m, merits two significant figures, i.e. $l = 4.7 \times 10^3$ m, and so on as summarised in Table A1.

Table A1

Distance l /m	Significant figures	Range of uncertainty	Precision of measurement
4700	unspecified	unspecified	unspecified
5×10^3	1	4.5 to 5.5	nearest 1000 m
4.7×10^3	2	4.65 to 4.75	nearest 100 m
4.70×10^3	3	4.695 to 4.705	nearest 10 m
4.700×10^3	4	4.6995 to 4.7005	nearest 1 m

Here is an Exercise to see if you can recognise the number of significant figures in a measured quantity.

EXERCISE A1

Answers on page 279

How many significant figures are in the following quantities?

a 2.54 g
b 2.205 g
c 1.1 g
d 14.0 cm^3
e 1.86×10^5 s

f 2.0070 g
g 9.993 g cm^{-3}
h 5070 m s^{-1}
i 127 000 kg

Now we look at significant figures in the results of combining uncertain values in calculations.

Addition and subtraction

After addition or subtraction, the answer should be rounded off to keep only the same number of decimal places as the **least** precise item. Here are some Worked Examples.

WORKED EXAMPLE Add the following quantities:

a

$$\begin{array}{r} 46.247\ cm^3 \\ 3.219\ cm^3 \\ 0.224\ cm^3 \\ \hline 49.690\ cm^3 \end{array}$$

Answer: 49.690 cm^3

Each volume to be added is expressed to the nearest 0.001 cm^3, so we can express the answer also to the nearest 0.001 cm^3.

b

$$\begin{array}{r} 26.6\ \ \ \ \ \ cm^3 \\ 0.0028\ cm^3 \\ 0.00002\ cm^3 \\ \hline 26.60282\ cm^3 \end{array}$$

Answer: 26.6 cm^3

The number 26.6 is expressed to one place past the decimal point so you cannot have the answer quoted to a greater accuracy than one place past the decimal point.

c

$$\begin{array}{l} 2.40\ \ \ \ cm^3 \\ 3.6584\ cm^3 \\ 0.029\ \ cm^3 \\ \hline 6.0874\ cm^3 \end{array}$$

Answer: 6.09 cm^3

The reasoning here is the same as in part **b**. The least accurate measurement is 2.40 cm^3 so, in the answer, the volume cannot be quoted to more than two places past the decimal point. In this case, however, we round up rather than round down.

The reasoning is the same for subtraction.

WORKED EXAMPLE Perform the following subtractions:

a

$$\begin{array}{r} 7.26\ g \\ -\ 0.2\ \ \ g \\ \hline 7.06\ g \end{array}$$

Answer: 7.1 g

b

$$\begin{array}{r} 539.27\ g \\ -\ 12.8\ \ \ g \\ \hline 526.47\ g \end{array}$$

Answer: 526.5 g

Try the following Exercises:

EXERCISE A2
Answers on page 279

Add the following, expressing your answer to the correct number of significant figures:

a

$$\begin{array}{rl} 203 & g \\ 4 & g \\ 0.77 & g \\ \hline \end{array}$$

b

$$\begin{array}{ll} 0.0034 & dm^3 \\ 0.094 & dm^3 \\ 0.552 & dm^3 \\ \hline \end{array}$$

EXERCISE A3
Answers on page 279

Perform the following subtractions:

a

$$\begin{array}{r} 4.0\ \ m \\ -\ 0.623\ m \\ \hline \end{array}$$

b

$$\begin{array}{rl} 76 & cm^3 \\ -\ 0.3 & cm^3 \\ \hline \end{array}$$

The rules for multiplication and division are even easier.

Multiplication and division

The result of multiplying or dividing can contain only as many significant figures as are contained in the factor with the least number of significant figures.

WORKED EXAMPLE Calculate the density of an object which weighs 17.32 g and has a volume of 2.4 cm^3.

Solution

$$\text{density} = \frac{\text{mass}}{\text{volume}} = \frac{17.32\ g}{2.4\ cm^3}$$

A calculator gives the result as 7.2166667 g cm^{-3}. But the volume has the smaller number of significant figures – two. So the result is rounded off to 7.2 g cm^{-3}.

Now do the next Exercise.

EXERCISE A4
Answers on page 279

Multiply the following, expressing your answer to the correct number of significant figures:
a 0.11 mol $dm^{-3} \times$ 0.0272 dm^3
b 2.43 mol $\times$ 27.9 g mol^{-1}

EXERCISE A5
Answers on page 279

Divide the following, expressing your answer to the correct number of significant figures:

a $$\frac{9.2\ g}{19.00\ g\ mol^{-1}}$$

b $$\frac{0.20\ g}{0.1\ cm^3}$$

ANSWERS

(Answers to questions from examination papers are provided by ILPAC and not by the examination boards.)

CHAPTER 1

EXERCISE 1.1

a

$$\text{time} = \frac{\text{number of molecules}}{\text{rate of removal}} = \frac{1.67 \times 10^{23}}{1\ \text{s}^{-1}}$$

$$= 1.67 \times 10^{23}\ \text{s}$$

b

$$1\ \text{y} = 365\ \text{dy} \times \frac{24\ \text{h}}{\text{dy}} \times \frac{60\text{min}}{\text{h}} \times \frac{60\ \text{s}}{\text{min}} = 3.15 \times 10^{7}\ \text{s}$$

$$\therefore\ 1\ \text{s} = \frac{1\ \text{y}}{3.15 \times 10^{7}}$$

$$\therefore\ \text{time taken} = 1.67 \times 10^{23}\ \text{s}$$

$$= 1.67 \times 10^{23} \times \frac{1\ \text{y}}{3.15 \times 10^{7}} = \mathbf{5.30 \times 10^{15}\ y}$$

EXERCISE 1.2

a **17.0 g mol^{-1}** 14.0 + 3(1.0) = 17.0
b **199.9 g mol^{-1}** 40.1 + 2(79.9) = 199.9
c **98.0 g mol^{-1}** 3(1.0) + 31.0 + 4(16.0) = 98.0
d **322.1 g mol^{-1}** 2(23.0) + 32.1 + 4(16.0) + 20(1.0) + 10(16.0) = 322.1

EXERCISE 1.3

a **35.5 g**
b **71.0 g**
c **31.0 g**
d **124.0 g**
e **126.9 g** (ignore the mass of the extra electrons)

EXERCISE 1.4

a Substituting into the expression

$$n = \frac{m}{M}$$

where m = 30.0 g and M = 32.0 g mol^{-1}

gives $$n = \frac{m}{M} = \frac{30.0\ \text{g}}{32.0\ \text{g mol}^{-1}} = \mathbf{0.938\ mol}$$

b $$n = \frac{m}{M} = \frac{31.0\ \text{g}}{124.0\ \text{g mol}^{-1}} = \mathbf{0.250\ mol}$$

c $$n = \frac{m}{M} = \frac{50.0\ \text{g}}{100.0\ \text{g mol}^{-1}} = \mathbf{0.500\ mol}$$

EXERCISE 1.5

a Substituting into the expression

$$n = \frac{m}{M} \text{ in the form } m = nM$$

where n = 1.00 mol and M = 2.00 g mol^{-1}
gives $m = nM$ = 1.00 mol × 2.00 g mol^{-1} = **2.00 g**

b $m = nM$ = 0.500 mol × 58.5 g mol^{-1} = **29.3 g**
c $m = nM$ = 0.250 mol × 44.0 g mol^{-1} = **11.0 g**

EXERCISE 1.6

a Substituting into the expression

$$n = \frac{m}{M}$$

where m = 1.00 g, M = 17.0 g mol^{-1}

gives $$n = \frac{m}{M} = \frac{1.00\ \text{g}}{17.0\ \text{g mol}^{-1}} = \mathbf{0.0588\ mol}$$

b If the number of molecules is to be the same, then the amount must be the same. For SO_2, M = 64.1 g mol^{-1} and from **a**, n = 0.0588 mol.
∴ substituting these values in the expression

$$n = \frac{m}{M} \text{ in the form } m = nM$$

gives $m = nM$ = 0.0588 mol × 64.1 g mol^{-1} = **3.77 g**

EXERCISE 1.7

a Substituting into the expression

$$n = \frac{m}{M}$$

where $m = 18.0$ g, $M = 12.0$ g mol^{-1}

gives $$n = \frac{m}{M} = \frac{18.0 \text{ g}}{12.0 \text{ g mol}^{-1}} = 1.50 \text{ mol}$$

Substituting into the expression

$$N = nL$$

where $n = 1.50$ mol, $L = 6.02 \times 10^{23}$ mol^{-1}
gives $N = nL = 1.50 \text{ mol} \times 6.02 \times 10^{23} \text{ mol}^{-1} = \mathbf{9.03 \times 10^{23}}$

Or, substituting $n = \frac{m}{M}$ into the expression $N = nL$

gives $$N = \frac{mL}{M} = \frac{18.0 \text{ g} \times 6.02 \times 10^{23} \text{ mol}^{-1}}{12.0 \text{ g mol}^{-1}} = \mathbf{9.03 \times 10^{23}}$$

b $$N = \frac{mL}{M} = \frac{18.0 \text{ g} \times 6.02 \times 10^{23} \text{ mol}^{-1}}{63.5 \text{ g mol}^{-1}} = \mathbf{1.71 \times 10^{23}}$$

c $$N = \frac{mL}{M} = \frac{7.20 \text{ g} \times 6.02 \times 10^{23} \text{ mol}^{-1}}{32.1 \text{ g mol}^{-1}} = \mathbf{1.35 \times 10^{23}}$$

Note that in **c** the number of **atoms** is the same whatever the molecular formula.

EXERCISE 1.8

a Substituting into the expression

$$n = \frac{m}{M}$$

where $m = 1.00$ g, $M = 17.0$ g mol^{-1}

gives $$n = \frac{1.00 \text{ g}}{17.0 \text{ g mol}^{-1}} = 0.0588 \text{ mol}$$

Substituting into the expression

$$N = nL$$

gives $N = 0.0588 \text{ mol} \times 6.02 \times 10^{23} \text{ mol}^{-1} = \mathbf{3.54 \times 10^{22}}$

Or, combining $n = \frac{m}{M}$ with $N = nL$ and substituting gives

$$N = \frac{mL}{M} = \frac{1.00 \text{ g} \times 6.02 \times 10^{23} \text{ mol}^{-1}}{17.0 \text{ g mol}^{-1}} = \mathbf{3.54 \times 10^{22}}$$

b $$N = \frac{mL}{M} = \frac{3.28 \text{ g} \times 6.02 \times 10^{23} \text{ mol}^{-1}}{64.1 \text{ g mol}^{-1}} = \mathbf{3.08 \times 10^{22}}$$

c $$N = \frac{mL}{M} = \frac{7.20 \text{ g} \times 6.02 \times 10^{23} \text{ mol}^{-1}}{8 \times 32.1 \text{ g mol}^{-1}} = \mathbf{1.69 \times 10^{22}}$$

EXERCISE 1.9

a 0.500 mol of NaCl contains 0.500 mol of Na^+ and 0.500 mol of Cl^-
∴ Total amount of ions, $n = 1.00$ mol
Substituting into the expression

$$N = nL$$

gives $N = 1.00 \text{ mol} \times 6.02 \times 10^{23} \text{ mol}^{-1} = \mathbf{6.02 \times 10^{23}}$

b The amount of NaCl is calculated by substituting into the expression

$$n = \frac{m}{M}$$

where $m = 14.6$ g and $M = 58.5$ g mol^{-1}

$$\therefore n = \frac{14.6 \text{ g}}{58.5 \text{ g mol}^{-1}} = 0.250 \text{ mol}$$

0.250 mol of NaCl contains 0.250 mol of Na^+ and 0.250 mol of Cl^-
∴ total amount of ions, $n = 0.500$ mol
Substituting into the expression

$$N = nL$$

gives $N = 0.500 \text{ mol} \times 6.02 \times 10^{23} \text{ mol}^{-1} = \mathbf{3.01 \times 10^{23}}$

c The amount of $CaCl_2$ is given by substituting in the expression

$$n = \frac{m}{M}$$

where $m = 18.5$ g and $M = 111.0$ g mol^{-1}

$$\therefore n = \frac{18.5 \text{ g}}{111.0 \text{ g mol}^{-1}} = 0.167 \text{ mol}$$

Since each mole of $CaCl_2$ contains 3 mol of ions (Ca^{2+}, Cl^-, Cl^-), the amount of ions,
$n = 3 \times 0.167 \text{ mol} = 0.501 \text{ mol}$
Substituting into the expression

$$N = nL$$

gives $N = 0.501 \text{ mol} \times 6.02 \times 10^{23} \text{ mol}^{-1} = \mathbf{3.02 \times 10^{23}}$

EXERCISE 1.10

a From the equation for the reaction we know that

amount of Mg = amount of S

The amount of S is found by using the expression

$$n = \frac{m}{M}$$

where $m = 16.0$ g and $M = 32.1$ g mol^{-1}

$$\therefore n = \frac{m}{M} = \frac{16.0 \text{ g}}{32.1 \text{ g mol}^{-1}} = 0.498 \text{ mol}$$

∴ the amount of Mg also = 0.498 mol.
The mass of Mg is found by using the expression

$$n = \frac{m}{M} \text{ in the form } m = nM$$

where $n = 0.498$ mol and $M = 24.3$ g mol^{-1},
$\therefore\ m = nM = 0.498\ \text{mol} \times 24.3\ \text{g mol}^{-1} =$ **12.1 g**

b For $NaNO_3$, $n = \frac{m}{M} = \frac{4.25\ \text{g}}{85.0\ \text{g mol}^{-1}} = 0.0500$ mol

But amount of O_2 = ½ × amount of $NaNO_3$
= ½ × 0.0500 mol = 0.0250 mol
For O_2, $m = nM$ = 0.0250 mol × 32.0 g mol^{-1} = **0.800 g**

EXERCISE 1.11

For P, substituting into the expression

$$n = \frac{m}{M}$$

where $m = 4.00$ g and $M = 31.0$ g mol^{-1}

gives $n = \frac{m}{M} = \frac{4.00\ \text{g}}{31.0\ \text{g mol}^{-1}} = 0.129$ mol

From the equation

$$\frac{\text{amount of } P_2O_5}{\text{amount of P}} = \frac{2}{4} = \frac{1}{2}$$

$\therefore$ amount of $P_2O_5 = \frac{1}{2} \times$ amount of P

For P_2O_5, $= \frac{1}{2} \times 0.129\ \text{mol} = 0.0645$ mol

substituting into the expression

$$n = \frac{m}{M} \text{ in the form } m = nM$$

where $n = 0.0645$ mol and $M = 142$ g mol^{-1}
gives $m = 0.0645\ \text{mol} \times 142\ \text{g mol}^{-1} =$ **9.16 g**

EXERCISE 1.12

The reacting amount of Al is given by substituting into the expression

$$n = \frac{m}{M}$$

where $m = 0.27$ g and $M = 27.0$ g mol^{-1}

$$\therefore\ n = \frac{0.27\ \text{g}}{27.0\ \text{g mol}^{-1}} = 0.010\ \text{mol}$$

The amount of Cu formed is given by substituting into the expression

$$n = \frac{m}{M}$$

where $m = 0.96$ g and $M = 63.5$ g mol^{-1}

$$\therefore\ n = \frac{0.96\ \text{g}}{63.5\ \text{g mol}^{-1}} = 0.015\ \text{mol}$$

$$\therefore\ \frac{\text{amount of Al}}{\text{amount of Cu}} = \frac{0.010\ \text{mol}}{0.015\ \text{mol}} = \frac{2}{3}$$

We can build up the equation from this ratio

$$2Al(s) + ?\ CuSO_4(aq) \rightarrow 3Cu(s) + ?\ Al_2(SO_4)_3(aq)$$

To equalise Cu atoms, the stoichiometric coefficient for $CuSO_4$ must be 3.
To equalise Al atoms, the stoichiometric coefficient for $Al_2(SO_4)_3$ must be 1.

$$\therefore\ \mathbf{2Al(s) + 3CuSO_4(aq) \rightarrow 3Cu(s) + Al_2(SO_4)_3(aq)}$$

EXERCISE 1.13

We must calculate the amount of each reagent to determine which limits the reaction.
For Fe, substituting into the expression

$$n = \frac{m}{M}$$

where $m = 2.8$ g and $M = 55.8$ g mol^{-1}

gives $n = \frac{2.8\ \text{g}}{55.8\ \text{g mol}^{-1}} = 0.050$ mol

For S, substituting into the expression

$$n = \frac{m}{M}$$

where $m = 2.0$ g and $M = 32.1$ g mol^{-1}

gives $n = \frac{2.0\ \text{g}}{32.1\ \text{g mol}^{-1}} = 0.062$ mol

From the equation, one mole of iron reacts with one mole of sulphur, so the amount of iron limits the amount of iron(II) sulphide formed

$\therefore$ amount of FeS = amount of Fe = 0.050 mol

Substituting into the expression

$$n = \frac{m}{M} \text{ in the form } m = nM$$

where $n = 0.050$ mol and $M = 87.9$ g mol^{-1}
gives $m = 0.050\ \text{mol} \times 87.9\ \text{g mol}^{-1} =$ **4.4 g**

EXERCISE 1.14

The equation tells us that 1 mol of ethanol produces 1 mol of iodoethane. Using the expression:

$$\text{density} = \frac{\text{mass}}{\text{volume}}$$

mass of ethanol = 0.79 g cm^{-3} × 5.0 cm^3 = 3.95 g.

Using $n = \frac{m}{M}$, amount of ethanol $= \frac{3.95\ \text{g}}{46\ \text{g mol}^{-1}} = 0.086$ mol

So, we should expect to produce 0.086 mol of iodoethane.
Maximum mass of iodoethane from amount using $m = nM$ gives $m = 0.086\ \text{mol} \times 156\ \text{g mol}^{-1} = 13.4$ g.
Substitution into the expression

$$\%\ \text{yield} = \frac{\text{actual mass of product}}{\text{maximum mass of product}} \times 100 = \frac{5.1\ \text{g}}{13.4\ \text{g}} \times 100 = \mathbf{38\%}$$

EXERCISE 1.15

Amount of 2-chloro-2-methylpropane produced using

$$n = \frac{m}{M} = \frac{5.5 \text{ g}}{92.5 \text{ g mol}^{-1}} = 0.059 \text{ mol}$$

From the equation, 1 mol of 2-chloro-2-methylpropane is produced from 1 mol of 2-methylpropan-2-ol. For a 100% yield the amount of 2-methylpropan-2-ol required would also be 0.059 mol.

For a 62% yield we should need more reactant to compensate for the loss of the product, so

$$\text{the amount of 2-methylpropan-2-ol} = \frac{100}{62} \times 0.059 \text{ mol} = 0.095 \text{ mol}$$

Converting this to mass gives $m = 0.095 \text{ mol} \times 74 \text{ g mol}^{-1} = 7.0 \text{ g}$

Converting this to volume using

$$\text{volume} = \frac{7.0 \text{ g}}{0.79 \text{ g cm}^{-3}} = \mathbf{9.0\ cm^3}$$ **of 2-methylpropan-1-ol**

We should also expect to use 0.095 mol HCl. Converting this to mass gives $m = 0.095 \text{ mol} \times 36.5 \text{ g mol}^{-1} = 3.5 \text{ g}$.

Since the hydrochloric acid is 32% acid, in order to obtain 3.5 g of hydrochloric acid we shall need to use $(100/32) \times 3.5 \text{ g} = 10.9 \text{ g}$ of the given acid solution. Convert this mass to volume from the given density, so

$$\text{volume} \approx \frac{10.9 \text{ g}}{1.16 \text{ g cm}^{-3}} = \mathbf{9.4\ cm^3}$$ **of hydrochloric acid**

This volume of acid represents the minimum volume required. Often more than this is used to ensure all of the alcohol reacts.

EXERCISE 1.16

	Co	S	O	H_2O
Mass/g	2.10	1.14	2.28	4.50
Molar mass/g mol^{-1}	58.9	32.1	16.0	18.0
Amount/mol	$\frac{2.10}{58.9} = 0.0357$	$\frac{1.14}{32.1} = 0.0355$	$\frac{2.28}{16.0} = 0.143$	$\frac{4.50}{18.0} = 0.250$
Amount/ smallest amount = relative amount	$\frac{0.0357}{0.0355} = 1.01$	$\frac{0.0355}{0.0355} = 1.00$	$\frac{0.143}{0.0355} = 4.03$	$\frac{0.250}{0.0355} = 7.04$
Simplest ratio	1	1	4	7

The formula is $\mathbf{CoSO_4 \cdot 7H_2O}$.

EXERCISE 1.17

	$BaCl_2$	H_2O
Mass/g	8.53	1.47
Molar mass/g mol^{-1}	208.2	18.0
Amount/mol	$\frac{8.53}{208.2} = 0.0410$	$\frac{1.47}{18.0} = 0.0817$
Amount/ smallest amount = relative amount	$\frac{0.0410}{0.0410} = 1.00$	$\frac{0.0817}{0.0410} = 1.99$
Simplest ratio	1	2

The formula is $BaCl_2 \cdot 2H_2O$, i.e. $\boldsymbol{x = 2}$.

EXERCISE 1.18

The mass of water removed = 0.585 g − 0.535 g = 0.050 g

	$UO(C_2O_4) \cdot 6H_2O$	H_2O (removed)
Mass/g	0.585	0.050
Molar mass/g mol^{-1}	450	18.0
Amount/mol	$\frac{0.585}{450} = 0.00130$	$\frac{0.050}{18.0} = 0.0028$
Amount/ smallest amount = relative amount	$\frac{0.00130}{0.00130} = 1.00$	$\frac{0.0028}{0.0013} = 2.2$
Simplest ratio	1	2

Thus, the ratio of the amount of original compound to the removed water is 1 : 2. This means that for every 1 mol of compound, 2 mol of water were removed. The resulting substance would therefore have the formula $\mathbf{UO(C_2O_4) \cdot 4H_2O}$.

EXERCISE 1.19

	C	H	O
Mass/g	40.0	6.6	53.4
Molar mass/g mol^{-1}	12.0	1.0	16.0
Amount/mol	$\frac{40.0}{12.0} = 3.3$	$\frac{6.6}{1.0} = 6.6$	$\frac{53.4}{16.0} = 3.34$
Amount/ smallest amount = relative amount	$\frac{3.33}{3.33} = 1.00$	$\frac{6.6}{3.33} = 2.0$	$\frac{3.34}{3.33} = 1.00$
Simplest ratio	1	2	1

The empirical formula is $\mathbf{CH_2O}$.

EXERCISE 1.20

	Na	Al	Si	O	H_2O
Mass/g	12.1	14.2	22.1	42.1	9.48
Molar mass/ g mol^{-1}	23.0	27.0	28.1	16.0	18.0
Amount/mol	$\frac{12.1}{23.0} = 0.526$	$\frac{14.2}{27.0} = 0.526$	$\frac{22.1}{28.1} = 0.786$	$\frac{42.1}{16.0} = 2.63$	$\frac{9.48}{18.0} = 0.527$
Amount/ smallest amount = relative amount	$\frac{0.526}{0.526} = 1.00$	$\frac{0.526}{0.526} = 1.00$	$\frac{0.786}{0.526} = 1.49$	$\frac{2.63}{0.526} = 5.00$	$\frac{0.527}{0.526} = 1.00$
Simplest ratio	2	2	3	10	2

Note: Since 1.49 is very close to 1.5, we are justified in rounding up.
The empirical formula is $\mathbf{Na_2Al_2Si_3O_{10}{\cdot}2H_2O}$.

EXERCISE 1.21

	C	H	O
Mass/g	3.91	0.87	5.22
Molar mass/g mol^{-1}	12.0	1.0	16.0
Amount/mol	0.326	0.87	0.326
Amount/ smallest amount = relative amount	$\frac{0.326}{0.326} = 1.00$	$\frac{0.87}{0.326} = 2.67$	$\frac{0.326}{0.326} = 1.00$
Simplest ratio	3	8	3

Note: We are not justified in rounding off 2.67 to 3. The number 2.67 is the decimal equivalent of 8/3, so we treble the relative amounts, thus converting 2.67 to 8. The resulting empirical formula is $\mathbf{C_3H_8O_3}$.

EXERCISE 1.22

Amount of CO_2, $n = \frac{m}{M} = \frac{0.66\ \text{g}}{44.0\ \text{g mol}^{-1}} = 0.015$ mol

$\therefore$ amount of C = 0.015 mol

amount of H_2O, $n = \frac{m}{M} = \frac{0.36\ \text{g}}{18.0\ \text{g mol}^{-1}} = 0.020$ mol

$\therefore$ amount of H = 0.020 mol × 2 = 0.040 mol

	C	H
Amount/mol	0.015	0.040
$\frac{\text{Amount}}{\text{Smallest amount}}$	$\frac{0.015}{0.015} = 1.0$	$\frac{0.040}{0.015} = 8/3$
Simplest ratio	3	8

Empirical formula = $\mathbf{C_3H_8}$.

EXERCISE 1.23

Amount of CO_2, $n = \frac{m}{M} = \frac{0.374\text{ g}}{44.0\text{ g mol}^{-1}} = 8.50 \times 10^{-3}$ mol

$\therefore$ amount of C = 8.50×10^{-3} mol

Amount of H_2O, $n = \frac{m}{M} = \frac{0.154\text{ g}}{18.0\text{ g mol}^{-1}} = 8.56 \times 10^{-3}$ mol

$\therefore$ amount of H = 2 × amount of water = 0.0171 mol
Mass of carbon, $m = nM = 8.50 \times 10^{-3}\text{ mol} \times 12.0\text{ g mol}^{-1} = 0.102$ g
Mass of hydrogen, $m = nM = 0.0171\text{ mol} \times 1.0\text{ g mol}^{-1} = 0.0171$ g
Mass of oxygen = mass of sample − (mass of C + mass of H)
= 0.146 g − (0.102 + 0.0171) g = 0.0269 g

	C	H	O
Mass/g			0.0269
Molar mass/g mol^{-1}			16.0
Amount/mol	8.50×10^{-3}	0.0171	1.68×10^{-3}
$\frac{\text{Amount}}{\text{Smallest amount}}$	$\frac{8.50 \times 10^{-3}}{1.68 \times 10^{-3}} = 5.06$	$\frac{0.0171}{1.68 \times 10^{-3}} = 10.2$	$\frac{1.68 \times 10^{-3}}{1.68 \times 10^{-3}} = 1.0$
Simplest ratio of relative amounts	5	10	1

Empirical formula = **$C_5H_{10}O$**.

EXERCISE 1.24

Amount of CO_2, $n = \frac{m}{M} = \frac{0.3771\text{ g}}{44.0\text{ g mol}^{-1}} = 8.57 \times 10^{-3}$ mol

$\therefore$ amount of C = 8.57×10^{-3} mol

Amount of H_2O, $n = \frac{m}{M} = \frac{0.0643\text{ g}}{18.0\text{ g mol}^{-1}} = 3.57 \times 10^{-3}$ mol

$\therefore$ amount of H = $2 \times 3.57 \times 10^{-3}$ mol = 7.14×10^{-3} mol

Amount of AgBr, $n = \frac{m}{M} = \frac{0.2685\text{ g}}{187.8\text{ g mol}^{-1}} = 1.43 \times 10^{-3}$ mol

$\therefore$ amount of Br = 1.43×10^{-3} mol

	C	H	Br
Amount/mol	8.57×10^{-3}	7.14×10^{-3}	1.43×10^{-3}
$\frac{\text{Amount}}{\text{Smallest amount}}$	$\frac{8.57}{1.43} = 5.99$	$\frac{7.14}{1.43} = 4.99$	$\frac{1.43}{1.43} = 1.00$
Simplest ratio of relative amounts	6	5	1

Empirical formula = **C_6H_5Br**.

EXERCISE 1.25

a For a sample of 100 g:

	C	H	O
Mass/g	74.2	7.9	17.9
Molar mass/g mol^{-1}	12.0	1.0	16.0
Amount/mol	6.18	7.90	1.12
$\frac{\text{Amount}}{\text{Smallest amount}}$	$\frac{6.18}{1.12} = 5.52$	$\frac{7.90}{1.12} = 7.05$	$\frac{1.12}{1.12} = 1$
Simplest ratio of relative amounts	11	14	2

Empirical formula of A = $\mathbf{C_{11}H_{14}O_2}$.

b $M_r(C_{11}H_{14}O_2) = (11 \times 12) + (14 \times 1) + (2 \times 16) = 178$
$\therefore$ molecular formula = empirical formula = $\mathbf{C_{11}H_{14}O_2}$

EXERCISE 1.26

a For a sample of 100 g of compound:

	C	H	O
Mass/g	60.0	13.3	26.7
Molar mass/g mol^{-1}	12.0	1.0	16.0
Amount/mol	5.00	13.3	1.67
$\frac{\text{Amount}}{\text{Smallest amount}}$	$\frac{5.00}{1.67} = 2.99$	$\frac{13.3}{1.67} = 7.96$	$\frac{1.67}{1.67} = 1.00$
Simplest ratio of relative amounts	3	8	1

Empirical formula = $\mathbf{C_3H_8O}$.

b $M_r(C_3H_8O) = (3 \times 12.0) + (8 \times 1.0) + (1 + 16.0) = 60.0$

$\therefore$ molecular formula = empirical formula = $\mathbf{C_3H_8O}$

EXERCISE 1.27

a Substituting into the expression where n = 0.100 mol and V

$c = \frac{n}{V}$ = (2000/1000) dm^3

gives $c = \frac{0.100 \text{ mol}}{2.00 \text{ dm}^3} = \mathbf{0.0500 \text{ mol dm}^{-3}}$

b $c = \frac{n}{V} = \frac{0.100 \text{ mol}}{1.00 \text{ dm}^3} = \mathbf{0.100 \text{ mol dm}^{-3}}$

c $c = \frac{n}{V} = \frac{0.100 \text{ mol}}{0.500 \text{ dm}^3} = \mathbf{0.200 \text{ mol dm}^{-3}}$

d $c = \frac{n}{V} = \frac{0.100 \text{ mol}}{0.250 \text{ dm}^3} = \mathbf{0.400 \text{ mol dm}^{-3}}$

e $c = \frac{n}{V} = \frac{0.100 \text{ mol}}{0.100 \text{ dm}^3} = \mathbf{1.00 \text{ mol dm}^{-3}}$

EXERCISE 1.28

a Substituting into the expression

$$n = \frac{m}{M}$$

where m = 8.50 g and M = 169.9 g mol^{-1}

gives $$n = \frac{8.50\ \text{g}}{169.9\ \text{g mol}^{-1}} = 0.0500\ \text{mol}$$

Substituting into the expression

$$c = \frac{n}{V}$$

where n = 0.0500 mol and V = 1.00 dm^3

gives $$c = \frac{0.0500\ \text{mol}}{1.00\ \text{dm}^3} = \mathbf{0.0500\ mol\ dm^{-3}}$$

Or, substituting $n = \frac{m}{M}$ into $c = \frac{n}{V}$

$$c = \frac{m}{MV} = \frac{8.50\ \text{g}}{169.9\ \text{g mol}^{-1} \times 1.00\ \text{dm}^3} = \mathbf{0.0500\ mol\ dm^{-3}}$$

b $$c = \frac{m}{MV} = \frac{10.7\ \text{g}}{214.0\ \text{g mol}^{-1} \times 0.250\ \text{dm}^3} = \mathbf{0.200\ mol\ dm^{-3}}$$

c $$c = \frac{m}{MV} = \frac{11.2\ \text{g}}{331.2\ \text{g mol}^{-1} \times 0.050\ \text{dm}^3} = \mathbf{0.676\ mol\ dm^{-3}}$$

d $$c = \frac{m}{MV} = \frac{14.3\ \text{g}}{294.1\ \text{g mol}^{-1} \times 0.250\ \text{dm}^3} = \mathbf{0.194\ mol\ dm^{-3}}$$

e $$c = \frac{m}{MV} = \frac{11.9\ \text{g}}{249.6\ \text{g mol}^{-1} \times 0.500\ \text{dm}^3} = \mathbf{0.0954\ mol\ dm^{-3}}$$

EXERCISE 1.29

a Substituting into the expression

$$c = \frac{n}{V} \text{ in the form } n = cV$$

where c = 5.00 mol dm^{-3} and V = 4.00 dm^3
gives n = 5.00 mol dm^{-3} × 4.00 dm^3 = **20.0 mol**

b $n = cV$ = 2.50 mol dm^{-3} × 1.00 dm^3 = **2.50 mol**

c $n = cV$ = 0.439 mol dm^{-3} × 0.020 dm^3 = **8.78 × 10^{-3} mol**

EXERCISE 1.30

a Substituting into the expression

$$c = \frac{n}{V} \text{ in the form } n = cV$$

where c = 0.100 mol dm^{-3} and V = 1.00 dm^3
gives n = 0.100 mol dm^{-3} × 1.00 dm^3 = 0.100 mol
Substituting into the expression

$$n = \frac{m}{M} \text{ in the form } m = nM$$

where n = 0.100 mol and M = 58.4 g mol^{-1}
gives m = 0.100 mol × 58.4 g mol^{-1} = **5.84 g**

Or, combining $c = \frac{n}{V}$ with $n = \frac{m}{M}$ and substituting

$m = nM = cVM$ = 0.100 mol dm^{-3} × 1.00 dm^3 × 58.4 g mol^{-1} = **5.84 g**

b $m = cVM$ = 1.00 mol dm^{-3} × 0.500 dm^3 × 110.9 g mol^{-1} = **55.5 g**

c $m = cVM$ = 0.200 mol dm^{-3} × 0.250 dm^3 × 158.0 g mol^{-1} = **7.90 g**

d $m = cVM$ = 0.117 mol dm^{-3} × 0.200 dm^3 × 40.0 g mol^{-1} = **0.936 g**

End-of-chapter questions

1.1 **a** 160 g
b 1.07 g
c 865 g
d 37.0 g

1.2 **a** 0.250 mol
b 0.0166 mol
c 1.25 mol
d 15.0 mol

1.3 132 g mol^{-1}

1.4 430 g

1.5 **a** 0.0500 mol dm^{-3}
b 0.100 mol dm^{-3}

1.6 **a** 0.0500 mol dm^{-3}
b 0.150 mol dm^{-3}

1.7 x = 18

1.8 **d** 7.84 g

1.9 C_2H_6O

1.10 $C_2H_4O_3$

1.11 C_6H_6

1.12 $C_5H_{11}NO$

1.13 C_4H_8O

1.14 butan-1-ol = 12.7 g (≈13 g to 2 significant figures)
ethanoic acid = 10.3 g (≈10 g to 2 significant figures)

1.15 **a** 10% yield

CHAPTER 2

EXERCISE 2.1

a $Ba(OH)_2(aq) + 2HCl(aq) \rightarrow BaCl_2(aq) + 2H_2O(l)$

Let A refer to HCl and B to $Ba(OH)_2$. Substituting into the expression

$$\frac{c_A V_A}{c_B V_B} = \frac{a}{b}$$

where $c_A = 0.0600 \text{ mol dm}^{-3}$, $V_A = 25.0 \text{ cm}^3$, $a = 2$; $c_B = ?$, $V_B = 20.0 \text{ cm}^3$, $b = 1$

gives $$\frac{0.0600 \text{ mol dm}^{-3} \times 25.0 \text{ cm}^3}{c_B \times 20.0 \text{ cm}^3} = \frac{2}{1}$$

Solving for c_B gives

$$c_B = \frac{0.0600 \text{ mol dm}^{-3} \times 25.0 \text{ cm}^3}{2 \times 20.0 \text{ cm}^3} = \mathbf{0.0375 \text{ mol dm}^{-3}}$$

b It is not necessary to convert from cm^3 to dm^3 because the units of volume cancel in the final expression.

EXERCISE 2.2

a $NaOH(aq) + HNO_3(aq) \rightarrow NaNO_3(aq) + H_2O(l)$

Let A refer to NaOH and B to HNO_3.
Substituting into the expression

$$\frac{c_A V_A}{c_B V_B} = \frac{a}{b}$$

where $c_A = 0.500 \text{ mol dm}^{-3}$, $V_A = ?$; $c_B = 0.100 \text{ mol dm}^{-3}$, $V_B = 50.0 \text{ cm}^3$

gives $$\frac{0.500 \text{ mol dm}^{-3} \times V_A}{0.100 \text{ mol dm}^{-3} \times 50.0 \text{ cm}^3} = \frac{1}{1}$$

$$\therefore V_A = \frac{0.100 \text{ mol dm}^{-3} \times 50.0 \text{ cm}^3}{0.500 \text{ mol dm}^{-3}} = \mathbf{10.0 \text{ cm}^3}$$

b $$\frac{c_A V_A}{c_B V_B} = \frac{a}{b}$$ (A refers to NaOH, B to H_2SO_4)

$$\therefore \frac{0.500 \text{ mol dm}^{-3} \times V_A}{0.262 \text{ mol dm}^{-3} \times 22.5 \text{ cm}^3} = \frac{2}{1}$$

and $$V_A = \frac{2 \times 0.262 \times 22.5 \text{ cm}^3}{0.500} = \mathbf{23.6 \text{ cm}^3}$$

EXERCISE 2.3

Substituting into the expression:

$$\frac{c_A V_A}{c_B V_B} = \frac{a}{b}$$

where $c_A = 0.50 \text{ mol dm}^{-3}$, $V_A = 25.0 \text{ cm}^3$; $c_B = 0.20 \text{ mol dm}^{-3}$, $V_B = 31.3 \text{ cm}^3$

$$\frac{a}{b} = \frac{0.50 \text{ mol dm}^{-3} \times 25.0 \text{ cm}^3}{0.20 \text{ mol dm}^{-3} \times 31.3 \text{ cm}^3} = \frac{2}{1}$$

$\therefore$ ***a* = 2, *b* = 1.**

EXERCISE 2.4

a **+1**
b **−2**
c **+3**
d **+1**
e **−1**

EXERCISE 2.5

a KCl is ionic $K^+ + Cl^-$ ***Ox*(K) = +1** ***Ox*(Cl) = −1**
b Na_2O is ionic $2Na^+ + O^{2-}$ ***Ox*(Na) = +1** ***Ox*(O) = −2**
c BaF_2 is ionic $Ba^{2+} + 2F^-$ ***Ox*(Ba) = +2** ***Ox*(F) = −1**
d Na_3P is ionic $3Na^+ + P^{3-}$ ***Ox*(Na) = +1** ***Ox*(P) = −3**
e Mg_3N_2 is ionic $3Mg^{2+} + 2N^{3-}$ ***Ox*(Mg) = +2** ***Ox*(N) = −3**
f CsH is ionic $Cs^+ + H^-$ ***Ox*(Cs) = +1** ***Ox*(H) = −1**

EXERCISE 2.6

a N is more electronegative than H.

```
δ+  δ−  δ+
H — N — H          imagined as        H+   N3-   H+
    |
    H                                       H+
    δ+
```

$\therefore$ ***Ox*(N) = −3, *Ox*(H) = +1**

b O is more electronegative than Cl.

```
δ+  δ−  δ+
Cl—O—Cl        imagined as  Cl+   O2−   Cl+
```

$\therefore$ ***Ox*(O) = −2, *Ox*(Cl) = +1**

EXERCISE 2.7

a The O—O bond is non-polar, i.e. neither oxygen atom can be regarded as having gained or lost an electron with respect to the other. However, the O—H bonds are polar, and each oxygen atom is regarded as having gained an electron from a hydrogen atom.

$\overset{\delta+}{H}$—$\overset{\delta-}{O}$—$\overset{\delta-}{O}$—$\overset{\delta+}{H}$ imagined as H^+ O^- O^- H^+

∴ ***Ox*(O) = −1**, ***Ox*(H) = +1**

b Fluorine is more electronegative than oxygen.

$\overset{\delta-}{F}$—$\overset{\delta+}{O}$—$\overset{\delta-}{F}$ imagined as F^- O^{2+} F^-

∴ ***Ox*(O) = +2**, ***Ox*(F) = −1**

EXERCISE 2.8

a *Ox*(As) + *Ox*(all other atoms) = 0
∴ *Ox*(As) = −*Ox*(all other atoms) = −(3*Ox*(H) + 4*Ox*(O))
= −(3(+1) + 4(−2)) = −3 + 8 = **+5**

b 2*Ox*(Cr) + *Ox*(all other atoms) = −1
∴ 2*Ox*(Cr) = −1 −*Ox*(all other atoms) = −1 −(*Ox*(H) + 7*Ox*(O))
= −1 −(+1 +7(−2)) = −1 −(−13)
= +12
∴ *Ox*(Cr) = **+6**

EXERCISE 2.9

a 2*Ox*(Cr) + *Ox*(all other atoms) = 0
∴ *Ox*(Cr) = −½*Ox*(all other atoms)
= −½(2*Ox*(K) + 7*Ox*(O)) = −½(2(+1) + 7(−2)) = **+6**

b $PbSO_4$ contains the SO_4^{2-} ion
∴ *Ox*(S) + 4*Ox*(O) = −2
∴ *Ox*(S) = −2 −4*Ox*(O) = −2 −4(−2) = −2 + 8 = **+6**

c *Ox*(N) = −(*Ox*(H) + 3*Ox*(O)) = −(+1 +3(−2)) = −(1 − 6) = **+5**

d *Ox*(S) = −½(2*Ox*(Na) + 3*Ox*(O)) = −½(2(+1) + 3(−2)) = −½(2 − 6)
= **+2**

e *Ox*(C) = − (*Ox*(Ca) + 3*Ox*(O)) = −(+2 +3(−2)) = −(2 − 6) = **+4**

f *Ox*(Ta) = −⅙(8*Ox*(Na) + 19*Ox*(O)) = −⅙(8(+1) + 19(−2))
= −⅙(8 − 38) = **+5**

g There are two methods, depending on whether we consider the formula as one entity or as two ions.

i) Considering the whole formula:
Ox(N) = −½(4*Ox*(H) + 3*Ox*(O)) = −½(4(+1) + 3(−2)) = ½(4 − 6)
= **+1**

ii) Considering the ion NH_4^+:
Ox(N) + 4*Ox*(H) = +1
∴ *Ox*(N) = +1 − 4*Ox*(H) = +1 − 4(+1) = **−3**
Considering the ion NO_3^-:
Ox(N) + 3*Ox*(O) = −1
∴ *Ox*(N) = −1 − 3*Ox*(O) = −1 − 3(−2) = **+5**

Since the oxidation number concept is merely a 'book-keeping' device, both methods and, therefore, both answers may be considered correct. However, since the substance is known to be ionic, the second method is preferred. Note that the first answer is the mean of the second pair of answers (+1 = ½(−3 + 5)).

h *Ox*(S) = −¼(2*Ox*(Na) + 6*Ox*(O)) = −¼(2(+1)+6(−2)) = −¼(2 − 12)
= **2½**

Note that fractional oxidation numbers sometimes appear in these calculations. This is usually because some atoms of the element concerned have one integral oxidation number and other atoms have another integral oxidation number.

Only if the structure is known can these integral oxidation numbers be allotted. In this case, the structure of the $S_4O_6^{2-}$ ion, with individual oxidation numbers added, is

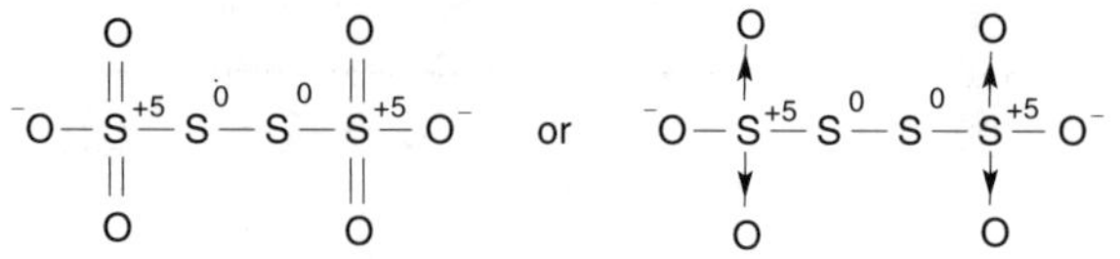

i *Ox*(Fe) = −⅓(4*Ox*(O)) = − ⅓(4(−2)) = ⁸⁄₃

This fraction arises because Fe_3O_4 contains both Fe^{2+} and Fe^{3+} ions – Fe^{2+}, $2Fe^{3+}$, $4O^{2-}$.

The sum of the oxidation numbers for the three iron atoms is +2 +3 +3 = 8. The mean *Ox*(Fe) is therefore **⁸⁄₃**.

EXERCISE 2.10

a $\overset{0}{Fe}(s) + \overset{+2}{Cu}SO_4(aq) \rightarrow \overset{+2}{Fe}SO_4(aq) + \overset{0}{Cu}(s)$
Iron is oxidised from Fe(0) to Fe(II)
Copper is reduced from Cu(II) to Cu(0)
∴ **Reaction is redox**

b $2\overset{+1}{K}\overset{-1}{Br}(aq) + \overset{0}{Cl_2}(aq) \rightarrow 2\overset{+1}{K}\overset{-1}{Cl}(aq) + \overset{0}{Br_2}(aq)$
Bromine is oxidised from Br(−I) to Br(0)
Chlorine is reduced from Cl(0) to Cl(−I)
∴ **Reaction is redox**

c $\overset{+2}{Cu}SO_4(aq) + \overset{+2}{Pb}(NO_3)_2(aq) \rightarrow \overset{+2}{Cu}(NO_3)_2(aq) + \overset{+2}{Pb}SO_4(s)$
No oxidation or reduction, i.e. **reaction is not redox**

d $\overset{-2}{C_2}\overset{+1}{H_4}(g) + \overset{0}{H_2}(g) \rightarrow \overset{-3}{C_2}\overset{+1}{H_6}(g)$
Some hydrogen is oxidised from H(0) to H(+I)
Carbon is reduced from C(−II) to C(−III)
∴ **Reaction is redox**
Note that the oxidation number concept is not generally useful in organic chemistry except in the simplest compounds.

e $\overset{+7}{Mn}O_4^-(aq) + 5\overset{+2}{Fe}{}^{2+}(aq) + 8H^+(aq) \rightarrow \overset{+2}{Mn}{}^{2+}(aq) + 5\overset{+3}{Fe}{}^{3+}(aq) + 4H_2O(l)$
Iron is oxidised from Fe(II) to Fe(III)
Manganese is reduced from Mn(VII) to Mn(II)
∴ **Reaction is redox**

EXERCISE 2.11

a
i) $\overset{+1}{H_2}\overset{-2}{S}(aq) + \overset{0}{Br_2}(aq) \rightarrow 2\overset{+1}{H}\overset{-1}{Br}(aq) + \overset{0}{S}(s)$

ii) $\overset{0}{Al}(s) + 1\tfrac{1}{2}\overset{0}{I_2}(s) \rightarrow \overset{+3}{Al}\overset{-1}{I_3}(s)$

iii) $\overset{0}{Cl_2}(g) + \overset{+1}{H_2}\overset{-2}{O}(l) \rightarrow \overset{+1}{H}\overset{+1}{Cl}\overset{-2}{O}(aq) + \overset{+1}{H}\overset{-1}{Cl}(aq)$

iv) $2\overset{+1}{Na_2}\overset{+2}{S_2}\overset{-2}{O_3}(aq) + \overset{0}{I_2}(aq) \rightarrow \overset{+1}{Na_2}\overset{+2.5}{S_4}\overset{-2}{O_6}(aq) + 2\overset{+1}{Na}\overset{-1}{I}(aq)$

b
i) S^{2-} is oxidised to S; Br_2 is reduced to Br^-.
ii) Al is oxidised to Al^{3+}; I_2 is reduced to I^-.
iii) Cl_2 is simultaneously oxidised to ClO^- and reduced to Cl^-.
iv) $S_2O_3^{2-}$ is oxidised to $S_4O_6^{2-}$; I_2 is reduced to I^-.

EXERCISE 2.12

a $\overset{+5}{I}O_3^-(aq) + \overset{-1}{I}^-(aq) + H^+(aq) \rightarrow \overset{0}{I}_2(aq) + H_2O(l)$

$\therefore$ oxidised species is I^- and reduced species is IO_3^-.

The balanced half-equations are:

$2IO_3^-(aq) + 12H^+(aq) + 10e^- \rightarrow I_2(aq) + 6H_2O(l)$

$2I^-(aq) \rightarrow I_2 + 2e^-$ (Multiply this equation by 5.)

Combining the two half-equations we get

$2IO_3^-(aq) + 12H^+(aq) + 10I^-(aq) \rightarrow 5I_2(aq) + I_2(aq) + 6H_2O(l)$

or $\mathbf{IO_3^-(aq) + 6H^+(aq) + 5I^-(aq) \rightarrow 3I_2(aq) + 3H_2O(l)}$

b $\overset{+2}{S}_2O_3^{2-}(aq) + \overset{0}{I}_2(aq) \rightarrow \overset{-1}{I}^-(aq) + \overset{+2.5}{S}_4O_6^{2-}(aq)$

$\therefore$ oxidized species is $S_2O_3^{2-}$ and reduced species is I_2.

The balanced half-equations are:

$2S_2O_3^{2-}(aq) \rightarrow S_4O_6^{2-}(aq) + 2e^-$

$I_2(aq) + 2e^- \rightarrow 2I^-(aq)$

Combining the two half-equations we get

$\mathbf{2S_2O_3^{2-}(aq) + I_2(aq) \rightarrow S_4O_6^{2-}(aq) + 2I^-(aq)}$

c $\overset{+7}{Mn}O_4^-(aq) + H^+(aq) + \overset{+2}{Fe}^{2+}(aq) \rightarrow \overset{+2}{Mn}^{2+}(aq) + \overset{+3}{Fe}^{3+}(aq) + H_2O(l)$

$MnO_4^-(aq) + 8H^+(aq) + 5e^- \rightarrow Mn^{2+}(aq) + 4H_2O(l)$

$Fe^{2+}(aq) \rightarrow Fe^{3+}(aq) + e^-$ (Multiply by 5 and add)

$\mathbf{MnO_4^-(aq) + 8H^+(aq) + 5Fe^{2+}(aq) \rightarrow Mn^{2+}(aq) + 4H_2O(l) + 5Fe^{3+}(aq)}$

d $\overset{+7}{Mn}O_4^-(aq) + H^+(aq) + \overset{+3}{C}_2O_4^{2-}(aq) \rightarrow \overset{+2}{Mn}^{2+}(aq) + 2\overset{+4}{C}O_2(g) + H_2O(l)$

$MnO_4^-(aq) + 8H^+(aq) + 5e^-(aq) \rightarrow Mn^{2+}(aq) + 4H_2O(l)$ (×2)

$C_2O_4^{2-}(aq) \rightarrow 2CO_2(g) + 2e^-$ (×5)

$\mathbf{2MnO_4^-(aq) + 16H^+(aq) + 5C_2O_4^{2-}(aq) \rightarrow 2Mn^{2+}(aq) + 8H_2O(l) + 10CO_2(g)}$

e $\overset{+7}{Mn}O_4^-(aq) + H^+(aq) + H_2\overset{-1}{O}_2(aq) \rightarrow \overset{+2}{Mn}^{2+}(aq) + \overset{0}{O}_2(g) + H_2\overset{-2}{O}(l)$

$MnO_4^-(aq) + 8H^+(aq) + 5e^- \rightarrow Mn^{2+}(aq) + 4H_2O(l)$ (×2)

$H_2O_2(aq) \rightarrow 2H^+(aq) + O_2(g) + 2e^-$ (×5)

$\mathbf{2MnO_4^-(aq) + 6H^+(aq) + 5H_2O_2(aq) \rightarrow 2Mn^{2+}(aq) + 8H_2O(l) + 5O_2(g)}$

(Note the simplification by deducting $10H^+$ from each side.)

EXERCISE 2.13

$M_r[FeSO_4{\cdot}(NH_4)_2SO_4{\cdot}6H_2O]$

= 55.8 + 2(32.1) + 14(16.0) + 2(14.0) + 20(1.01) = 392.2

$$\therefore c = \frac{n}{V} = \frac{9.80\ \text{g}/392.2\ \text{g mol}^{-1}}{0.250\ \text{dm}^3} = 0.100\ \text{mol dm}^{-3}$$

Equation: $5Fe^{2+}(aq) + MnO_4^-(aq) + 8H^+(aq) \rightarrow 5Fe^{3+}(aq) + Mn^{2+}(aq) + 4H_2O(l)$

Substituting into the expression:

$$\frac{c_A V_A}{c_B V_B} = \frac{a}{b}$$ where A refers to Fe^{2+} and B to MnO_4^-

$$\frac{0.100\ \text{mol dm}^{-3} \times 25.0\ \text{cm}^3}{c_B \times 24.6\ \text{cm}^3} = \frac{5}{1}$$

$$\therefore c_B = \frac{0.100\ \text{mol dm}^{-3} \times 25.0\ \text{cm}^3}{5 \times 24.6\ \text{cm}^3} = \mathbf{0.0203\ mol\ dm^{-3}}$$

EXERCISE 2.14

Equation: $H_2C_2O_4(aq) + 2NaOH(aq) \rightarrow Na_2C_2O_4(aq) + 2H_2O(l)$

$$\frac{c_A V_A}{c_B V_B} = \frac{a}{b}$$ where A refers to NaOH and B to $H_2C_2O_4$

$$\text{i.e.}\quad \frac{0.100\ \text{mol dm}^{-3} \times 14.8\ \text{cm}^3}{c_B \times 25.0\ \text{cm}^3} = \frac{2}{1}$$

$$\therefore c_B = \frac{0.100\ \text{mol dm}^{-3} \times 14.8\ \text{cm}^3}{2 \times 25.0\ \text{cm}^3} = 0.0296\ \text{mol dm}^{-3}$$

$M_r(H_2C_2O_4) = 2(1.01) + 2(12.0) + 4(16.0) = 90.0$

$\therefore$ mass of $H_2C_2O_4$ per dm^3 = 0.0296 mol dm^{-3} × 90.0 g mol^{-1}

= **2.66 g dm^{-3}**

Equation: $2\tfrac{1}{2}C_2O_4^{2-}(aq) + MnO_4^-(aq) + 8H^+(aq) \rightarrow 5CO_2(g) + Mn^{2+}(aq) + 4H_2O(l)$

$$\frac{c_A V_A}{c_B V_B} = \frac{a}{b}$$ where A refers to MnO_4^- and B to $C_2O_4^{2-}$

$$\text{i.e.}\quad \frac{0.0200\ \text{mol dm}^{-3} \times 29.5\ \text{cm}^3}{c_B \times 25.0\ \text{cm}^3} = \frac{1}{2.50}$$

$$\therefore c_B = \frac{0.0200\ \text{mol dm}^{-3} \times 29.5\ \text{cm}^3 \times 2.50}{25.0\ \text{cm}^3} = 0.0590\ \text{mol dm}^{-3}$$

$[C_2O_4^{2-}(aq)]$ from $Na_2C_2O_4$ = total $[C_2O_4^{2-}(aq)]$ − $[C_2O_4^-(aq)]$ from $H_2C_2O_4$

= 0.0590 mol dm^{-3} − 0.0296 mol dm^{-3}

= 0.0294 mol dm^{-3}

$M_r(Na_2C_2O_4) = 2(23.0) + 2(12.0) + 4(16.0) = 134$

$\therefore$ mass of $Na_2C_2O_4$ per dm^{-3} = 0.0294 mol dm^{-3} × 134 g mol^{-1}

= **3.94 g dm^{-3}**

EXERCISE 2.15

a Relative molecular mass of KIO_3 = 39 + 127 + 48 = 214

i.e. 1 dm^3 of (*M*/60) KIO_3 contains (214/60) g = 3.57 g

$\therefore$ 250 cm^3 requires (3.57/4) = **0.89 g**

b From the equations given,

1 mol of IO_3^- produces 3 mol of I_2 which react with 6 mol of $S_2O_3^{2-}$.

$\therefore$ 1 mol of IO_3^- is equivalent to 6 mol of $S_2O_3^{2-}$

Let A refer to IO_3^- and B refer to $S_2O_3^{2-}$

Substituting into the expression,

$$\frac{c_A V_A}{c_B V_B} = \frac{a}{b}$$

where c_A = 1/60 mol dm^{-3} $\quad c_B$ = ? $\quad a = 1$

V_A = 25.0 cm^3 $\quad V_B$ = 20.0 cm^3 $\quad b = 6$

$$\text{gives}\quad \frac{\tfrac{1}{60}\ \text{mol dm}^{-3} \times 25.0\ \text{cm}^3}{c_B \times 20.0\ \text{cm}^3} = \frac{1}{6}$$

$$\therefore c_B = \frac{\tfrac{1}{60}\ \text{mol dm}^{-3} \times 25.0\ \text{cm}^3 \times 6}{20.0\ \text{cm}^3} = \frac{2.5\ \text{mol dm}^{-3}}{20} = \mathbf{0.125\ mol\ dm^{-3}}$$

c To a 25 cm^3 aliquot of HCl(aq) add excess KI(aq) and KIO_3(aq). Titrate the liberated I_2 with sodium thiosulphate solution of known concentration to the usual 'decolorisation of starch indicator' end-point.

From the chemical equations, 1 mol of H^+ is equivalent to 1 mol of $S_2O_3^{2-}$. The concentration of H^+ can be calculated by substituting in the equation

$\dfrac{c_A V_A}{c_B V_B} = \dfrac{1}{1}$ where A refers to H^+ and B refers to $S_2O_3^{2-}$

d i) $S_2O_3^{2-}$: $Ox(S) = +2$ (Actually one atom is +4 and the other is 0.)
   ii) From the equations for the reactions,
        2 mol of $S_2O_3^{2-}$ reacts with 1 mol of I_2 released by 1 mol of Br_2
        $\therefore$ 20 cm^3 of 0.1 M $Na_2S_2O_3$ reacts with 10 cm^3 of 0.1 M I_2
        released by 10 cm^3 of 0.1 M Br_2
        If only 10 cm^3 of 0.1 M Br_2 remains, then $50 - 10 = 40$ cm^3 must
        have reacted, i.e.
        40 cm^3 of 0.1 M Br_2 reacts with 10 cm^3 of 0.1 M $S_2O_3^{2-}$
        $\therefore$ **4 mol of Br_2 reacts with 1 mol of $S_2O_3^{2-}$**
   iii) In reactants, $Ox(Br) = 0$ and $Ox(S) = +2$

In products, $Ox(Br) = -1$ (assumed to be Br^-)
8 Br atoms change oxidation number by -1 each. Total change
$= -8$
$\therefore$ total change for S atoms must be $+8$, i.e. $+4$ each for 2 atoms
$\therefore$ $Ox(S)$ in products $= +2 + (+4) = $ **+6**
   iv) The most likely product with $Ox(S) = 6$ is SO_4^{2-}
        $\therefore$ half-equations can be written:

$$4Br_2(aq) + 8e^- \rightarrow 8Br^-(aq)$$
$$S_2O_3^{2-}(aq) + 5H_2O(l) \rightarrow 2SO_4^{2-}(aq) + 10H^+(aq) + 8e^-$$

Combining these half-equations gives:

$S_2O_3^{2-}(aq) + 4Br_2(aq) + 5H_2O(l) \rightarrow 2SO_4^{2-}(aq) + 10H^+(aq) + 8Br^-(aq)$

EXERCISE 2.16

a Amount of Fe $= \dfrac{0.8 \text{ g}}{56 \text{ g mol}^{-1}} = \dfrac{1}{70}$ mol

Amount of $HNO_3 = \dfrac{0.9 \text{ g}}{63 \text{ g mol}^{-1}} = \dfrac{1}{70}$ mol

$\therefore$ molar ratio of Fe : HNO_3 is **1 : 1**

b
Gain of 3 electrons

$1\,Fe(s) + 1\,HNO_3(aq) \longrightarrow 1Fe^{3+}(aq) + \begin{bmatrix} N_2O \\ \text{or NO} \\ \text{or } N_2O_4 \end{bmatrix}$

Loss of 3 electrons

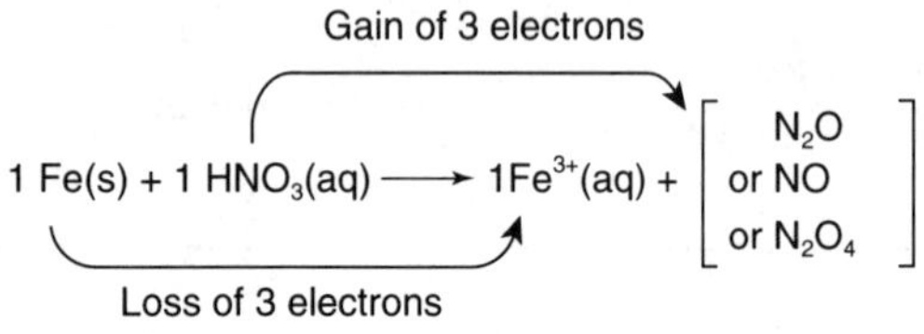

c Nitrogen in HNO_3 is in the +5 oxidation state. If the nitric acid gains
   three electrons, the final oxidation state of the nitrogen must be +2,
   $\therefore$ the oxide produced must be **NO**

d $1Fe(s) - 3e^- \rightarrow 1Fe^{3+}(aq)$
   $3H^+(aq) + 1HNO_3(aq) + 3e^- \rightarrow 1NO(g) + 2H_2O(l)$
   Addition of these two equations gives the overall equation:
   $Fe(s) + 3H^+(aq) + HNO_3(aq) \rightarrow Fe^{3+}(aq) + NO(g) + 2H_2O(l)$

EXERCISE 2.17

a $Ag^+(aq) + Cl^-(aq) \rightarrow AgCl(s)$
   Amount of Cl^- in 25 cm^3 of solution = amount of Ag^+ used $= cV$

   $= 0.100 \text{ mol dm}^{-3} \times \dfrac{18.4}{1000} \text{ dm}^3 = \mathbf{1.84 \times 10^{-3}}$ **mol**

b Amount of Cl^- from whole sample $= 1.84 \times 10^{-3} \text{ mol} \times \dfrac{250 \text{ cm}^3}{25.0 \text{ cm}^3}$

   $= \mathbf{1.84 \times 10^{-2}}$ **mol**

   Mass of Cl in whole sample $= n \times M$

   $\therefore$ mass of Cl in whole sample $= 1.84 \times 10^{-2} \text{ mol} \times 35.5 \text{ g mol}^{-1}$

   $= \mathbf{0.653}$ **g**

c Mass of P in the whole sample = (mass of sample) − (mass of Cl in sample)

   $= 0.767 \text{ g} - 0.653 \text{ g}$

   $= 0.114 \text{ g}$

c $\therefore$ amount of P in the whole sample $= \dfrac{m}{N}$

   $= \dfrac{0.114 \text{ g}}{31.0 \text{ g mol}^{-1}} = \mathbf{3.68 \times 10^{-3}}$ **mol**

d Amount of chlorine combined with one mole of phosphorus:
   3.68×10^{-3} mol of P combines with 1.84×10^{-2} mol of Cl^-

   $\therefore$ 1 mol of P combines with $\dfrac{1.84 \times 10^{-2} \text{ mol}}{3.68 \times 10^{-3}}$

   $= \mathbf{5.00}$ **mol of Cl atoms**

e The simplest formula of the chloride is **PCl_5**.

EXERCISE 2.18

a 1. $Ag^+(aq) + Cl^-(aq) \rightarrow AgCl(s)$
        Amount of Cl^- in 25 cm^3 of solution = amount of Ag^+ used $= cV$

        $= 0.100 \text{ mol dm}^{-3} \times \dfrac{21.0}{1000} \text{ dm}^3 = 2.10 \times 10^{-3}$ mol

   2. Amount of Cl^- from whole sample $= 2.10 \times 10^{-3} \text{ mol} \times \dfrac{250 \text{ cm}^3}{25.0 \text{ cm}^3}$

       $= 2.10 \times 10^{-2}$ mol

3. Mass of Cl in whole sample = $n \times M$
= 2.10×10^{-2} mol × 35.5 g mol^{-1}
= 0.746 g
4. Mass of S in the whole sample
= 1.420 g − 0.746 g
= 0.674 g
5. Amount of S in the whole sample $= \dfrac{m}{M}$

$$= \frac{0.674\ \text{g}}{32.1\ \text{g mol}^{-1}}$$

$$= 2.10 \times 10^{-2}\ \text{mol}$$

There are equal amounts of chloride ion and sulphur, therefore **one mole of chloride is combined with each mole of sulphur**.

b The molar mass of the chloride is 135.2.
Since S and Cl atoms occur in equal numbers the molecular formula is $(SCl)_n$.
For $n = 1$ molar mass = 67.6, for $n = 2$ molar mass = 135.2.
The molecular formula is $\mathbf{S_2Cl_2}$.

EXERCISE 2.19

First calculate the mass of Mg^{2+}, Cl^- and H_2O in the 0.203 g sample. The mass of Cl^- can be calculated from the titration data, the mass of H_2O from the percentage lost on dehydration and finally the mass of Mg^{2+} by difference.

Next, convert masses to relative amounts to give values for m and n. The procedure can be summarised in a flow-chart, as follows:

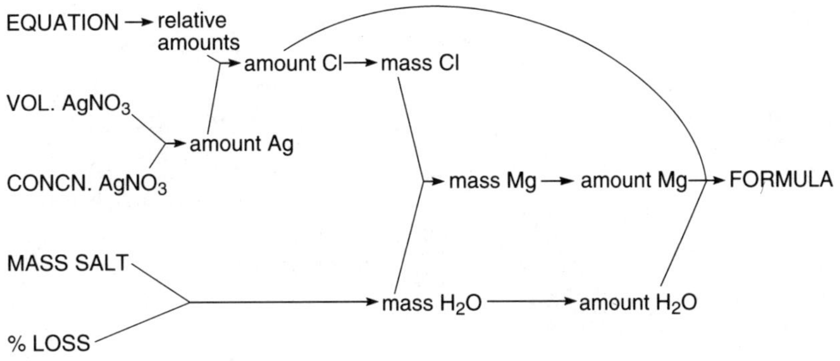

The detailed calculation follows, step by step.

1. Amount and mass of Cl

$$Ag^+(aq) + Cl^-(aq) \rightarrow AgCl(s)$$

∴ amount of Cl = amount of $AgNO_3$
$= cV = 0.100\ \text{mol dm}^{-3} \times 0.0200\ \text{dm}^3$
$= 2.00 \times 10^{-3}$ mol

∴ mass of Cl = $nM = 2.00 \times 10^{-3}\ \text{mol} \times 35.5\ \text{g mol}^{-1} = 0.0710$ g

2. Mass and amount of H_2O

mass of H_2O = loss in mass of sample

$$= \frac{53.2}{100} \times 0.203\ \text{g} = 0.108\ \text{g}$$

$$\therefore \text{amount of } H_2O = \frac{m}{M} = \frac{0.108\ \text{g}}{18.0\ \text{g mol}^{-1}} = 6.00 \times 10^{-3}\ \text{mol}$$

3. Mass and amount of Mg

mass of Mg = mass of salt − mass of Cl − mass of H_2O

= 0.203 g − 0.071 g − 0.108 g = 0.024 g

$$\therefore \text{amount of Mg} = \frac{m}{M} = \frac{0.024\ \text{g}}{24.0\ \text{g mol}^{-1}} = 1.00 \times 10^{-3}\ \text{mol}$$

4. Relative amounts

	Mg	**Cl**	**H_2O**
Amount/mol	1.0×10^{-3}	2.0×10^{-3}	6.0×10^{-3}
Amount/ smallest amount = relative amount	$\frac{1.0}{1.0} = 1.0$	$\frac{2.0}{1.0} = 2.0$	$\frac{6.0}{1.0} = 6.0$

∴ $m = 2$, $n = 6$ and the formula is $\mathbf{MgCl_2 \cdot 6H_2O}$.

End-of-chapter questions

2.1 9.19 mol dm^{-3}
2.2 0.628 mol dm^{-3}
2.3 **a** 0.0150 mol
b 7.5 cm^3
2.4 **a** i) 0.010 mol
ii) 0.025 mol
iii) 1 mol of P deposits 2.5 mol of Cu.
b i) The change in oxidation number is from +2 to 0.
ii) +5
iii) $n = 3$
c $P_4(s) + 10CuSO_4(aq) + 12H_2O(l) \rightarrow 10Cu(s) + 4HPO_3(aq) + 10H_2SO_4(aq)$
2.5 **a** i) −2
ii) +5
b i) $NH_3(aq) + H_2O(l) \rightleftharpoons NH_4^+(aq) + OH^-(aq)$
ii) $NH_3(aq) + HNO_3(aq) \rightarrow NH_4NO_3(aq)$
iii) 0.10 dm^3 or 100 cm^3
2.6 5.0×10^{-8} mol $litre^{-1}$ (mol dm^{-3})
2.7 **a** $SO_2(aq) + H_2O_2(aq) \rightarrow H_2SO_4(aq)$
b 2.45×10^{-4} g m^{-3}
2.8 0.500 g
2.9 $2SO_4^{2-}(aq) \rightarrow S_2O_8^{2-}(aq) + 2e^-$
If S is oxidised, *Ox*(O) remains at −2 so that average *Ox*(S) = ½(−2 − (−16)) = +7
If O is oxidised, *Ox*(S) remains at +6 so that average *Ox*(O) = ⅛(−2 − 12) = −1.75
∴ KI was in excess.
$S_2O_8^{2-}(aq) + 2I^-(aq) \rightarrow 2SO_4^{2-}(aq) + I_2(aq)$
2.10 **a** i) 0.010 ii) 0.18 g
b i) 0.010 ii) 0.96 g
c i) 0.0050 ii) 0.32 g
d i) 0.54 g ii) 0.030
e $Cu(NH_4)_2(SO_4)_2{\cdot}6H_2O$ or $CuSO_4{\cdot}(NH_4)_2SO_4{\cdot}6H_2O$
2.11 73.9% Fe^{2+} 26.1% Fe^{3+}
2.12 1.31 tonnes

CHAPTER 3

EXERCISE 3.1

a Height of lithium-6 peak = 0.50 cm
Height of lithium-7 peak = 6.30 cm
Substituting into the expression:

$$\% \text{ abundance} = \frac{\text{amount of isotope}}{\text{total amount of all isotopes}} \times 100$$

$$\% \text{ abundance of } {}^{6}\text{Li} = \frac{0.50}{6.30 + 0.50} \times 100 = 7.4\%$$

$$\% \text{ abundance of } {}^{7}\text{Li} = \frac{6.30}{6.30 + 0.50} \times 100 = 92.6\%$$

b Total mass of a hundred atoms = (6 × 7.4) + (7 × 92.6) amu
= 44.4 + 648.2 amu = 692.6 amu
= 693 amu (3 significant figures)

Substituting into the expression:

$$\text{average mass} = \frac{\text{total mass}}{\text{number of atoms}}$$

$$\text{average mass} = \frac{693}{100} \text{ amu} = 6.93 \text{ amu}$$

∴ relative atomic mass = **6.93**

EXERCISE 3.2

The mass/charge ratios refer to the singly-charged ions from the isotopes present in the sample: ${}^{20}_{10}Ne$, ${}^{21}_{10}Ne$, ${}^{22}_{10}Ne$.

The relative intensities show the relative abundance of each isotope in the sample.

Substituting into the expression:

$$\% \text{ abundance} = \frac{\text{amount of isotope}}{\text{total amount of all isotopes}} \times 100$$

$$\% \text{ abundance of } {}^{20}\text{Ne} = \frac{0.910}{0.910 + 0.0026 + 0.088} \times 100 = 91\%$$

$$\% \text{ abundance of } {}^{21}\text{Ne} = \frac{0.0026}{0.910 + 0.0026 + 0.088} \times 100 = 0.26\%$$

$$\% \text{ abundance of } {}^{22}\text{Ne} = \frac{0.088}{0.910 + 0.0026 + 0.088} \times 100 = 8.8\%$$

(Did you notice that the sum of the relative intensities is 1?)

Total mass of 100 atoms = (20 × 91) + (21 × 0.26) + (22 × 8.8)
= 1820 + 5.46 + 193.6 = 2.02×10^3 amu

$$\text{Average mass} = \frac{\text{total mass}}{\text{number of atoms}} = \frac{2.02 \times 10^3 \text{ amu}}{100} = 20.2 \text{ amu}$$

∴ relative atomic mass = **20.2**

EXERCISE 3.3

a $^{50}_{24}Cr$ – 4.31%; $^{52}_{24}Cr$ – 83.76%; $^{53}_{24}Cr$ – 9.55%; $^{54}_{24}Cr$ – 2.38%.

b–d

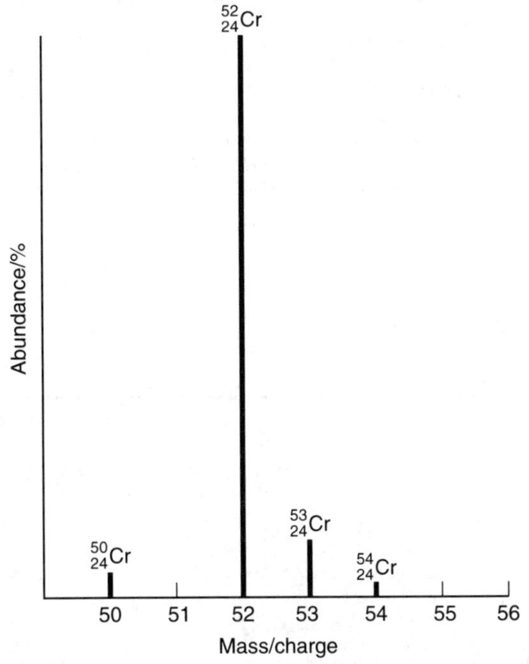

$$\text{Total mass of 100 atoms} = (50 \times 4.31) + (52 \times 83.76) + (53 \times 9.55) + (54 \times 2.38)$$

$$= 215.5 + 4355.2 + 506.15 + 128.52$$

$$= 5.21 \times 10^3 \text{ amu}$$

$$\therefore \text{ average mass} = \frac{5.21 \times 10^3 \text{ amu}}{100} = 52.1 \text{ amu}$$

and **relative atomic mass = 52.1**

EXERCISE 3.4

Mass/charge ratio	Species
35	$^{35}_{17}\mathbf{Cl}^+$
37	$^{37}_{17}\mathbf{Cl}^+$
70	$(^{35}_{17}\mathbf{Cl}{-}^{35}_{17}\mathbf{Cl})^+$
72	$(^{35}_{17}\mathbf{Cl}{-}^{37}_{17}\mathbf{Cl})^+$
74	$(^{37}_{17}\mathbf{Cl}{-}^{37}_{17}\mathbf{Cl})^+$

EXERCISE 3.5

a The bombardment of molecules of ethanol by a high-energy electron beam not only produces ions of the parent molecule $C_2H_5OH^+$ but also fragment ions which result from some of the molecules splitting up, e.g. $C_2H_5^+$, CH_2OH^+, etc.

b Peak **A** (m/e = 29) corresponds to an ion of relative molecular mass 29; therefore, **A = $(C_2H_5)^+$ or possibly $(COH)^+$**.
Peak **B** (m/e = 31) corresponds to an ion of relative molecular mass 31; therefore, **B = $(CH_2OH)^+$**.
Peak **C** (m/e = 45) corresponds to an ion of relative molecular mass 45; therefore, **C = $(C_2H_5O)^+$**.
Peak **D** (m/e = 46) corresponds to an ion of relative molecular mass 46; therefore, **D = parent molecular ion $(C_2H_5OH)^+$**.

EXERCISE 3.6

Empirical formula determination. For a sample of 100 g:

	C	H
Mass/g	92.4	7.6
Molar mass/g mol^{-1}	12.0	1.0
Amount/mol	$\frac{92.4}{12.0} = 7.7$	$\frac{7.6}{1.0} = 7.6$
Simplest ratio of relative amounts	1	1

Empirical formula of X = **CH**.

Substitute into the expression:

$$\frac{\text{height of } (M+1) \text{ peak}}{\text{height of } M \text{ peak}} = \frac{1.1\,N}{100}$$

$$N = \frac{1}{11.5} \times \frac{100}{1.1}$$

$$= 7.9$$

$= \mathbf{8}$ (to the nearest whole number)

$\therefore$ X contains 8 carbon atoms, and so the molecular formula is $\mathbf{C_8H_8}$.

EXERCISE 3.7

a For hydrocarbon A, the species of highest mass/charge ratio, 170, corresponds to an ion of the complete molecule being analysed. The isotopic peak due to molecular ions containing ^{13}C is too small to be visible.

For hydrocarbon B, the species of highest mass/charge ratio 191 corresponds to an ion of the complete molecule containing the ^{13}C isotope.

b The major peaks are caused by ion fragments resulting from breaking the weaker bonds in the molecule. Associated minor peaks are caused by a variety of other ion fragments, e.g.:

i) Major fragments as above, but with one, two or three more H atoms removed. These give smaller peaks at lower mass/charge ratio.

ii) Major fragments containing isotopes of different mass. For a hydrocarbon, the only significant isotope is ^{13}C and this would give very small peaks at higher mass/charge ratio.

iii) Impurities from air or residues from the previous sample. These can give small peaks anywhere along the spectrum.

c $\mathbf{C_8H_{17}^+}$ (from mass/charge ratio 113 and by comparison with labelled peaks).

EXERCISE 3.8

The two peaks with smaller mass/charge ratio than $C_5H_{11}^+$ are due to $\mathbf{C_5H_{10}^+}$ and $\mathbf{C_5H_9^+}$. The one with higher mass/charge ratio at about 5% of the intensity is due to $\mathbf{^{13}C^{12}C_4H_{11}^+}$.

EXERCISE 3.9

a Molecular ions of mass/charge ratios 112 and 114 are due to $C_6H_5{}^{35}Cl^+$ and $C_6H_5{}^{37}Cl^+$ respectively. The relative abundance of ^{35}Cl and ^{37}Cl means the $C_6H_5{}^{35}Cl^+$ peak will be three times as high as the $C_6H_5{}^{37}Cl^+$ peak.

b The peaks at mass/charge ratios 113 and 115 are due to some of the molecular ions from **a** containing the ^{13}C isotope.

c The fragment responsible for peak mass/charge ratio 77 is $\mathbf{C_6H_5^+}$ (as $M_r = (6 \times 12) + (5 \times 1) = 77$).

This would be produced when the parent molecule splits as shown below:

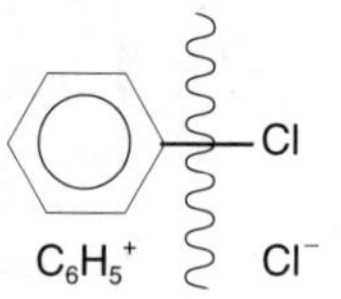

The Cl^- fragment may or may not then have electrons removed by bombardment to form a Cl^+ ion.

EXERCISE 3.10

There would be **three peaks due to** $\mathbf{CH_2{}^{79}Br_2^+}$, $\mathbf{CH_2{}^{81}Br_2^+}$ **and** $\mathbf{CH_2{}^{81}Br^{79}Br^+}$. Note that peaks due to ^{13}C would be scarcely visible because the molecule has only one carbon atom.

EXERCISE 3.11

a $^{212}_{84}Po \rightarrow {}^{4}_{2}He + {}^{208}_{82}Pb$

b $^{220}_{86}Rn \rightarrow {}^{4}_{2}He + {}^{216}_{84}Po$

EXERCISE 3.12

a $^{212}_{84}Po \rightarrow {}^{0}_{-1}e + {}^{212}_{85}At$

b $^{24}_{11}Na \rightarrow {}^{0}_{-1}e + {}^{24}_{12}Mg$

c $^{108}_{47}Ag \rightarrow {}^{0}_{-1}e + {}^{108}_{48}Cd$

EXERCISE 3.13

a $^{226}_{88}Ra \rightarrow {}^{4}_{2}He + {}^{222}_{86}Rn$

b $^{43}_{19}K \rightarrow {}^{0}_{-1}e + {}^{43}_{20}Ca$

c $^{43}_{20}Ca \rightarrow {}^{4}_{2}He + {}^{39}_{18}Ar$

End-of-chapter questions

3.1 65

3.2 **a** 63.6

b $^{220}_{86}Rn \rightarrow {}^{216}_{84}Po + {}^{4}_{2}He$

$^{214}_{82}Pb \rightarrow {}^{214}_{83}Bi + {}^{0}_{-1}e$

3.3 **a** Two peaks at *m/e* values 35 and 37, *m/e* = 35 being the more intense.

b Peaks at *m/e* values 70, 72 and 74 (due to $^{35}Cl-{}^{35}Cl^+$, $^{35}Cl-{}^{37}Cl^+$ and $^{37}Cl-{}^{37}Cl^+$, respectively). Most intense peak at *m/e* = 70.

3.4 **a** 79.99 (= 80.0 to 3 significant figures)

b ii)

Mass No.	158	160	162
Ion	$^{79}Br—{}^{79}Br^+$	$^{79}Br—{}^{81}Br^+$	$^{81}Br—{}^{81}Br^+$

3.5 **a** Molecular formula = $C_3H_6O_2$

b Empirical formula = $C_3H_6O_2$

c

74	$C_3H_6O_2^+$	29	$C_2H_5^+$ or COH^+
57	$C_2H_5CO^+$	28	$C_2H_4^+$ or CO^+
45	CO_2H^+	27	$C_2H_3^+$

CHAPTER 4

EXERCISE 4.1

a $C(s) + O_2(g) \rightarrow CO_2(g)$; $\Delta H^{\ominus} = -394\ \text{kJ mol}^{-1}$

b i) Part **a** shows the enthalpy change for 1.00 mol of carbon. For 10.0 mol the enthalpy change is

$10.0\ \text{mol} \times (-394\ \text{kJ mol}^{-1}) = \mathbf{-3940\ kJ}$

ii) For 0.25 mol of carbon

$\Delta H = 0.25\ \text{mol} \times (-394\ \text{kJ mol}^{-1}) = \mathbf{-98.5\ kJ}$

iii) $\text{Amount of carbon} = \frac{m}{M} = \frac{18.0\ \text{g}}{12.0\ \text{g mol}^{-1}} = 1.5\ \text{mol}$

$\therefore \Delta H = 1.5\ \text{mol} \times (-394\ \text{kJ mol}^{-1}) = \mathbf{-591\ kJ}$

c i) You may be able to see at a glance that 197 is half 394, so that the amount of carbon is ½ mol and the mass is 6 g. Otherwise, since the quantity of heat depends directly on the amount of carbon burnt:

$$\frac{1.00\ \text{mol of C}}{-394\ \text{kJ}} = \frac{x\ \text{mol of C}}{-197\ \text{kJ}}$$

$$\therefore x = \frac{197}{394} = 0.500$$

$\therefore$ mass of carbon = $0.500\ \text{mol} \times 12.0\ \text{g mol}^{-1} = \mathbf{6.00\ g}$

ii)

$$\frac{1.00\ \text{mol of C}}{-394\ \text{kJ}} = \frac{y\ \text{mol of C}}{-1000\ \text{kJ}}$$

$$\therefore y = \frac{1000}{394} = 2.54$$

$\therefore$ mass of carbon = $2.54\ \text{mol} \times 12.0\ \text{g mol}^{-1} = \mathbf{30.5\ g}$

EXERCISE 4.2

a

b

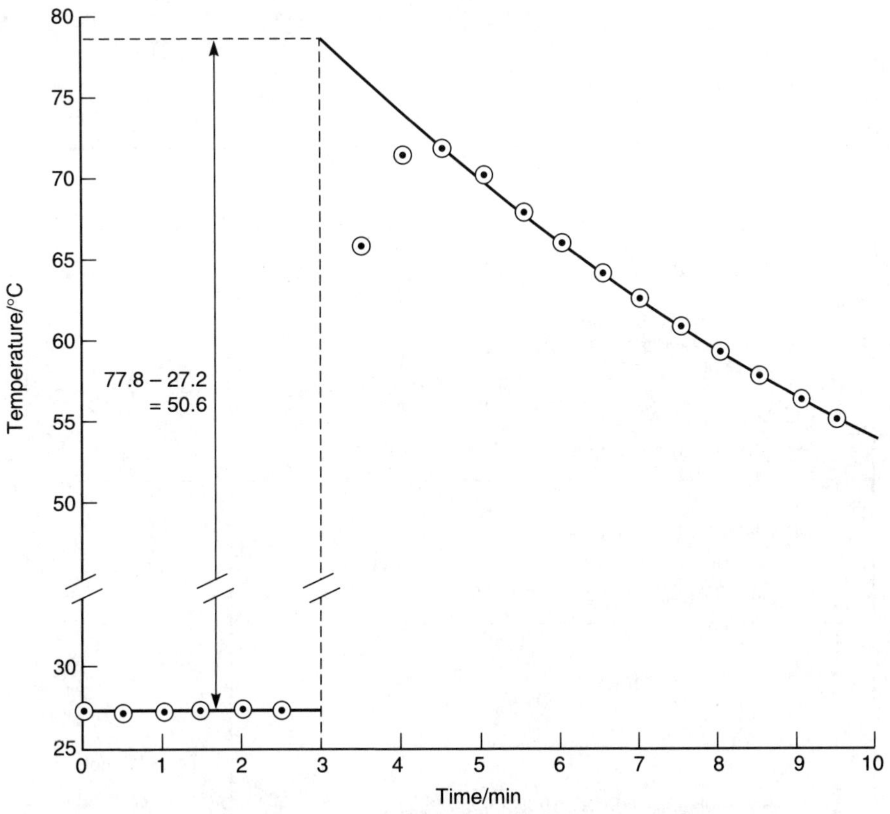

c

$$\begin{bmatrix}\text{enthalpy change due to}\\ \text{reaction (at constant } T)\end{bmatrix} = \begin{bmatrix}\text{enthalpy change in restoring}\\ \text{solution to original temperature}\end{bmatrix}$$

i.e. $\Delta H \quad = \quad mc_p\Delta T$

$\therefore \Delta H = mc_p\Delta T = 0.0250\ \text{kg} \times 4.18\ \text{kJ kg}^{-1}\ \text{K}^{-1} \times (-50.6\ \text{K}) = \mathbf{-5.29\ kJ}$

d Amount of Cu^{2+} used = $cV = 1.00\ \text{mol dm}^{-3} \times 0.0250\ \text{dm}^3$
= 0.0250 mol
Scaling up to the amount in the equation, 1 mol,

$$\Delta H = -5.29\ \text{kJ} \times \frac{1}{0.0250} = \mathbf{-212\ kJ}$$

$\mathbf{Zn(s) + Cu^{2+}(aq) \rightarrow Cu(s) + Zn^{2+}(aq);\ \Delta H^{\ominus} = -212\ kJ\ mol^{-1}}$

EXERCISE 4.3

ΔT = final temperature – initial temperature = 22.7 °C – 24.5 °C = –1.8 K

$$\begin{bmatrix}\text{enthalpy change on} \\ \text{dissolving } NH_4Cl\end{bmatrix} = \begin{bmatrix}\text{enthalpy change in restoring} \\ \text{solution to original temperature}\end{bmatrix}$$

i.e. $\Delta H = mc_p\Delta T$

$\therefore \Delta H = mc_p\Delta T = 0.090\ \text{kg} \times 4.18\ \text{kJ kg}^{-1}\ \text{K}^{-1} \times 1.8\ \text{K} = +0.68\ \text{kJ}$

Scaling up from 0.050 mol to 1 mol.

$$\Delta H = +0.68\ \text{kJ} \times \frac{1}{0.050} = +14\ \text{kJ}$$

$NH_4Cl(s) + 100H_2O(l) \rightarrow NH_4Cl(aq, 100H_2O)$; $\Delta H^\ominus$ = **+14 kJ mol^{-1}**

Note that only two significant figures are justified here because ΔT is small.

EXERCISE 4.4

a

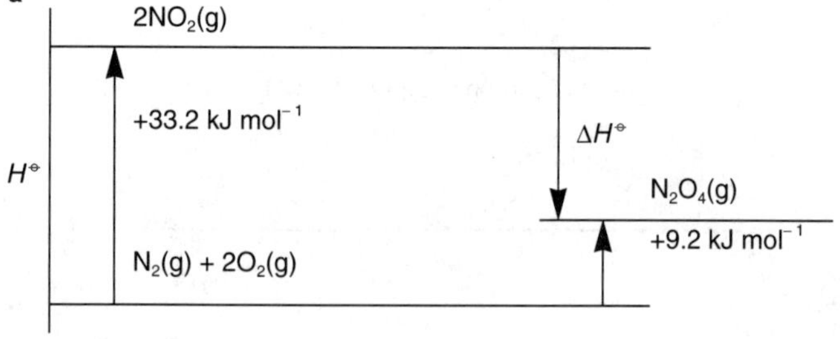

$+33.2\ \text{kJ mol}^{-1} = +9.2\ \text{kJ mol}^{-1} + (-\Delta H^\ominus)$

$\therefore \Delta H^\ominus = (9.2 - 33.2)\ \text{kJ mol}^{-1} =$ **–24.0 kJ mol^{-1}**

b

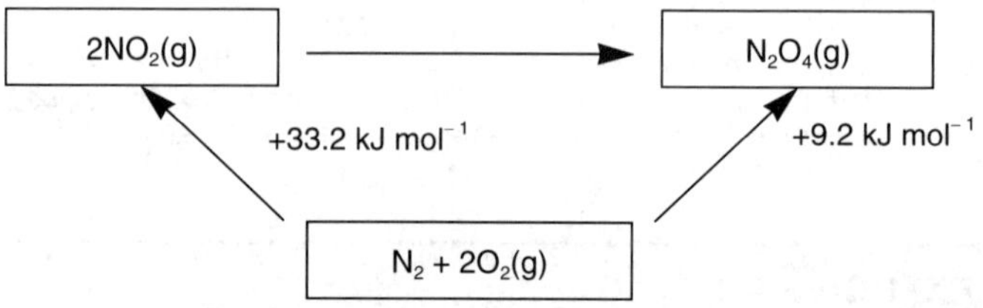

$+33.2\ \text{kJ mol}^{-1} + \Delta H^\ominus = +9.2\ \text{kJ mol}^{-1}$

$\therefore \Delta H^\ominus = (9.2 - 33.2)\ \text{kJ mol}^{-1} =$ **–24.0 kJ mol^{-1}**

EXERCISE 4.5

a $NH_3(g) + HCl(g) +$ ~~$NH_4Cl(s)$~~ $+ 200H_2O(l)$
$\rightarrow$ ~~$NH_4Cl(s)$~~ $+ NH_4Cl(aq, 200H_2O)$; $\Delta H^\ominus = (-175.3 + 16.3)\ \text{kJ mol}^{-1}$
$NH_3(g) + HCl(g) + 200H_2O(l) \rightarrow NH_4Cl(aq, 200H_2O)$; $\Delta H^\ominus$ = **–159.0 kJ mol^{-1}**

b $NH_3(g) + 100H_2O(l) + HCl(g) + 100H_2O(l) + NH_3$~~$(aq, 100H_2O)$~~ $+ HCl$~~$(aq, 100H_2O)$~~ $\rightarrow NH_3(aq, 100H_2O) + H$~~$Cl(aq, 100H_2$~~$O) + NH_4Cl(aq, 200H_2O)$; $\Delta H^\ominus = (-35.6 - 73.2 - 50.2)\ \text{kJ mol}^{-1}$

$NH_3(g) + HCl(g) + 200H_2O(l) \rightarrow NH_4Cl(aq, 200H_2O)$; $\Delta H^\ominus$ = **–159.0 kJ mol^{-1}**

c The enthalpy change for the overall process is the same by either Method 1 or Method 2.

d

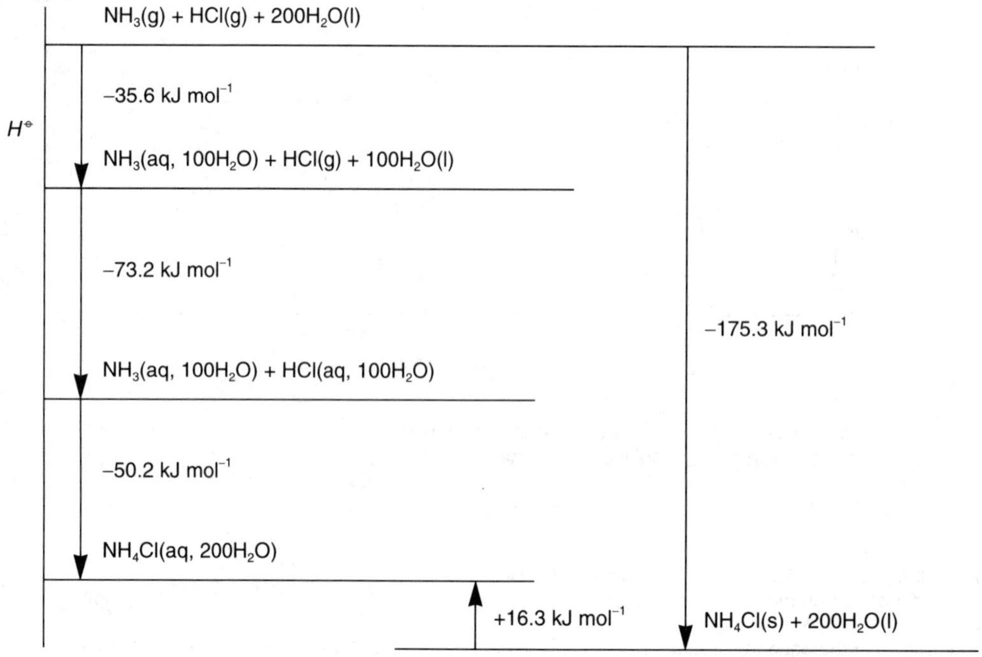

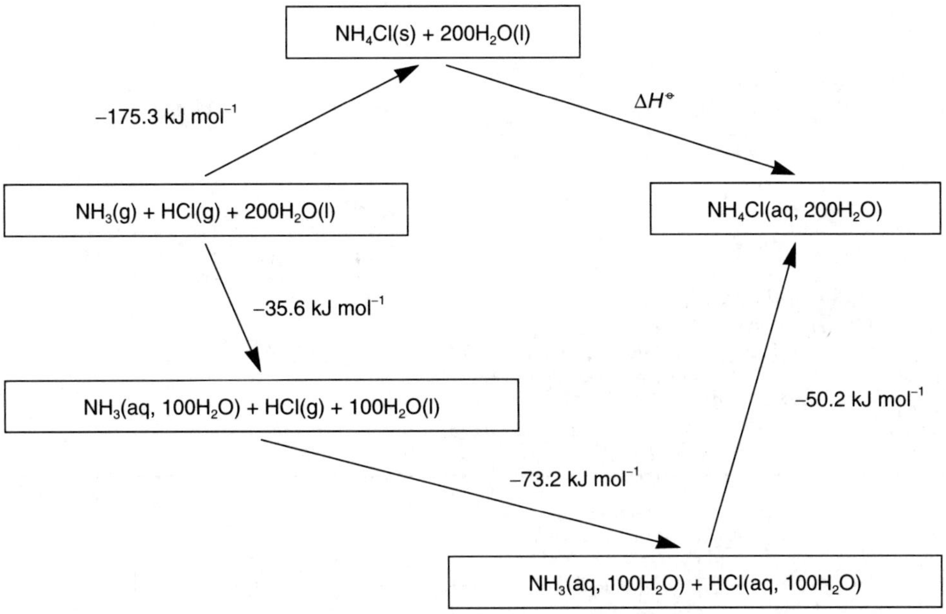

EXERCISE 4.6

a $\Delta H = mc_p\Delta T = 0.0450\ \text{kg} \times 4.18\ \text{kJ kg}^{-1}\ \text{K}^{-1} \times (-11.3\ \text{K}) = -2.13\ \text{kJ}$

Scaling up, $\Delta H = -2.13\ \text{kJ} \times \dfrac{1\ \text{mol}}{0.0250\ \text{mol}} = -85.0\ \text{kJ}$

Thus $MgSO_4(s) + 100H_2O(l) \rightarrow MgSO_4(aq, 100H_2O)$; $\Delta H^\ominus = $ **-85.0 kJ mol^{-1}**

b $\Delta H = mc_p\Delta T = 0.450\ \text{kg} \times 4.18\ \text{kJ kg}^{-1}\ \text{K}^{-1} \times 1.4\ \text{K} = +0.26\ \text{kJ}$

Note that the **total** amount of water includes the water of crystallisation and is therefore the same as in **a**.

Scaling up, $\Delta H = +0.26\ \text{kJ} \times \dfrac{1\ \text{mol}}{0.0250\ \text{mol}} = +10.5\ \text{kJ}$

Thus $MgSO_4{\cdot}7H_2O(s) + 93H_2O(l) \rightarrow MgSO_4(aq, 100H_2O)$; $\Delta H^\ominus = $ **$+10.5$ kJ mol^{-1}**

c

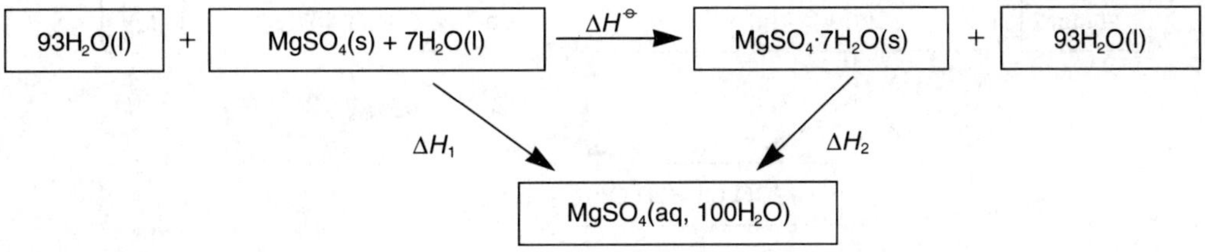

$\Delta H^\ominus = \Delta H_1 - \Delta H_2 = (-85.0 - 10.5)\ \text{kJ mol}^{-1} = $ **-95.5 kJ mol^{-1}**

d

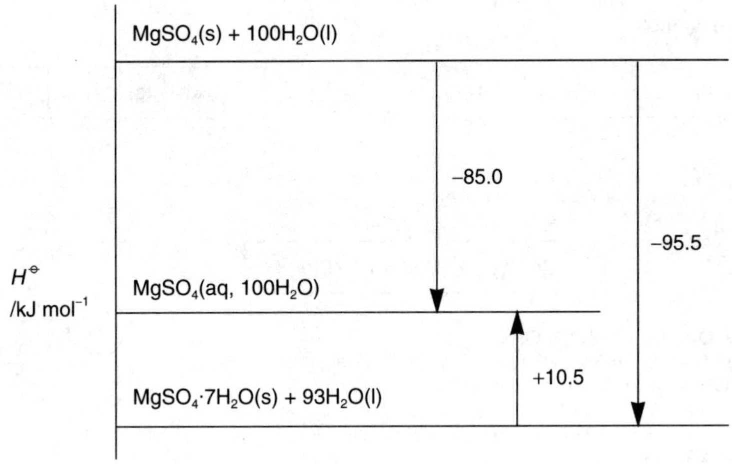

EXERCISE 4.7

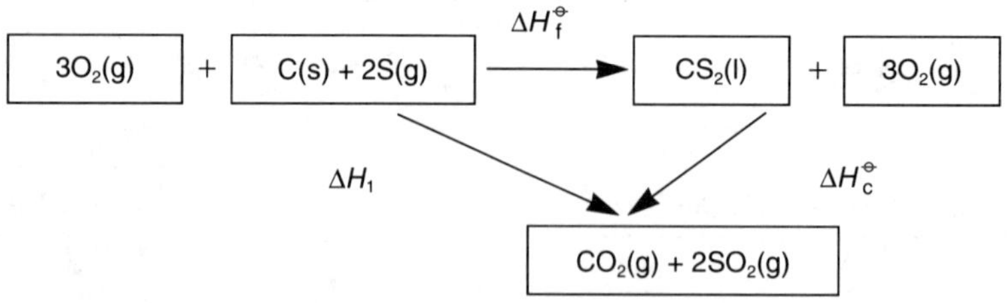

$\Delta H^{\ominus}_{f}[CS_2(l)] = \Delta H_1 - \Delta H^{\ominus}_{c}[CS_2(l)]$
$\Delta H_1 = \Delta H^{\ominus}_{f}[CO_2(g)] + 2\Delta H^{\ominus}_{f}[SO_2(g)]$
$\therefore\ \Delta H^{\ominus}_{f}[CS_2(l)] = \Delta H^{\ominus}_{f}[CO_2(g)] + 2\Delta H^{\ominus}_{f}[SO_2(g)] - \Delta H^{\ominus}_{c}[CS_2(l)]$
$= (-393.5 + 2(-296.9) + 1075.2)\ \text{kJ mol}^{-1}$
$\Delta H^{\ominus}_{f}[CS_2(l)] =$ **$+87.9\ \text{kJ mol}^{-1}$**

EXERCISE 4.8

a

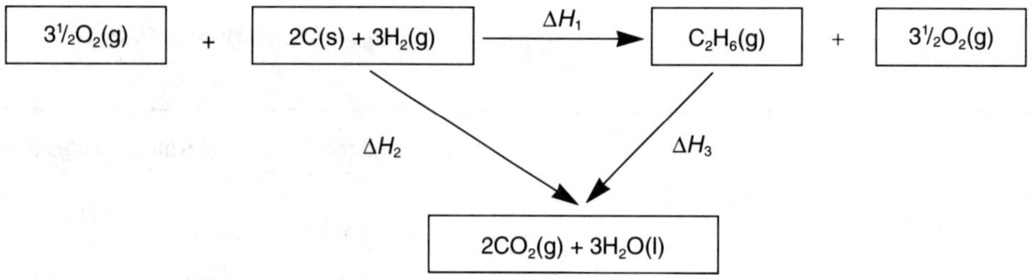

$\Delta H_1 + \Delta H_3 = \Delta H_2$ or $\quad \Delta H_1 = \Delta H_2 - \Delta H_3$
$\Delta H_2 = 2\Delta H^{\ominus}_{f}[CO_2(g)] + 3\Delta H^{\ominus}_{f}[H_2O(l)]$
$= (2(-393.5) + 3(-285.9))\ \text{kJ mol}^{-1} = -1644.7\ \text{kJ mol}^{-1}$
$\Delta H_3 = \Delta H^{\ominus}_{c}[C_2H_6(g)] = -1559.8\ \text{kJ mol}^{-1}$
$\therefore\ \Delta H_1 = (-1644.7 - (-1559.8))\ \text{kJ mol}^{-1} =$ **$-84.9\ \text{kJ mol}^{-1}$**

b

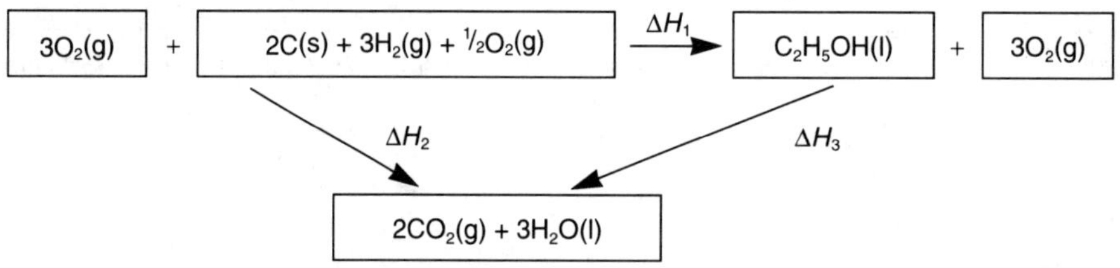

$\Delta H_2 = 2\Delta H^{\ominus}_{f}[(CO_2(g)] + 3\Delta H^{\ominus}_{f}[H_2O(l)]$
$= (2(-393.5) + 3(-285.9))\ \text{kJ mol}^{-1} = -1644.7\ \text{kJ mol}^{-1}$
$\Delta H_3 = \Delta H^{\ominus}_{c}[C_2H_5OH(l)] = -1366.7\ \text{kJ mol}^{-1}$
$\Delta H_1 = \Delta H_2 - \Delta H_3 = (-1644.7 - (-1366.7))\ \text{kJ mol}^{-1}$
$=$ **$-278.0\ \text{kJ mol}^{-1}$**

c

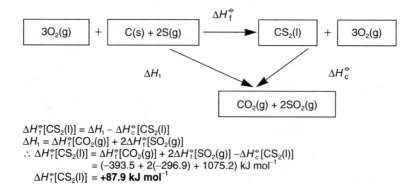

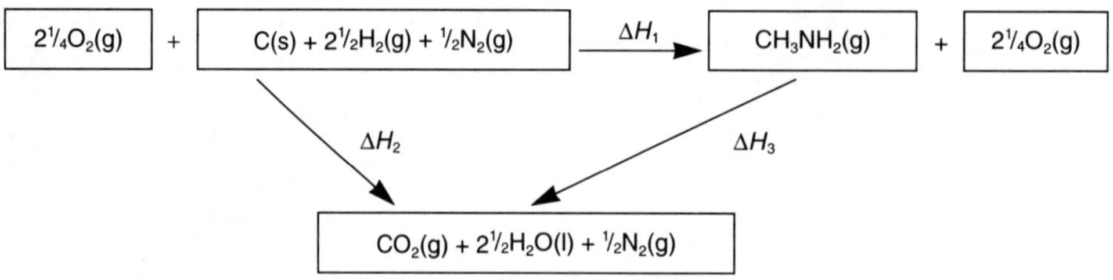

$\Delta H_2 = \Delta H^{\ominus}_{f}[CO_2(g)] + 2\frac{1}{2}\Delta H^{\ominus}_{f}[H_2O(l)]$
$= (-393.5 + 2\frac{1}{2}(-285.9))\ \text{kJ mol}^{-1} = -1108.3\ \text{kJ mol}^{-1}$
$\Delta H_3 = \Delta H^{\ominus}_{f}[CH_3NH_2(g)] = -1079.9\ \text{kJ mol}^{-1}$
$\Delta H_1 + \Delta H_3 = \Delta H_2$
$\therefore\ \Delta H_1 = \Delta H_2 - \Delta H_3 = (-1108.3 - (-1079.9))\ \text{kJ mol}^{-1}$
$=$ **$-28.4\ \text{kJ mol}^{-1}$**

EXERCISE 4.9

a $CH_3OH(l) + 1\frac{1}{2}O_2(g) \rightarrow CO_2(g) + 2H_2O(l)$

$\Delta H_f^\ominus$/kJ mol^{-1} −238.9   0   −393.5   −2(285.9)

$\Delta H^\ominus = (-571.8 - 393.5 - (-238.9))$ kJ mol^{-1} = **−726.4 kJ mol^{-1}**

b $2CO(g) + O_2(g) \rightarrow 2CO_2(g)$

$\Delta H_f^\ominus$/kJ mol^{-1} −2(110.5)   0   −2(393.5)

$\Delta H^\ominus = (-787.0 - (-221.0))$ kJ mol^{-1} = **−566.0 kJ mol^{-1}**

c $ZnCO_3(s) \rightarrow ZnO(s) + CO_2(g)$

$\Delta H_f^\ominus$/kJ mol^{-1} −812.5   −348.0   −393.5

$\Delta H^\ominus = (-393.5 - 348.0 - (-812.5))$ kJ mol^{-1} = **+71.0 kJ mol^{-1}**

d $2Al(s) + Fe_2O_3(s) \rightarrow 2Fe(s) + Al_2O_3(s)$

$\Delta H_f^\ominus$/kJ mol^{-1} 0   −822.2   0   −1675.7

$\Delta H^\ominus = (-1675.7 - (-822.2))$ kJ mol^{-1} = **−853.5 kJ mol^{-1}**

[The very large quantity of heat released is sufficient to melt the iron produced and other iron in contact with the reaction mixture. This is the basis of the 'Thermit' process which is still sometimes used to weld steel rails *in situ* on railway tracks.]

EXERCISE 4.10

$CH_4(g) + H_2O(g) \rightarrow CO(g) + 3H_2(g)$

$\Delta H_f^\ominus$/kJ mol^{-1} −75   −242   −110   0

$\Delta H^\ominus = (-110 - (-242) - (-75))$ kJ mol^{-1} = **+207 kJ mol^{-1}**

EXERCISE 4.11

$\Delta H_r^\ominus = \Sigma\Delta H_f^\ominus(\text{products}) - \Sigma\Delta H_f^\ominus(\text{reactants})$

$= \Delta H_f^\ominus(Si(g)) + 2\Delta H_f^\ominus(O(g)) - \Delta H_f^\ominus(SiO_2(s))$

$= (455 + (2 \times 249) - (-910))$ kJ mol^{-1} = **+1863 kJ mol^{-1}**

(Note that a much-used data book gives a false value for $\Delta H_f^\ominus(Si(g))$.)

EXERCISE 4.12

For the reaction:

$2Al_2O_3 + 3C \rightarrow 4Al + 3CO_2$

$\Delta H^\ominus_{reaction} = 3\Delta H_f^\ominus(CO_2) + 4\Delta H_f^\ominus(Al) - 2\Delta H_f^\ominus(Al_2O_3) - 3\Delta H_f^\ominus(C)$

$= (3 \times -395 \text{ kJ mol}^{-1}) + 0 - (2 \times -1676 \text{ kJ mol}^{-1}) - 0$

= **+2167 kJ mol^{-1}**

This very high positive value suggests that direct reduction with carbon is not feasible. If you have studied $\Delta G^\ominus$ and Ellingham diagrams, you might realise that the reaction is feasible at very high temperatures (above 2000 °C). However, at this temperature aluminium combines with carbon to form aluminium carbide, Al_4C_3.

EXERCISE 4.13

a Adding the reactions in Table 4.2 gives the overall reaction:

$CH_4(g) \rightarrow C(g) + 4H(g)$; ΔH = **+1662 kJ mol^{-1}**

b $\bar{E}$(C—H) = (1662 kJ mol^{-1}/4) = **416 kJ mol^{-1}**

c Each step represents a bond dissociation energy.

d Each bond dissociation energy is different because the environment in each step is different; i.e. the C—H bond broken in

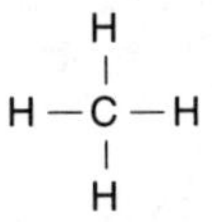

is different from the ones broken in

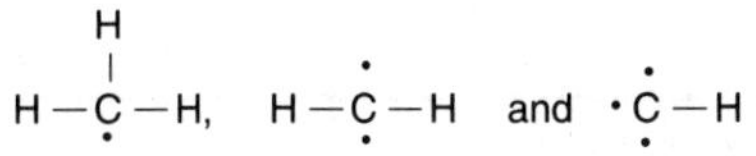

EXERCISE 4.14

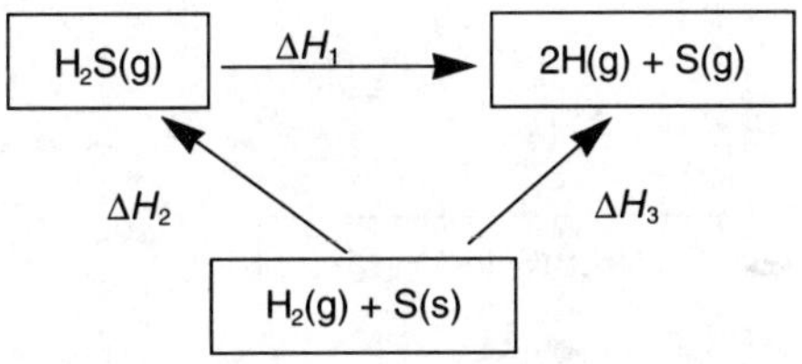

$\Delta H_3 = 2\Delta H_{at}^\ominus[H(g)] + \Delta H_{at}^\ominus[S(g)] = (2(218.0) + 238.1))$ kJ mol^{-1}

$= 674.1$ kJ mol^{-1}

$\Delta H_2 = \Delta H_f^\ominus[H_2S(g)] = -20.6$ kJ mol^{-1}

$\Delta H_1 = -\Delta H_2 + \Delta H_3 = (20.6 + 674.1)$ kJ mol^{-1} = 694.7 kJ mol^{-1}

$\therefore$ $\bar{E}$(H—S) = 694.7/2 kJ mol^{-1} = **347.4 kJ mol^{-1}**

EXERCISE 4.15

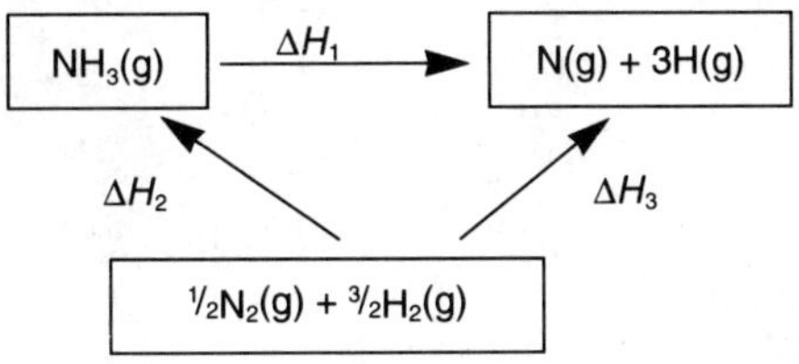

$\Delta H_3 = \Delta H_{at}^\ominus[N(g)] + 3\Delta H_{at}^\ominus[H(g)] = (473 + 3(218))$ kJ mol^{-1}

$= 1127$ kJ mol^{-1}

$\Delta H_2 = -46$ kJ mol^{-1}

$\therefore$ $\Delta H_1 = -\Delta H_2 + \Delta H_3 = (46 + 1127)$ kJ mol^{-1} = 1173 kJ mol^{-1}

$\bar{E}$(N—H) = 1173/3 kJ mol^{-1} = **391 kJ mol^{-1}**

EXERCISE 4.16

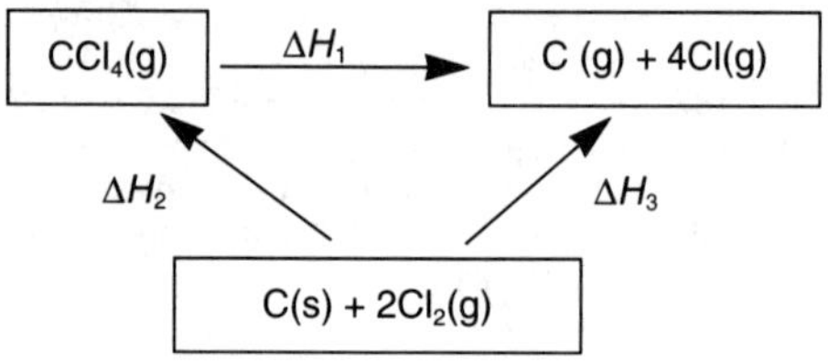

$\Delta H_2 = \Delta H_f^{\ominus}[CCl_4 \text{ (l)}] + \Delta H_{vap}^{\ominus}[CCl_4 \text{ (l)}] = (-135.5 + 30.5) \text{ kJ mol}^{-1}$
$= -105.0 \text{ kJ mol}^{-1}$
$\Delta H_3 = \Delta H_{at}^{\ominus}[C \text{ (g)}] + 4\Delta H_{at}^{\ominus}[Cl \text{ (g)}] = (715.0 + 4(121.1)) \text{ kJ mol}^{-1}$
$= 1199.4 \text{ kJ mol}^{-1}$
$\Delta H_1 = -\Delta H_2 + \Delta H_3 = (105.0 + 1199.4) \text{ kJ mol}^{-1} = 1304.4 \text{ kJ mol}^{-1}$
$\therefore \bar{E}(\text{C—Cl}) = 1304.4/4 \text{ kJ mol}^{-1} = \textbf{326 kJ mol}^{-1}$

EXERCISE 4.17

Let x kJ mol$^{-1} = \bar{E}(\text{C—C})$ and y kJ mol$^{-1} = \bar{E}(\text{C—H})$
$C_6H_{14}(g) \rightarrow 6C(g) + 14H(g); \quad \Delta H^{\ominus} = +7512 \text{ kJ mol}^{-1}$
$C_7H_{16}(g) \rightarrow 7C(g) + 16H(g); \quad \Delta H^{\ominus} = +8684 \text{ kJ mol}^{-1}$
$\therefore 5x + 14y = 7512 \hspace{2cm} (1)$
and $6x + 16y = 8684 \hspace{2cm} (2)$
Multiplying equation (1) by 6 and equation (2) by 5 and subtracting:

$$30x + 84y = 45072$$
$$\underline{30x + 80y = 43420}$$
$$4y = 1652 \quad \therefore y = 413$$

Substituting in equation (1)

$$5x + 5782 = 7512 \quad \therefore x = \frac{7512 - 5782}{5} = 346$$

i.e. $\bar{E}(\text{C—C}) = \textbf{346 kJ mol}^{-1}$ and $\bar{E}(\text{C—H}) = \textbf{413 kJ mol}^{-1}$

EXERCISE 4.18

a $H_2(g) + Cl_2(g) \rightarrow 2H\text{—}Cl(g)$
Bonds broken: $\bar{E}(\text{H—H}) + \bar{E}(\text{Cl—Cl}) = (436 + 242) \text{ kJ mol}^{-1}$
$= 678 \text{ kJ mol}^{-1}$
Bonds made: $-2\bar{E}(\text{H—Cl}) = -2(431) \text{ kJ mol}^{-1} = -862 \text{ kJ mol}^{-1}$
$\therefore \Delta H^{\ominus} = (678 - 862) \text{ kJ mol}^{-1} = \textbf{−184 kJ mol}^{-1}$

b $N_2(g) + 3H_2(g) \rightarrow 2NH_3(g)$
Bonds broken: $\bar{E}(\text{N}\equiv\text{N}) + 3\bar{E}(\text{H—H}) = (945 + 3(436)) \text{ kJ mol}^{-1}$
$= 2253 \text{ kJ mol}^{-1}$
Bonds made: $-6\bar{E}(\text{N—H}) = -6(389) \text{ kJ mol}^{-1} = -2334 \text{ kJ mol}^{-1}$
$\therefore \Delta H^{\ominus} = (2253 - 2334) \text{ kJ mol}^{-1} = \textbf{−81 kJ mol}^{-1}$

EXERCISE 4.19

a Bond energy term is the average value of the enthalpy changes for the dissociation of a particular type of bond.

b i) For the reaction

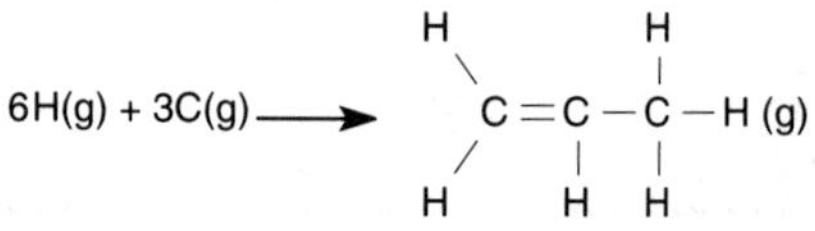

$$6H(g) + 3C(g) \longrightarrow \begin{array}{c} H H \\ \backslash | \\ C = C - C - H \text{ (g)} \\ / | | \\ H H H \end{array}$$

Bonds formed: six (C—H); $\quad -6\bar{E}(\text{C—H}) = -6(415) \text{ kJ mol}^{-1}$
one (C=C); $\quad -\bar{E}(\text{C=C}) = -598 \text{ kJ mol}^{-1}$
one (C—C); $\quad -\bar{E}(\text{C—C}) = -356 \text{ kJ mol}^{-1}$
$\therefore H^{\ominus} = -(6(415) + 598 + 356) \text{ kJ mol}^{-1} = \textbf{−3444 kJ mol}^{-1}$

ii)

$$3C(g) + 6H(g) + 2Br(g) \longrightarrow \begin{array}{c} H Br H \\ | | | \\ H - C - C - C - H \text{ (g)} \\ | | | \\ Br H H \end{array}$$

Bonds formed: six (C—H);$-6\bar{E}(\text{C—H}) = -6(415) \text{ kJ mol}^{-1}$
two (C—Br)$-\bar{E}(\text{C—Br}) = -2(284) \text{ kJ mol}^{-1}$
two (C—C);$- 2\bar{E}(\text{C—C}) = -2(356) \text{ kJ mol}^{-1}$
$\therefore \Delta H^{\ominus} = -(6(415) + 2(284) + 2(356)) \text{ kJ mol}^{-1} = \textbf{−3770 kJ mol}^{-1}$

c

$\Delta H_2 = -3444 \text{ kJ mol}^{-1} - \bar{E}(\text{Br—Br}) = -(3444 + 193) = -3637 \text{ kJ mol}^{-1}$
$\Delta H_1 = -\Delta H_2 + \Delta H_3 = (3637 - 3770) \text{ kJ mol}^{-1} = -133 \text{ kJ mol}^{-1}$

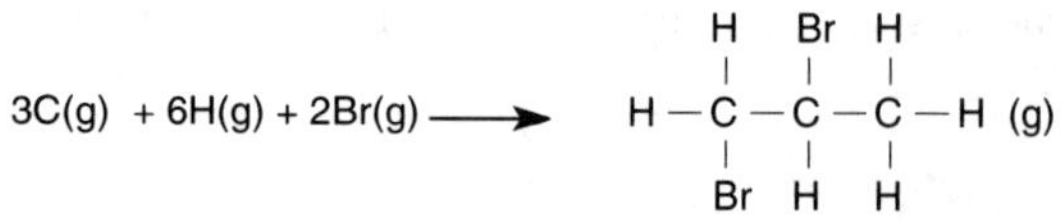

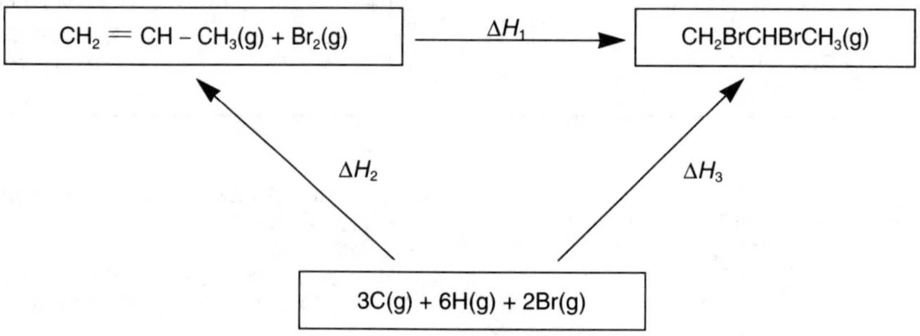

EXERCISE 4.20

a

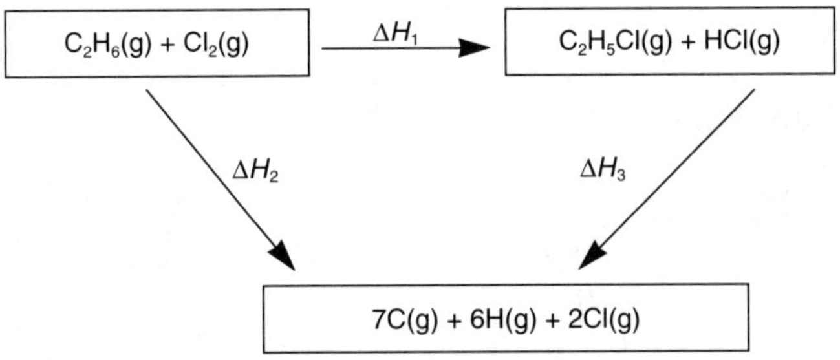

$\Delta H_2 = \bar{E}(C—C) + 6\bar{E}(C—H) + \bar{E}(Cl—Cl)$
$= (346 + 6(413) + 242)\ kJ\ mol^{-1} = 3066\ kJ\ mol^{-1}$
$\Delta H_3 = \bar{E}(C—C) + 5\bar{E}(C—H) + \bar{E}(C—Cl) + \bar{E}(H—Cl)$
$= (346 + 5(413) + 339 + 431)\ kJ\ mol^{-1} = 3181\ kJ\ mol^{-1}$
$\Delta H_1 = \Delta H_2 - \Delta H_3 = (3066 - 3181)\ kJ\ mol^{-1} = \mathbf{-115\ kJ\ mol^{-1}}$

Or, more simply,

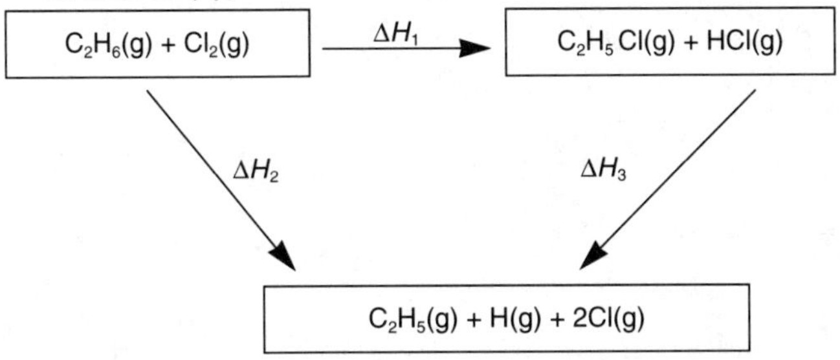

$\Delta H_2 = \bar{E}(C—H) + \bar{E}(Cl—Cl) = (413 + 242)\ kJ\ mol^{-1} = 655\ kJ\ mol^{-1}$
$\Delta H_3 = \bar{E}(C—Cl) + \bar{E}(H—Cl) = (339 + 431)\ kJ\ mol^{-1} = 770\ kJ\ mol^{-1}$
$\Delta H_1 = \Delta H_2 - \Delta H_3 = (655 - 770)\ kJ\ mol^{-1} = \mathbf{-115\ kJ\ mol^{-1}}$

b

	$C_2H_6(g)$ +	$Cl_2(g)$ →	$C_2H_5Cl(g)$ +	$HCl(g)$
$\Delta H_f^{\ominus}/kJ\ mol^{-1}$	−84.6	0	−136.5	−92.3

$\Delta H^{\ominus} = (-136.5 - 92.3 - (-84.6))\ kJ\ mol^{-1} = \mathbf{-144.2\ kJ\ mol^{-1}}$

c The bond energy terms $\bar{E}$(C—H) and $\bar{E}$(C—Cl) used in the first calculation are average values and are not quite the same as the bond dissociation energies for the particular compounds C_2H_6 and C_2H_5Cl. If the bond dissociation energies were known and used in the calculation they would give a more accurate result, much closer to −144.2 kJ mol^{-1}. Even then, however, the two answers might not quite agree because there is some uncertainty in data book values for $\Delta H_f^{\ominus}$ and bond dissociation energies.

EXERCISE 4.21

a

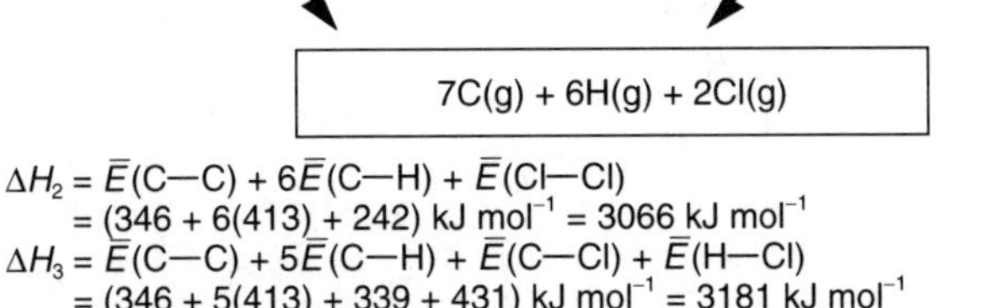

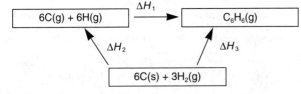

$\Delta H_1 = -\Delta H_2 + \Delta H_3 = -(6\Delta H_{at}^{\ominus}(C) + 6\Delta H_{at}^{\ominus}(H)) + \Delta H_f^{\ominus}$(benzene **vapour**)
$= (-(6 \times 715) - (6 \times 218) + 82.9)\ kJ\ mol^{-1} = \mathbf{-5515\ kJ\ mol^{-1}}$

b $\Delta H^{\ominus} = -(6\bar{E}(C—H) + 3\bar{E}(C—C) + 3\bar{E}(C{=}C))$
$= -((6 \times 412) + (3 \times 348) + (3 \times 612))\ kJ\ mol^{-1}$
$= \mathbf{-5352\ kJ\ mol^{-1}}$

c The formation of the Kekulé structure would release less energy than the formation of the actual structure. The actual structure is, therefore, at a lower energy level and is energetically more stable.
The stabilisation energy is (5515 − 5352) kJ mol^{-1} = **163 kJ mol^{-1}**

d 6C(g) + 6H(g) → 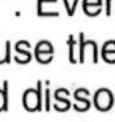 (g); $\Delta H^{\ominus} = -5515\ kJ\ mol^{-1}$

But also $\Delta H^{\ominus} = -6\bar{E}(C—H) - 6\bar{E}(C{\cdots}C)$
$\therefore \bar{E}(C{\cdots}C) = \frac{1}{6}(-\Delta H^{\ominus} - 6\bar{E}(C—H))$
$= \frac{1}{6}(5515 - (6 \times 412))\ kJ\ mol^{-1} = \mathbf{507\ kJ\ mol^{-1}}$
This value is intermediate between $\bar{E}$(C—C) and $\bar{E}$(C=C), 348 and 612 kJ mol^{-1}.

EXERCISE 4.22

a $-\Delta H^{\ominus} \simeq \bar{E}(C—C) + 6\bar{E}(C—H)$
$\therefore \bar{E}(C—C) = -\Delta H^{\ominus} - 6\bar{E}(C—H)$
$= 2820\ kJ\ mol^{-1} - (6 \times 412)\ kJ\ mol^{-1} = \mathbf{348\ kJ\ mol^{-1}}$

b $-\Delta H^{\ominus} \simeq \bar{E}(C{=}C) + 4\bar{E}(C—H)$
$\therefore \bar{E}(C{=}C) = -\Delta H^{\ominus} - 4\bar{E}(C—H)$
$= 2260\ kJ\ mol^{-1} - (4 \times 412)\ kJ\ mol^{-1} = \mathbf{612\ kJ\ mol^{-1}}$

c $\bar{E}(C{=}C) = \bar{E}(\sigma) + \bar{E}(\pi)$
$\therefore \bar{E}(\pi) = \bar{E}(C{=}C) - \bar{E}(\sigma)$
$= 612\ kJ\ mol^{-1} - 348\ kJ\ mol^{-1} = \mathbf{264\ kJ\ mol^{-1}}$
This is about $^3/_4$ of $\bar{E}(\sigma)$.

d It requires less energy to break the π-bond in ethene than the σ-bond in ethane. This suggests that ethene might be more reactive than ethane.

EXERCISE 4.23

a i)

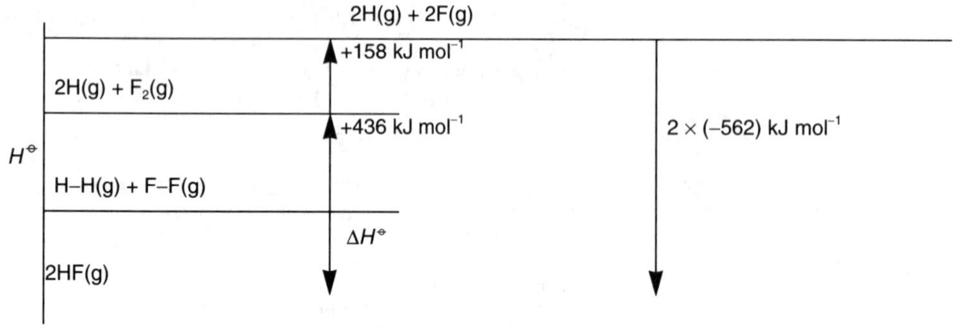

$\Delta H^{\ominus} = 2\,(-562) + 158 + 436 =$ **-530 kJ mol^{-1}**

ii)

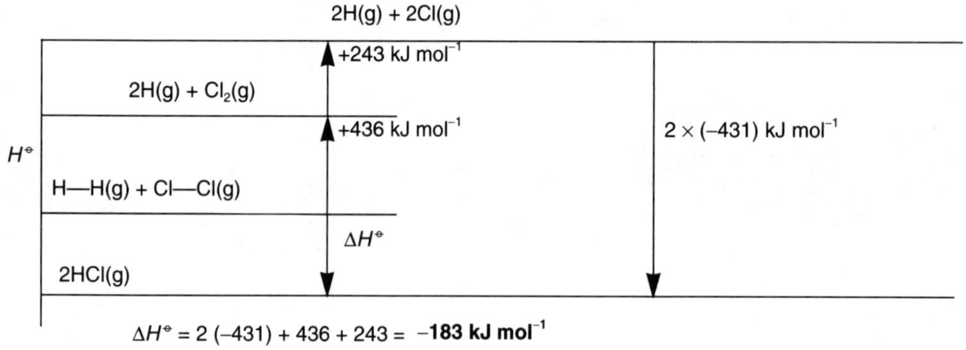

$\Delta H^{\ominus} = 2\,(-431) + 436 + 243 =$ **-183 kJ mol^{-1}**

EXERCISE 4.24

a There are six covalent P—P bonds in each P_4 molecule.

b i) $4P(g) \rightarrow P_4(g)$ $\quad \Delta H^{\ominus} = -(6 \times 201)$ kJ mol^{-1} = **-1206 kJ mol^{-1}**

ii) $4N(g) \rightarrow N_4(g)$ $\quad \Delta H^{\ominus} = -(6 \times 163)$ kJ mol^{-1} = **-978 kJ mol^{-1}**

iii) $4P(g) \rightarrow 2P{\equiv}P(g)$ $\quad \Delta H^{\ominus} = -(2 \times 488)$ kJ mol^{-1} = **-976 kJ mol^{-1}**

iv) $4N(g) \rightarrow 2N{\equiv}N(g)$ $\quad \Delta H^{\ominus} = -(2 \times 945)$ kJ mol^{-1} = **-1890 kJ mol^{-1}**

c The calculations above show that the formation of P_4 and N_2 molecules is energetically more favourable than the formation of P_2 and N_4 molecules. Similar calculations show that As_4 and Sb_4 are more likely than As_2 and Sb_2.

P_2 molecules would be likely to convert to P_4 molecules:

$$2P_2(g) \rightarrow P_4(g)$$
$$\Delta H^{\ominus} = (-1206 + 976)\ \text{kJ mol}^{-1} = -230\ \text{kJ mol}^{-1}$$

N_4 molecules would be even more likely to convert to N_2 molecules:

$$N_4(g) \rightarrow 2N_2(g)$$
$$\Delta H^{\ominus} = (-1890 + 978)\ \text{kJ mol}^{-1} = -912\ \text{kJ mol}^{-1}$$

EXERCISE 4.25

a i) $S_8(g) \rightarrow 4S_2(g)$

Bonds broken: $8\bar{E}(S{-}S) = (8 \times 264)$ kJ mol^{-1} = 2112 kJ mol^{-1}

Bonds made: $4\bar{E}(S{=}S) = (4 \times 431)$ kJ mol^{-1} = 1724 kJ mol^{-1}

$\therefore \Delta H^{\ominus} = (+2112 - 1724)$ kJ mol^{-1} = **$+388$ kJ mol^{-1}**

ii) $O_8(g) \rightarrow 4O_2(g)$

Bonds broken: $8\bar{E}(O{-}O) = (8 \times 142)$ kJ mol^{-1} = 1136 kJ mol^{-1}

Bonds made: $4\bar{E}(O{=}O) = (4 \times 498)$ kJ mol^{-1} = 1992 kJ mol^{-1}

$\therefore \Delta H^{\ominus} = (+1136 - 1992)$ kJ mol^{-1} = **-856 kJ mol^{-1}**

b The positive $\Delta H^{\ominus}$ value for the change in **a** i) suggests that the single bonds in the S_8 molecule are not likely to break in order to form S=S double bonds. The high negative $\Delta H^{\ominus}$ value for the change in **a** ii) suggests that single bonds in a hypothetical O_8 molecule would be very likely to break in order to form O=O double bonds. Thus, O_2 molecules are energetically favourable with respect to O_8 molecules, and S_8 molecules are energetically favourable with respect to S_2 molecules.

EXERCISE 4.26

a

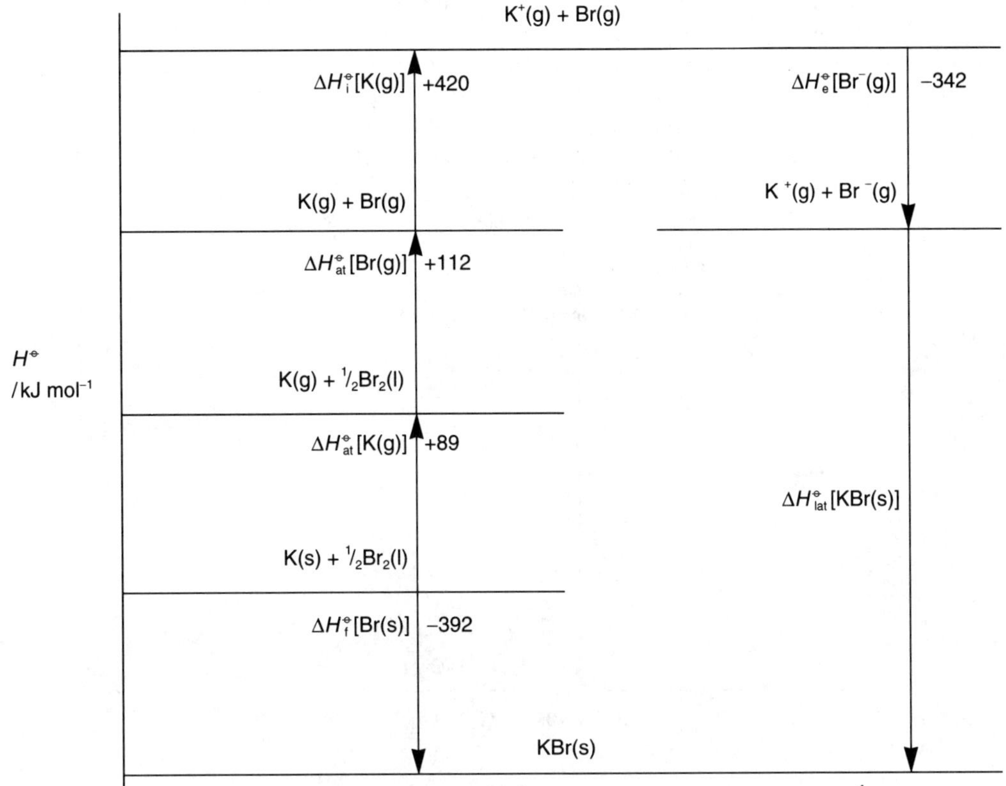

$\Delta H^{\ominus}_{lat}$[KBr(s)] = – [392 + 89 + 112 + 420 – 342] kJ mol^{-1} = **–671 kJ mol^{-1}**

b

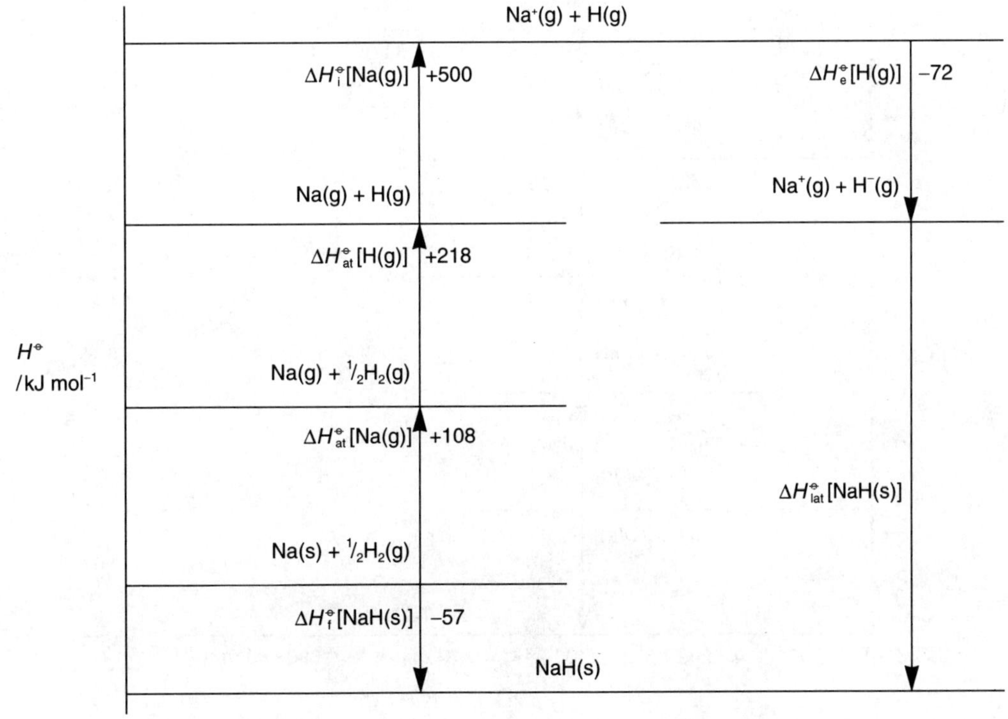

$\Delta H^{\ominus}_{lat}$[NaH(s)] = –[57 + 108 + 218 + 500 – 72] kJ mol^{-1} = **–811 kJ mol^{-1}**

EXERCISE 4.27

a

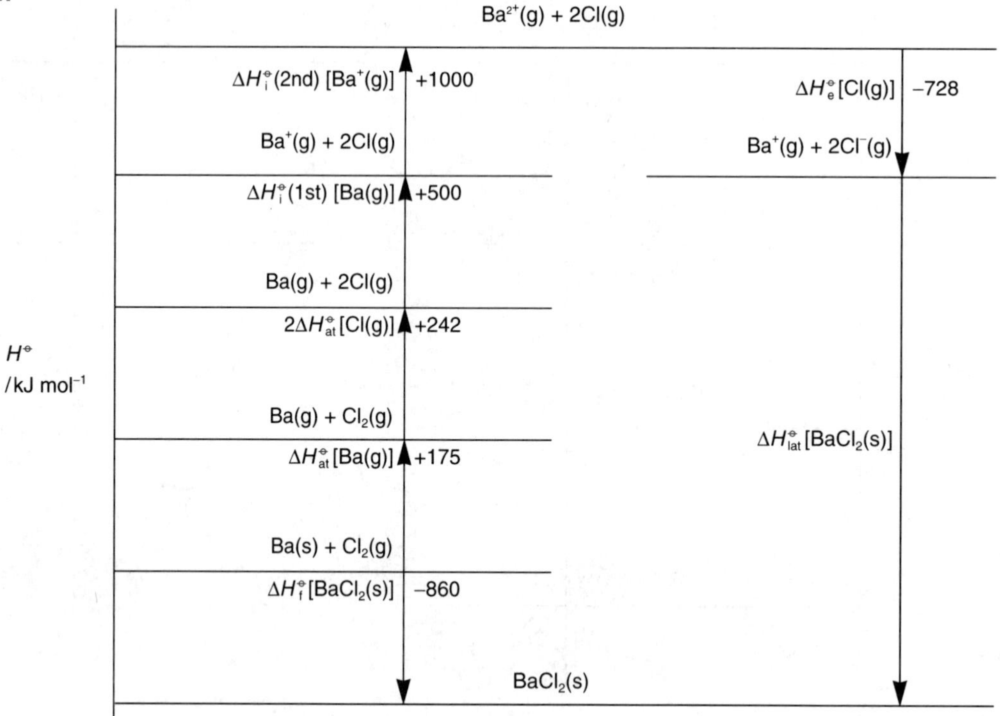

$\Delta H^{\ominus}_{lat}[BaCl_2(s)] = -[860 + 175 + 242 + 500 + 1000 - 728]\ \text{kJ mol}^{-1} = \mathbf{-2049\ kJ\ mol^{-1}}$

b

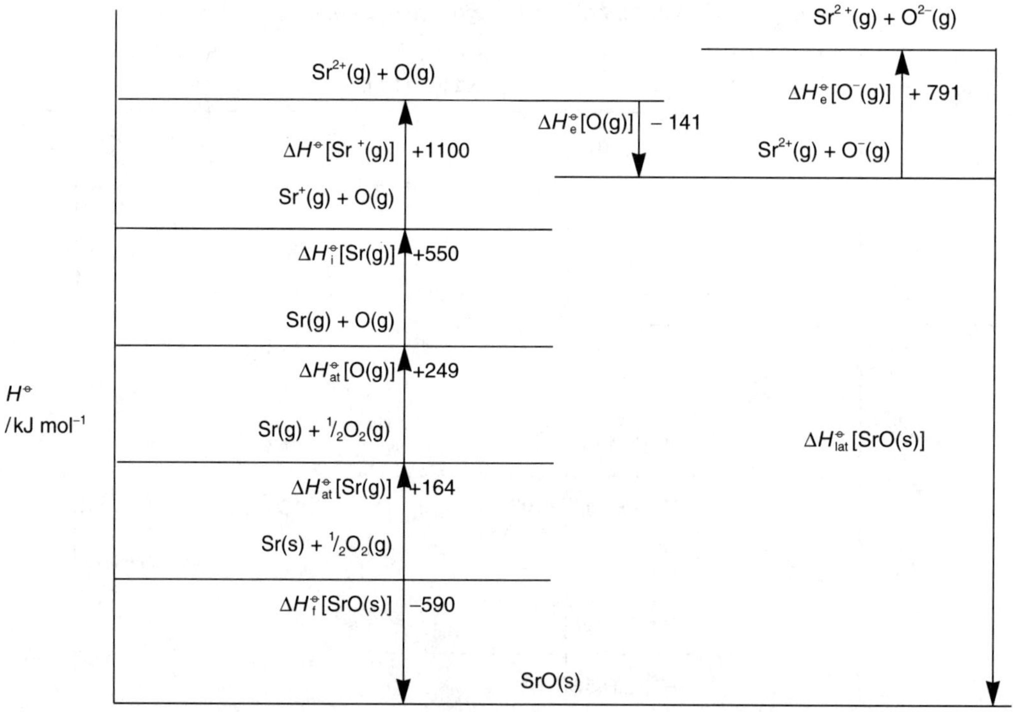

$\Delta H^{\ominus}_{lat}[SrO(s)] = -[590 + 164 + 249 + 550 + 1100 - 141 + 791]\ \text{kJ mol}^{-1} = \mathbf{-3303\ kJ\ mol^{-1}}$

EXERCISE 4.28

a

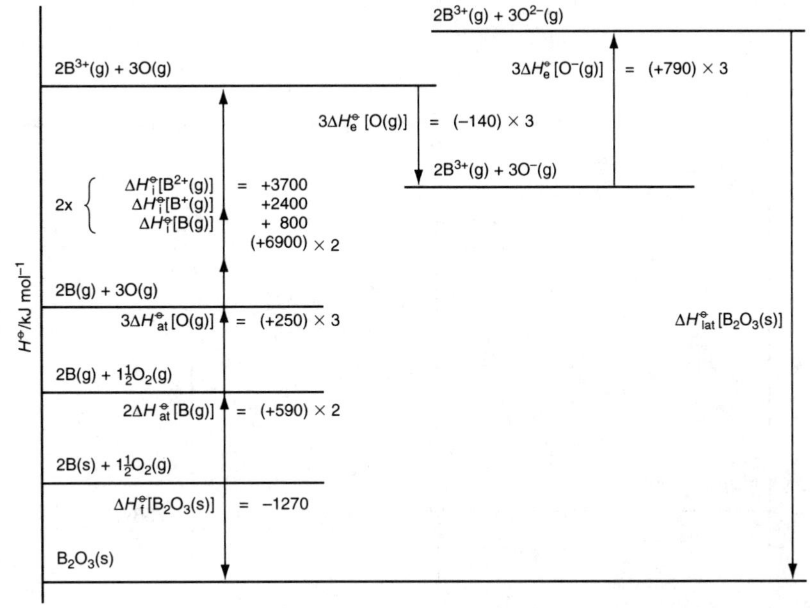

$$\Delta H_{lat}^\ominus[B_2O_3(s)] = -(1270 + 1180 + 750 + 13\,800 - 420 + 2370)\text{ kJ mol}^{-1}$$
$$= \mathbf{-18\,950\text{ kJ mol}^{-1}}$$

b There is poor agreement because the theoretical derivation assumes that the lattice consists of spherical ions. B^{3+} is so small and highly charged that it causes extensive distortion of O^{2-} ions leading to a considerable degree of covalent bonding for which the calculation is, at best, an approximation.

EXERCISE 4.29

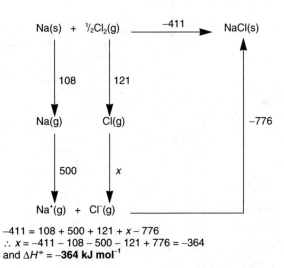

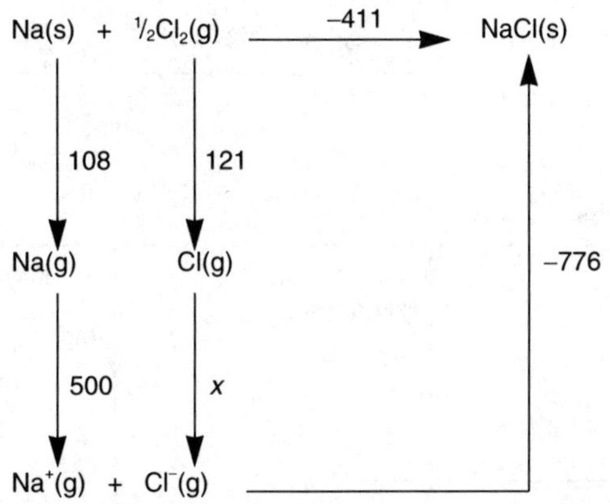

$-411 = 108 + 500 + 121 + x - 776$

$\therefore x = -411 - 108 - 500 - 121 + 776 = -364$

and $\Delta H^\ominus$ = **−364 kJ mol^{-1}**

EXERCISE 4.30

a i) B is the sum of the first and second ionisation energies of calcium.

ii) C is twice the enthalpy change of atomisation of chlorine.

iii) E is the lattice energy of $CaCl_2(s)$.

iv) F is the enthalpy change of formation of $CaCl_2(s)$.

b F = A + B + C + D + E

$\therefore$ D = F − A − B − C − E = (−795 − 177 − 1690 − 242 + 2197) kJ mol^{-1}

= **−707 kJ mol^{-1}**

(This is twice the electron affinity of chlorine.)

EXERCISE 4.31

a

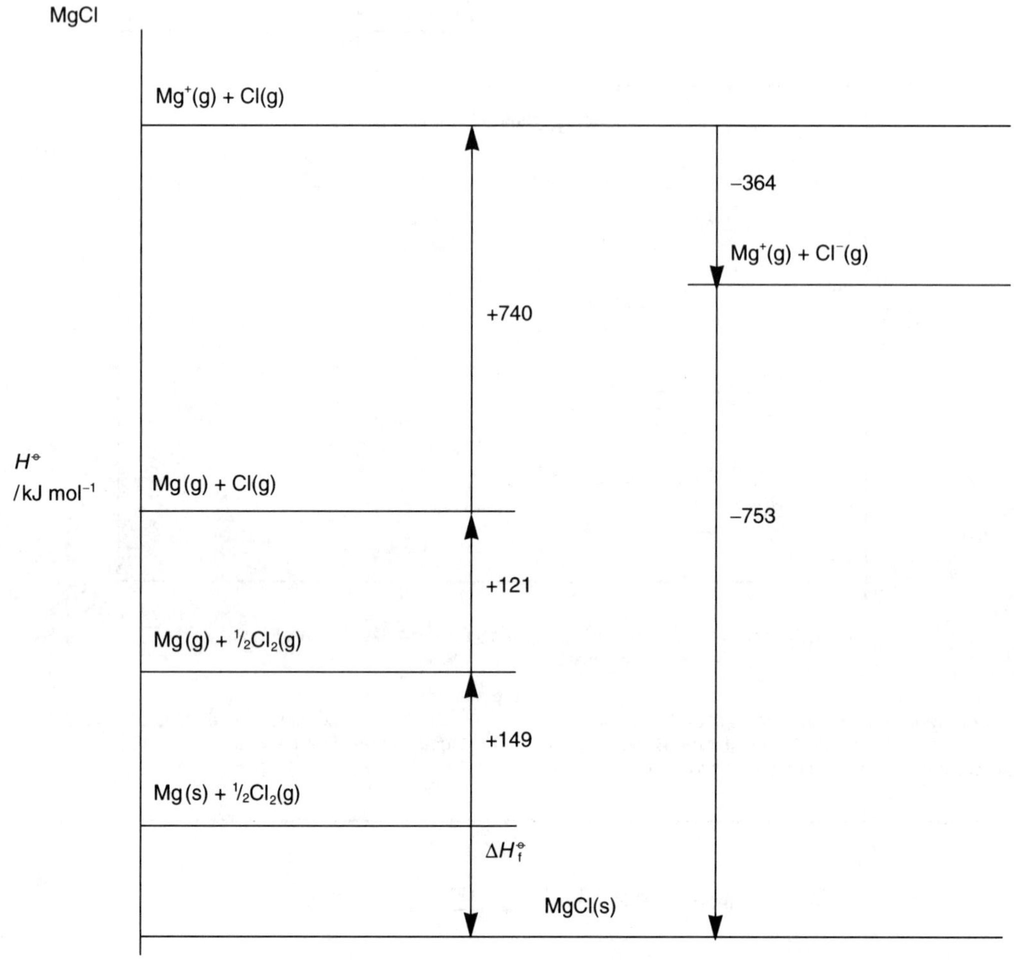

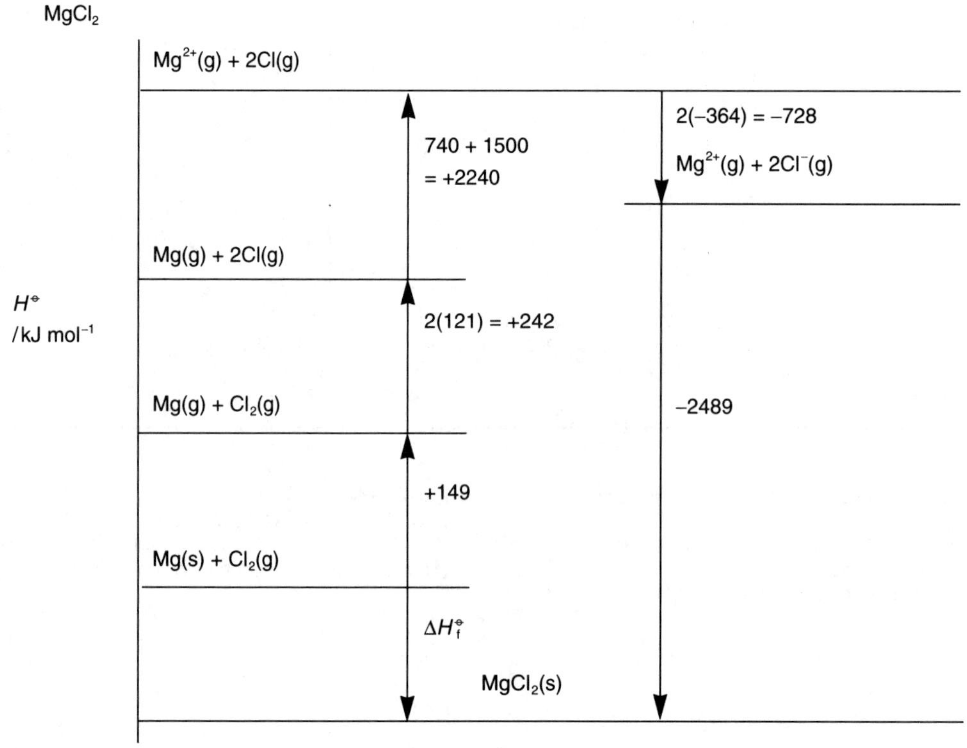

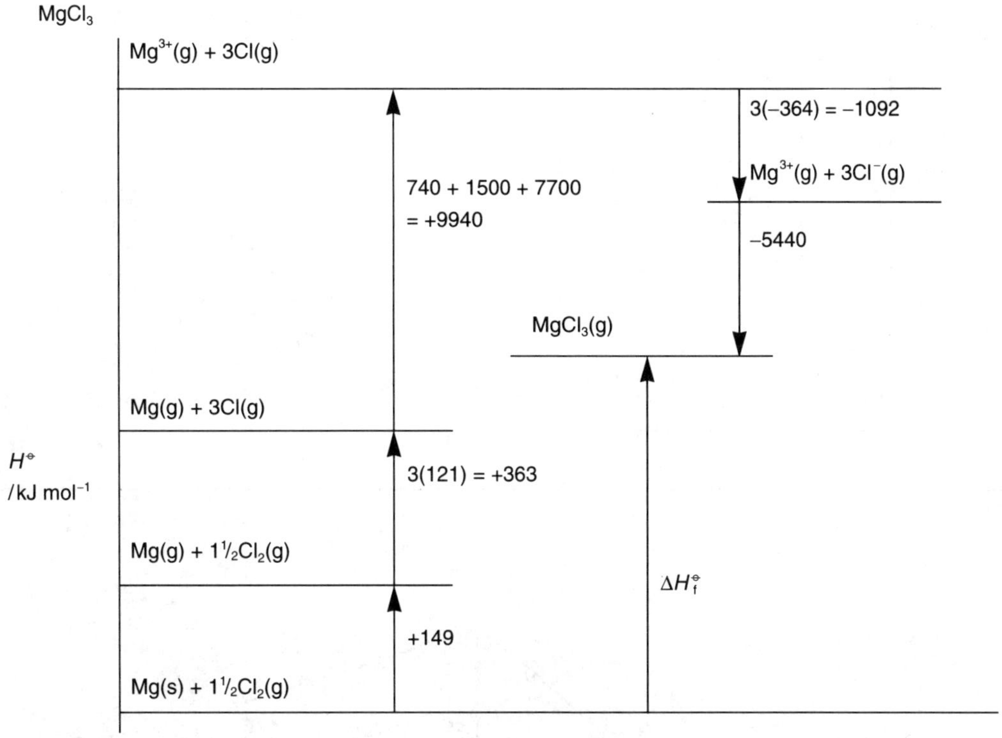

b $\Delta H^{\ominus}_{f}$[MgCl] = (149 + 121 + 740 – 364 – 753) kJ mol^{-1}
= **–107 kJ mol^{-1}**.
$\Delta H^{\ominus}_{f}$[$MgCl_2$] = (149 + 242 + 2240 – 728 – 2489) kJ mol^{-1}
= **–586 kJ mol^{-1}**.
$\Delta H^{\ominus}_{f}$[$MgCl_3$] = (149 + 363 + 9940 – 1092 – 5440) kJ mol^{-1}
= **+3920 kJ mol^{-1}**.

c MgCl and $MgCl_2$ are both energetically stable with respect to the elements, but $MgCl_2$ is more stable than MgCl.

d 2MgCl(s) → $MgCl_2$(s) + Mg(s)

	2MgCl(s)	$MgCl_2$(s)	Mg(s)
$\Delta H^{\ominus}_{f}$/kJ mol^{-1}	–214	–586	0

$\Delta H^{\ominus}$ = $\Sigma\Delta H^{\ominus}_{f}$[products] – $\Sigma\Delta H^{\ominus}_{f}$[reactants]
= (–586 + 214) kJ mol^{-1} = **–372 kJ mol^{-1}**

e MgCl is unstable relative to $MgCl_2$ and Mg. This explains why MgCl is unknown; as soon as it forms, it would be converted into $MgCl_2$ and Mg.

EXERCISE 4.32

a NaCl(s) → Na^+(g) + Cl^-(g); $\Delta H^{\ominus} = -\Delta H^{\ominus}_{lat}$ = +780 kJ mol^{-1}.

b Na^+(g) + aq → Na^+(aq); $\Delta H^{\ominus} = \Delta H^{\ominus}_{hyd}(Na^+)$ = –406 kJ mol^{-1}.
Cl^-(g) + aq → Cl^-(aq); $\Delta H^{\ominus} = \Delta H^{\ominus}_{hyd}(Cl^-)$= –364 kJ mol^{-1}.

c NaCl(s) → Na^+(g) + Cl^-(g); $\Delta H^{\ominus}$ = +780 kJ mol^{-1}.
Na^+(g) + aq → Na^+(aq); $\Delta H^{\ominus}$ = –406 kJ mol^{-1}.
Cl^-(g) + aq → Cl^-(aq); $\Delta H^{\ominus}$ = –364 kJ mol^{-1}.

Overall: NaCl(s) + aq → Na^+(aq) + Cl^-(aq);
$\Delta H^{\ominus}$ = (+780 – 406 – 364) kJ mol^{-1} = **+10 kJ mol^{-1}**.
The overall process is the dissolving of sodium chloride and the enthalpy change is $\Delta H_{solution}$.

EXERCISE 4.33

a *A* is the lattice enthalpy of calcium chloride.
B is the enthalpy of solution of calcium chloride.
C is the enthalpy of hydration of calcium chloride.

b $B = C - A = \Delta H^{\ominus}_{hyd}(Ca^{2+}) + 2\Delta H^{\ominus}_{hyd}(Cl^-) - \Delta H^{\ominus}_{lat}(CaCl_2)$
= {–1561 – 2(384) – (–2197)} kJ mol^{-1}
= **–132 kJ mol^{-1}**

EXERCISE 4.34

a

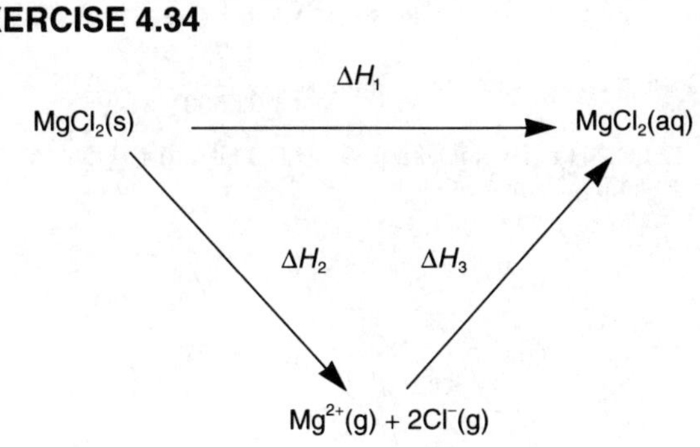

$\Delta H_1 = \Delta H_2 + \Delta H_3$
i.e. $\Delta H_{sol} = -\Delta H_{lat} + \Delta H_{hyd}$
= +2526 kJ mol^{-1} + –1920 kJ mol^{-1} + –2(364) kJ mol^{-1}
= **–122 kJ mol^{-1}**

b

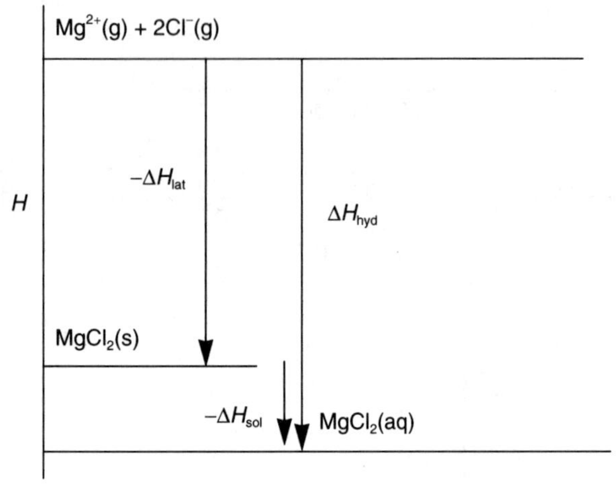

c Since ΔH_{sol} is negative and fairly large, anhydrous magnesium chloride would be expected to be soluble in water.

EXERCISE 4.35

a and **b**

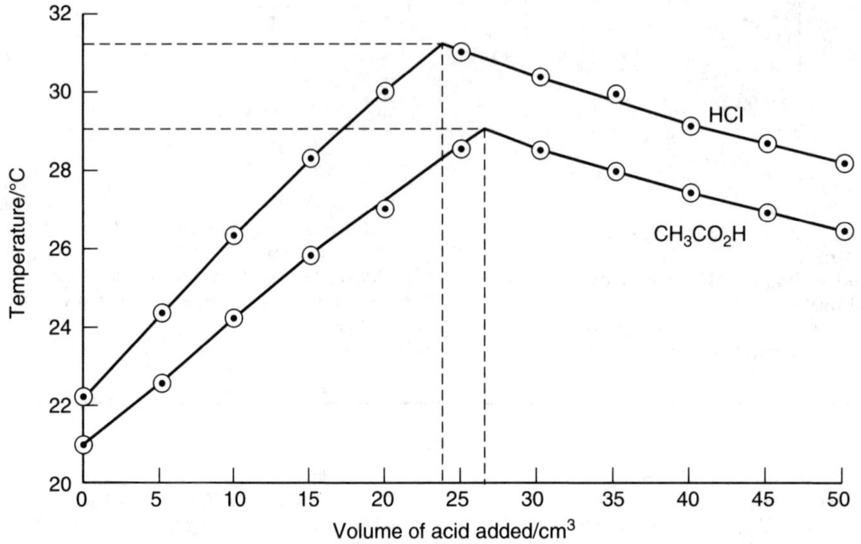

c i) $HCl(aq) + NaOH(aq) \rightarrow NaCl(aq) + H_2O(l)$
Amount of HCl = amount of NaOH

$$c \times \frac{23.5}{1000}\ dm^3 = 1.00\ mol\ dm^{-3} \times \frac{50.0}{1000} dm^3$$

$$\therefore c = 1.00\ mol\ dm^{-3} \times \frac{50.0}{23.5} = \mathbf{2.13\ mol\ dm^{-3}}$$

ii) $CH_3CO_2H(aq) + NaOH(aq) \rightarrow CH_3CO_2Na(aq) + H_2O(l)$
Amount of CH_3CO_2H = amount of NaOH

$$c \times \frac{26.5}{1000}\ dm^3 = 1.00\ mol\ dm^{-3} \times \frac{50.0}{1000} dm^3$$

$$\therefore c = 1.00\ mol\ dm^{-3} \times \frac{50.0}{26.5} = \mathbf{1.89\ mol\ dm^{-3}}$$

d i) Volume of mixture when reaction is complete = (50.0 + 23.5) cm^3 = 73.5 cm^3
ΔT = (31.3 – 22.2) K = 9.1 K

$\Delta H = -mc_p\Delta T = -0.0735\ kg \times 4.18\ kJ\ kg^{-1}\ K^{-1} \times 9.1\ K = -2.80\ kJ$
Amount of NaOH used = cV = 1.00 mol dm^{-3} × 0.0500 dm^3
= 0.0500 mol

$$\text{Scaling up to 1 mol, } \Delta H = -2.80\ kJ \times \frac{1}{0.0500} = -56.0\ kJ$$

$\therefore HCl(aq) + NaOH(aq) \rightarrow NaCl(aq) + H_2O(l)$; $\Delta H^{\ominus}$
= **–56.0 kJ mol^{-1}**

ii) Volume of mixture when reaction is complete = (50.0 + 26.5) cm^3
= 76.5 cm^3
ΔT = (29.1 – 21.0) K = 8.1 K
$\Delta H = -mc_p\Delta T = -0.0765\ kg \times 4.18\ kJ\ kg^{-1}\ K^{-1} \times 8.1\ K = -2.59\ kJ$

$$\text{Scaling up to 1 mol, } \Delta H = -2.59\ kJ \times \frac{1}{0.0500} = -51.8\ kJ$$

$\therefore CH_3CO_2H(aq) + NaOH(aq) \rightarrow CH_3CO_2Na(aq) + H_2O(l)$; $\Delta H^{\ominus}$
= **–51.8 kJ mol^{-1}**

EXERCISE 4.36

a $CH_3OH(l) + 1\frac{1}{2}O_2(g) \rightarrow CO_2(g) + 2H_2O(l)$

$\Delta G_f^\ominus$/kJ mol^{-1} −166.7 0 −394.4 −2(237.2)

$\Delta G^\ominus$ =(−394.4 − 474.4 −(−166.7)) kJ mol^{-1} = **−702.1 kJ mol^{-1}**

($\Delta H^\ominus$ = −726.4 kJ mol^{-1})

b $2CO(g) + O_2(g) \rightarrow 2CO_2(g)$

$\Delta G_f^\ominus$/kJ mol^{-1} −2(137.3) 0 −2(394.4)

$\Delta G^\ominus$ = (−788.8 −(−274.6)) kJ mol^{-1} = **−514.2 kJ mol^{-1}**

($\Delta H^\ominus$ = −566.0 kJ mol^{-1})

c $ZnCO_3(s) \rightarrow ZnO(s) + CO_2(g)$

$\Delta\Delta G_f^\ominus$/kJ mol^{-1} −731.4 −318.2 −394.4

$\Delta G_f^\ominus$ = (−318.2 − 394.4 −(−731.4)) kJ mol^{-1} = **+18.8 kJ mol^{-1}**

($\Delta H^\ominus$ = +71.0 kJ mol^{-1})

d $2Al(s) + Fe_2O_3(s) \rightarrow 2Fe(s) + Al_2O_3(s)$

$\Delta G_f^\ominus$/kJ mol^{-1} 0 −741.0 0 −1582.4

$\Delta G^\ominus$ = (−1582.4 −(−741.0)) kJ mol^{-1} = **−841.4 kJ mol^{-1}**

($\Delta H^\ominus$ = −853.5 kJ mol^{-1})

In these examples (and many others) the values of $\Delta H^\ominus$ and $\Delta G^\ominus$ are not greatly different and the use of $\Delta H^\ominus$ values to predict the feasibility of reactions would give the correct result.

Note, however, that this only applies to standard conditions, especially T = 298 K. ZnO reacts with CO_2 at 298 K, whereas $ZnCO_3$ decomposes readily at higher temperatures. You will see how $\Delta G^\ominus$ varies with temperature later.

EXERCISE 4.37

a $H_2(g) + C_2H_4(g) \rightarrow C_2H_6(g)$

$S^\ominus$/J K^{-1} mol^{-1} 130.6 219.5 229.5

$\Delta S^\ominus$ = (229.5 − (130.6 + 219.5)) J K^{-1} mol^{-1} = **−120.6 J K^{-1} mol^{-1}**

b $N_2(g) + 3H_2(g) \rightarrow 2NH_3(g)$

$S^\ominus$/J K^{-1} mol^{-1} 191.4 3 × 130.6 2 × 192.5

$\Delta S^\ominus$ = (385.0 − (191.4 + 391.8)) J K^{-1} mol^{-1} = **−198.2 J K^{-1} mol^{-1}**

c $2NaNO_3(s) \rightarrow 2NaNO_2(s) + O_2(g)$

$S^\ominus$/J K^{-1} mol^{-1} 2 × 116.3 2 × 120.5 204.9

$\Delta S^\ominus$ = (204.9 + 241.0 − 232.6) J K^{-1} mol^{-1} = **+213.3 J K^{-1} mol^{-1}**

EXERCISE 4.38

a Negative. $\Delta S^\ominus$ = (289 − (198 + 223)) J K^{-1} mol^{-1}
= **−132 J K^{-1} mol^{-1}**

b Positive. $\Delta S^\ominus$ = ((2 × 70.0) + 204.9 − (2 × 102)) J K^{-1} mol^{-1}
= **+140.9 J K^{-1} mol^{-1}**

c Negative. $\Delta S^\ominus$ = ((2 × 213.6 + 70.0) − 219.5 − (3 × 204.9)) J K^{-1} mol^{-1}
= **−337.0 J K^{-1} mol^{-1}**

d Positive. $\Delta S^\ominus$ = ((2 × 82.7) + (3 × 204.9)) − (2 × 143.0)) J K^{-1} mol^{-1}
= **+494.1 J K^{-1} mol^{-1}**

e Positive. $\Delta S^\ominus$ = ((2 × 197.9) − (2 × 5.7) −204.9) J K^{-1} mol^{-1}
= **+179.5 J K^{-1} mol^{-1}**

EXERCISE 4.39

a $2NO(g) + O_2(g) \rightarrow N_2O_4(g)$

$\Delta H_f^\ominus$/kJ mol^{-1} 2 × 90.4 0 9.2

$S^\ominus$/J K^{-1} mol^{-1} 2 × 210.5 204.9 304.2

$\Delta H^\ominus$ = (9.2 − 180.8 − 0) kJ mol^{-1} = −171.6 kJ mol^{-1}

$\Delta S^\ominus$ = (304.2 − 421 − 204.9) J K^{-1} mol^{-1} = −321.7 J K^{-1} mol^{-1}
= −0.322 kJ K^{-1} mol^{-1}

$\Delta G^\ominus = \Delta H^\ominus - T\Delta S^\ominus$ = −171.6 kJ mol^{-1} −(298 K × (−0.322 kJ K^{-1} mol^{-1}))
= **−75.6 kJ mol^{-1}**

b $NH_3(g) + HCl(g) \rightarrow NH_4Cl(s)$

$\Delta H_f^\ominus$/kJ mol^{-1} −46.0 −92.3 −315.5

$S^\ominus$/J K^{-1} mol^{-1} 192.5 186.7 94.6

$\Delta H^\ominus$ = (−315.5 −(−46.0 − 92.3)) kJ mol^{-1} = −177.2 kJ mol^{-1}

$\Delta S^\ominus$ = (94.6 − 186.7 − 192.5) J K^{-1} mol^{-1} = −284.6 J K^{-1} mol^{-1}
= −0.285 kJ K^{-1} mol^{-1}

$\Delta G^\ominus = \Delta H^\ominus - T\Delta S^\ominus$ = −177.2 kJ mol^{-1} −(298 K × (−0.285 kJ K^{-1}mol^{-1}))
= **−92.3 kJmol^{-1}**

c $H_2O(l) \rightarrow H_2O(g)$

$\Delta H_f^\ominus$/kJ mol^{-1} −285.9 −241.8

$S^\ominus$/J K^{-1} mol^{-1} 70.0 188.7

$\Delta H^\ominus$ = (−241.8 −(−285.9)) kJ mol^{-1} = +44.1 kJ mol^{-1}

$\Delta S^\ominus$ = (188.7 − 70.0) J K^{-1} mol^{-1} = 118.7 J K^{-1} mol^{-1}
= + 0.119 kJ K^{-1} mol^{-1}

$\Delta G^\ominus = \Delta H^\ominus - T\Delta S^\ominus$ = 44.1 kJ mol^{-1} − (298 K × 0.119 kJ K^{-1} mol^{-1})
= **+8.6 kJ mol^{-1}**

EXERCISE 4.40

a As in Exercise 4.39, $\Delta H^\ominus$ = −171.6 kJ mol^{-1} and
$\Delta S^\ominus$ = −0.322 kJ K^{-1} mol^{-1}

$\Delta G^\ominus = \Delta H^\ominus - T\Delta S^\ominus$
= −171.6 kJ mol^{-1} − (1000 K × (−0.322 kJ K^{-1} mol^{-1}))
= (−171.6 + 322) kJ mol^{-1} = **+150 kJ mol^{-1}**

This reaction is feasible at 298 K ($\Delta G^\ominus$ negative) but not at 1000 K ($\Delta G^\ominus$ positive).

b As in Exercise 4.39, $\Delta H^\ominus$ = −177.2 kJ mol^{-1} and $\Delta S^\ominus$
= −0.285 kJ K^{-1} mol^{-1}

$\Delta G^\ominus = \Delta H^\ominus - T\Delta S^\ominus$
= −177.2 kJ mol^{-1} −(1000 K × (−0.285 kJ K^{-1} mol^{-1}))
= (−177.2 + 285) kJ mol^{-1} = **+108 kJ mol^{-1}**

This reaction is feasible at 298 K ($\Delta G^\ominus$ negative) but not at 1000 K.

c As in Exercise 4.39, $\Delta H^\ominus$ = +44.1 kJ mol^{-1} and $\Delta S^\ominus$
= +0.119 kJ K^{-1} mol^{-1}

$\Delta G^\ominus = \Delta H^\ominus - T\Delta S^\ominus$ = +44.1 kJ mol^{-1} −(1000 K × 0.119 kJ K^{-1} mol^{-1})
= (44.1 − 119) kJ mol^{-1} = **−75 kJ mol^{-1}**

At 298 K water vaporises only to a small extent ($\Delta G^\ominus$ small and positive) but at 1000 K vaporisation is complete ($\Delta G^\ominus$ substantially negative).

d $CaCO_3(s) \rightarrow CaO(s) + CO_2(g)$

$\Delta H_f^\ominus$/kJ mol^{-1} −1206.9 −635.5 −393.5

$S^\ominus$/J K^{-1} mol^{-1} 92.9 39.7 213.6

$\Delta H^\ominus$ = (−635.5 − 393.5 + 1206.9) kJ mol^{-1} = 177.9 kJ mol^{-1}

$\Delta S^\ominus$ = (213.6 + 39.7 −92.9) J K^{-1} mol^{-1} = 160.4 J K^{-1} mol^{-1}

$\Delta G^\ominus = \Delta H^\ominus - T\Delta S^\ominus$
= 177.9 kJ mol^{-1} −(1000 K × 0.1604 kJ K^{-1} mol^{-1})
= (177.9 − 160.4) kJ mol^{-1} = **+17.5 kJ mol^{-1}**

This reaction is not feasible at 298 K ($\Delta G^\ominus$ substantially positive) but can proceed to an equilibrium mixture of reactants and products at 1000 K ($\Delta G^\ominus$ small and positive).

End-of-chapter questions

4.1 −42 kJ mol^{-1}
4.2 −484 kJ mol^{-1}
4.3 −1031 kJ mol^{-1}
4.4 **a** $C_4H_{10}(g) + 6\frac{1}{2}O_2(g) \rightarrow 4CO_2(g) + 5H_2O(l)$
b 357 g
4.5 −487 kJ mol^{-1}
4.6 **a** $\Delta H^\circ = +109.5$ kJ mol^{-1} $\Delta G^\circ = -11.2$ kJ mol^{-1}
b $\Delta H^\circ = +52.3$ kJ mol^{-1} $\Delta G^\circ = -16.0$ kJ mol^{-1}
4.7 +412 kJ mol^{-1}
4.8 +391 kJ mol^{-1}
4.9 **a** $C(s) + O_2(g) \rightarrow CO_2(g)$
$C_2H_5OH(l) + 3O_2(g) \rightarrow 2CO_2(g) + 3H_2O(l)$
$H_2(g) + \frac{1}{2}O_2(g) \rightarrow H_2O(l)$
b i) −394 kJ mol^{-1}
ii) −1368 kJ mol^{-1}
c −283 kJ mol^{-1}
d i) coke 32.8 kJ g^{-1}; ethanol 29.7 kJ g^{-1}; hydrogen 141.5 kJ g^{-1}

4.10 **b** −3325(±10) kJ mol^{-1}
c −279 kJ mol^{-1}
d −1830(±5) kJ mol^{-1}
4.11 **a** +376 kJ mol^{-1}
b −411 kJ mol^{-1}
c $\Delta H_f^\circ(CaCl_3) = +1371$ kJ mol^{-1} (endothermic, ∴ energetically unstable)
$\Delta H_f^\circ(CaCl_2) = -762$ kJ mol^{-1} (exothermic, ∴ energetically stable)
4.12 −57 kJ mol^{-1}
4.13 **a** 2884 J
b $\Delta H^\circ_{neut} = -57.7$ kJ mol^{-1}
4.14 **a** $\Delta S^\circ = -142$ kJ mol^{-1}; $\Delta G^\circ = -70$ kJ mol^{-1}
b i) Yes (ΔG° negative)
ii) No (ΔG° positive above 789 K)

CHAPTER 5

EXERCISE 5.1

a $p_1V_1 = p_2V_2$ (Boyle's Law)

$$\therefore V_2 = V_1 \times \frac{p_1}{p_2}$$

$$= 25.0\ \text{dm}^3 \times \frac{150\ \text{atm}}{1.00\ \text{atm}} = \mathbf{3.75 \times 10^3\ dm^3}$$

b $\dfrac{V_1}{T_1} = \dfrac{V_2}{T_2}$ (Charles' Law)

$$\therefore T_2 = T_1 \times \frac{V_2}{V_1}$$

$$= 288\ \text{K} \times \frac{4.00 \times 10^3\ \text{dm}^3}{3.75 \times 10^3\ \text{dm}^3} = \mathbf{307\ K\ (or\ 34\ °C)}$$

EXERCISE 5.2

$$p_2 = p_1 \times \frac{V_1}{V_2} = 1.00\ \text{atm} \times \frac{100\ \text{cm}^3}{37.0\ \text{cm}^3} = \mathbf{2.70\ atm}$$

EXERCISE 5.3

$$V_2 = V_1 \times \frac{p_1}{p_2} = 73.0\ \text{cm}^3 \times \frac{1456\ \text{mmHg}}{760\ \text{mmHg}} = \mathbf{140\ cm^3}$$

EXERCISE 5.4

Combining Boyle's Law and Charles' Law gives:

$$V_2 = V_1 \times \frac{p_1}{p_2} \times \frac{T_2}{T_1}$$

$$= 38.2\ \text{cm}^3 \times \frac{765\ \text{mmHg}}{760\ \text{mmHg}} \times \frac{273\ \text{K}}{(273 + 18)\ \text{K}} = \mathbf{36.1\ cm^3}$$

EXERCISE 5.5

$$V_2 = V_1 \times \frac{p_1}{p_2} \times \frac{T_2}{T_1}$$

$$= 79.0\ \text{cm}^3 \times \frac{756\ \text{mm}}{760\ \text{mm}} \times \frac{273\ \text{K}}{294\ \text{K}} = \mathbf{73.0\ cm^3}$$

EXERCISE 5.6

In each case, $\frac{p_1V_1}{T_1} = \frac{p_2V_2}{T_2}$

or $V_2 = V_1 \times \frac{p_1}{p_2} \times \frac{T_2}{T_1}$

a $V_2 = 29.2\ \text{cm}^3 \times \frac{762}{760} \times \frac{273}{298} = \mathbf{26.8\ cm^3}$

b $V_2 = 5.13\ \text{dm}^3 \times \frac{1.02}{1.00} \times \frac{273}{335} = \mathbf{4.26\ dm^3}$

c $V_2 = 132\ \text{cm}^3 \times \frac{760}{758} \times \frac{298}{273} = \mathbf{144\ cm^3}$

d $V_2 = 42.1\ \text{cm}^3 \times \frac{752}{200} \times \frac{293}{304} = \mathbf{153\ cm^3}$

EXERCISE 5.7

$T_2 = T_1 \times \frac{p_2}{p_1} \times \frac{V_2}{V_1} = 300\ \text{K} \times \frac{2}{1} \times \frac{2}{1} = \mathbf{1.20 \times 10^3\ K}$

EXERCISE 5.8

$p_2 = p_1 \times \frac{V_1}{V_2} \times \frac{T_2}{T_1} = (28.5 + 14.7)\ \text{lb in}^{-2} \times \frac{21.0}{21.6} \times \frac{315}{288} = 45.9\ \text{lb in}^{-2}$

$\therefore$ gauge reads $(45.9 - 14.7)\ \text{lb in}^{-2} = \mathbf{31.2\ lb\ in^{-2}}$

EXERCISE 5.9

a

$2N_2O(g)$	$\rightarrow 2N_2(g)$	$+\ O_2(g)$
2 mol	2 mol	1 mol
2 volumes	2 volumes	1 volume (equal amounts occupy equal volumes)

$\therefore$ volume of N_2O = volume of N_2 = **15 cm³**
and volume of $O_2 = \frac{1}{2}$ volume of N_2 = **7.5 cm³**

b

$4NH_3(g)$	$+\ 3O_2(g)$	$\rightarrow 2N_2(g)$	$+\ 6H_2O(g)$
4 mol	3 mol	2 mol	+ 6 mol

$\therefore$ 4 volumes + 3 volumes $\rightarrow$ 8 volumes (since equal amounts occupy equal volumes)

or $\frac{4}{8}$ volumes + $\frac{3}{8}$ volumes $\rightarrow$ 1 volume

$\therefore$ volume of $NH_3 = \frac{4}{8} \times$ volume of products

$= \frac{4}{8} \times 128\ \text{cm}^3 = \mathbf{64.0\ cm^3}$

and volume of $O_2 = \frac{3}{8} \times 128\ \text{cm}^3 = \mathbf{48.0\ cm^3}$

EXERCISE 5.10

$H_2(g) + Cl_2(g) \rightarrow 2HCl(g)$
$\therefore$ volume of HCl = 2 × volume of H_2 = $2 \times 75\ \text{cm}^3 = \mathbf{150\ cm^3}$

EXERCISE 5.11

Volume of $NH_3 = 15\ \text{cm}^3 \times \frac{4}{5} = \mathbf{12\ cm^3}$

Volume of products $= 15\ \text{cm}^3 \times \frac{(4+6)}{5} = \mathbf{30\ cm^3}$

Percentage change $= \frac{\text{volume increase}}{\text{volume at start}} \times 100$

$= \frac{30\ \text{cm}^3 - 27\ \text{cm}^3}{27\ \text{cm}^3} \times 100 = \mathbf{11\%}$

EXERCISE 5.12

$C_2H_5OH(g) + 3O_2(g) \rightarrow 2CO_2(g) + 3H_2O(g)$
Volume of O_2 = 3 × volume of C_2H_5OH
$= 3 \times 78.5\ \text{cm}^3 = \mathbf{236\ cm^3}$

Volume of products $= (3 + 2) \times 78.5\ \text{cm}^3 = \mathbf{393\ cm^3}$

EXERCISE 5.13

Volume of oxygen used = starting volume − volume after combustion
$= 60\ \text{cm}^3 - 15\ \text{cm}^3$
$= 45\ \text{cm}^3$
Volume of carbon dioxide formed = volume absorbed by NaOH
$= 45\ \text{cm}^3 - 15\ \text{cm}^3$
$= 30\ \text{cm}^3$

Equation	$C_xH_y(g)$	$+\ (x+\frac{y}{4})O_2(g)$	$\rightarrow\ xCO_2(g) + \frac{y}{2}H_2O(l)$
Volume, cm³	15	45	30
$\frac{\text{Volume}}{15\ \text{cm}^3}$	1	3	2

$\therefore\ x = 2$ and $x + \frac{y}{4} = 3$

$\therefore\ y = 4$

$\therefore$ formula of hydrocarbon = $\mathbf{C_2H_4}$

EXERCISE 5.14

Completing and balancing equations for combustion of each hydrocarbon:

A $C_3H_8(g) + 5O_2(g) \rightarrow 3CO_2(g) + 4H_2O(g)$
B $C_3H_6(g) + 4.5O_2(g) \rightarrow 3CO_2(g) + 3H_2O(g)$
C $C_3H_7OH(g) + 4.5O_2(g) \rightarrow 3CO_2(g) + 4H_2O(g)$
D $C_2H_5CHO(g) + 4O_2(g) \rightarrow 3CO_2(g) + 3H_2O(g)$

10 cm^3 of X requires 45 cm^3 of O_2 for combustion.
$\therefore$ 1 cm^3 of X requires 4.5 cm^3 of O_2 for combustion.
$\therefore$ X could be **B or C**.

EXERCISE 5.15

Volume of CO_2 = 150 cm^3 − 70 cm^3 = 80 cm^3 = 4 × volume of C_xH_y
Volume of O_2 = 200 cm^3 − 70 cm^3 = 130 cm^3 = 6½ × volume of C_xH_y
$C_xH_y(g) + 6\frac{1}{2}O_2(g) \rightarrow 4CO_2(g) + zH_2O(l)$

C atoms: $x = 4$ O atoms: $13 = 8 + z\ \therefore\ z = 5$
H atoms: $y = 2z = 10\ \therefore$ formula is $\mathbf{C_4H_{10}}$

EXERCISE 5.16

Volume of H_2O = 90 cm^3 = 3 × volume of C_3H_x
$C_3H_x(g) + yO_2(g) \rightarrow zCO_2(g) + 3H_2O(l)$
C atoms: $3 = z$ O atoms: $2y = 2z + 3\ \therefore\ y = 4\frac{1}{2}$

Volume of O_2 = 4½ × volume of C_3H_x = 4½ × 30 cm^3 = **135 cm^3**
H atoms: $x = 3 \times 2 = 6$
$\therefore\ \mathbf{x = 6}$

EXERCISE 5.17

a The volume of 0.122 g of H_2 at s.t.p. is given by:

$$V_2 = V_1 \times \frac{p_1}{p_2} \times \frac{T_2}{T_1}$$

$$= 0.211\ \text{dm}^3 \times \frac{7.00\ \text{atm}}{1.00\ \text{atm}} \times \frac{273\ \text{K}}{(273 + 20)\ \text{K}} = 1.38\ \text{dm}^3$$

The amount of H_2 is given by:

$$n = \frac{m}{M} = \frac{0.122\ \text{g}}{2.02\ \text{g mol}^{-1}} = 0.0604\ \text{mol}$$

Also $n = \frac{V}{V_m}$

$$\therefore\ V_m = \frac{V}{n} = \frac{1.38\ \text{dm}^3}{0.0604\ \text{mol}} = \mathbf{22.8\ dm^3\ mol^{-1}}$$

b The volume of 1.10 g of C_4H_{10} at s.t.p. is given by:

$$V_2 = V_1 \times \frac{p_1}{p_2} \times \frac{T_2}{T_1}$$

$$= 34.4\ \text{dm}^3 \times \frac{30.0\ \text{mmHg}}{760\ \text{mmHg}} \times \frac{273\ \text{K}}{(273 + 600)\ \text{K}} = 0.425\ \text{dm}^3$$

The amount of C_4H_{10} is given by:

$$n = \frac{m}{M} = \frac{1.10\ \text{g}}{58.1\ \text{g mol}^{-1}} = 0.0189\ \text{mol}$$

Also, $n = \frac{V}{V_m}$

$$\therefore\ V_m = \frac{V}{n} = \frac{0.425\ \text{dm}^3}{0.0189\ \text{mol}} = \mathbf{22.5\ dm^3}$$

c The volume of 2.00 g of O_2 at s.t.p. is given by:

$$V_2 = V_1 \times \frac{p_1}{p_2} \times \frac{T_2}{T_1}$$

$$= 51.9\ \text{dm}^3 \times \frac{5.00\ \text{kPa}}{100\ \text{kPa}} \times \frac{273\ \text{K}}{500\ \text{K}} = 1.42\ \text{dm}^3$$

The amount of O_2 is given by:

$$n = \frac{m}{M} = \frac{2.00\ \text{g}}{32.0\ \text{g mol}^{-1}} = 0.0625\ \text{mol}$$

Also, $n = \frac{V}{V_m}$

$$\therefore\ V_m = \frac{V}{n} = \frac{1.42\ \text{dm}^3}{0.0625\ \text{mol}} = \mathbf{22.7\ dm^3}$$

EXERCISE 5.18

Let the volume of the gas at s.t.p. be V_2. Then:

$$V_2 = V_1 \times \frac{p_1}{p_2} \times \frac{T_2}{T_1}$$

$$= 10.0\ \text{dm}^3 \times \frac{350\ \text{mmHg}}{760\ \text{mmHg}} \times \frac{273\ \text{K}}{(273 + 27)\ \text{K}} = 4.19\ \text{dm}^3$$

$$\text{Then } n = \frac{V_2}{V_m} = \frac{4.19\ \text{dm}^3}{22.4\ \text{dm}^3\ \text{mol}^{-1}} = \mathbf{0.187\ mol}$$

EXERCISE 5.19

Let the molar volume at 23 °C and 0.200 atm be V_2. Then:

$$V_2 = V_1 \times \frac{p_1}{p_2} \times \frac{T_2}{T_1}$$

$$= 22.4\ \text{dm}^3\ \text{mol}^{-1} \times \frac{1.00\ \text{atm}}{0.200\ \text{atm}} \times \frac{(273 + 23)\ \text{K}}{273\ \text{K}} = 121\ \text{dm}^3\ \text{mol}^{-1}$$

The amount, n, of oxygen = $\frac{m}{M}$, but also, $n = \frac{V}{V_m}$

$$\therefore \frac{m}{M} = \frac{V}{V_m}, \text{ and}$$

$$V = \frac{m}{M} V_m = \frac{8.00\ \text{g}}{32.0\ \text{g mol}^{-1}} \times 121\ \text{dm}^3\ \text{mol}^{-1} = \mathbf{30.3\ dm^3}$$

EXERCISE 5.20

We have two molar volumes at different temperatures but constant pressure. Applying Charles' Law:

$$\frac{V_1}{T_1} = \frac{V_2}{T_2}$$

$$\therefore T_2 = T_1 \times \frac{V_2}{V_1} = 273\ \text{K} \times \frac{24.0\ \text{dm}^3\ \text{mol}^{-1}}{22.4\ \text{dm}^3\ \text{mol}^{-1}} = \mathbf{293\ K\ or\ 20\ °C}$$

EXERCISE 5.21

a $V = 86.1\ \text{cm}^3 \times \frac{1.25}{1.00} \times \frac{273}{290} = 101\ \text{cm}^3 = 0.101\ \text{dm}^3$

$$V_m = \frac{V}{n} = \frac{0.101\ \text{dm}^3}{0.0780\ \text{g}/17.0\ \text{g mol}^{-1}} = \mathbf{22.0\ dm^3\ mol^{-1}}$$

b $V = 90.2\ \text{cm}^3 \times \frac{298}{273} = 98.5\ \text{cm}^3 = 0.0985\ \text{dm}^3$

$$V_m = \frac{V}{n} = \frac{0.0985\ \text{dm}^3}{0.0591\ \text{g}/16.0\ \text{g mol}^{-1}} = \mathbf{26.7\ dm^3\ mol^{-1}}$$

EXERCISE 5.22

$pV = nRT$

$$\therefore R = \frac{pV}{nT} = \frac{1.00\ \text{atm} \times 22.4\ \text{dm}^3}{1.00\ \text{mol} \times 273\ \text{K}} = \mathbf{0.0821\ atm\ dm^3\ K^{-1}\ mol^{-1}}$$

EXERCISE 5.23

a $R = 0.0821\ \text{atm dm}^3\ \text{K}^{-1}\ \text{mol}^{-1}$
$= 0.0821\ (760\ \text{mmHg})\ \text{dm}^3\ \text{K}^{-1}\ \text{mol}^{-1}$
$= \mathbf{62.4\ dm^3\ mmHg\ K^{-1}\ mol^{-1}}$

b The value of p to use in the calculation is found by making successive substitutions as follows:
$p = 1.00\ \text{atm} = 100\ \text{kN m}^{-2} = 100\ (10^3\ \text{N})\ \text{m}^{-2} = 1.00 \times 10^5\ \text{N m}^{-2}$
$= 1.00 \times 10^5\ (\text{N m})\ \text{m}^{-3} = 1.00 \times 10^5\ \text{J m}^{-3}$
$= 1.00 \times 10^5\ \text{J}\ (10\ \text{dm})^{-3} = 1.00 \times 10^2\ \text{J dm}^{-3}$

Now $pV = nRT$

$$\therefore R = \frac{pV}{nT} = \frac{1.00 \times 10^2\ \text{J dm}^{-3} \times 22.4\ \text{dm}^3}{1.00\ \text{mol} \times 273\ \text{K}} = \mathbf{8.21\ J\ mol^{-1}\ K^{-1}}$$

EXERCISE 5.24

a $pV = nRT$

$$\therefore V = \frac{nRT}{p} = \frac{0.500\ \text{mol} \times 0.0821\ \text{atm dm}^3\ \text{K}^{-1}\ \text{mol}^{-1} \times 273\ \text{K}}{1.00\ \text{atm}}$$

$$= \mathbf{11.2\ dm^3}$$

b $pV = nRT$ or $V = \frac{nRT}{p}$

where n, the amount of H_2, $= \frac{m}{M} = \frac{1.50\ \text{g}}{2.02\ \text{g mol}^{-1}} = 0.743\ \text{mol}$

and $p = 750\ \text{mmHg} = \frac{750}{760}\ \text{atm} = 0.987\ \text{atm}$

$$\therefore V = \frac{0.743\ \text{mol} \times 0.0821\ \text{atm dm}^3\ \text{K}^{-1}\ \text{mol}^{-1} \times 288\ \text{K}}{0.987\ \text{atm}} = \mathbf{17.8\ dm^3}$$

c $pV = nRT$ or $T = \frac{pV}{nR}$

where $p = 760\ \text{mmHg} = 1.00\ \text{atm}$ and

$$n = \frac{4.71\ \text{g}}{28.0\ \text{g mol}^{-1}} = 0.168\ \text{mol}$$

$$\therefore T = \frac{pV}{nR} = \frac{1.00\ \text{atm} \times 12.0\ \text{dm}^3}{0.168\ \text{mol} \times 0.0821\ \text{atm dm}^3\ \text{K}^{-1}\ \text{mol}^{-1}}$$

$$= \mathbf{870\ K\ or\ 597\ °C}$$

EXERCISE 5.25

$Zn(s) + 2H^+(aq) \rightarrow Zn^{2+}(aq) + H_2(g)$

$\therefore$ the amount of Zn = the amount of H_2

The amount, n, of H_2 is given by:

$pV = nRT$ in the form $n = \frac{pV}{RT}$

where $p = \frac{755}{760}$ atm = 0.993 atm

$$\therefore n = \frac{0.993 \text{ atm} \times 2.00 \text{ dm}^3}{0.0821 \text{ atm dm}^3 \text{ K}^{-1} \text{ mol}^{-1} \times 288 \text{ K}} = 0.0840 \text{ mol}$$

mass of zinc = amount of zinc × molar mass
= 0.0840 mol × 65.4 g mol^{-1} = **5.49 g**

EXERCISE 5.26

a $C_7H_5N_3O_6(s) + 5\frac{1}{4}O_2(g) = 7CO_2(g) + 2\frac{1}{2}H_2O(g) + 1\frac{1}{2}N_2(g)$
or $4C_7H_5N_3O_6(s) + 21O_2(g) = 28CO_2(g) + 10H_2O(g) + 6N_2(g)$

b From the equation, one mole of TNT gives $7 + 2\frac{1}{2} + 1\frac{1}{2} = 11$ mol of gas

$pV = nRT$ or $V = \frac{nRT}{p}$

$$\therefore V = \frac{11 \text{ mol} \times 0.0821 \text{ atm dm}^3 \text{ mol}^{-1} \text{ K}^{-1} \times 673 \text{ K}}{1.00 \text{ atm}} = \mathbf{608 \text{ dm}^3}$$

c Ratio of volumes = $\frac{608 \text{ dm}^3}{0.500 \text{ dm}^3} = \frac{1216}{1}$

As a percentage this is:
1216 × 100 = **1.22 × 10^5%**

d $pV = nRT$ or $p = \frac{nRT}{V}$

$$\therefore p = \frac{11.0 \text{ mol} \times 0.0821 \text{ atm dm}^3 \text{ mol}^{-1} \text{ K}^{-1} \times 873 \text{ K}}{2.00 \text{ dm}^3} = \mathbf{394 \text{ atm}}$$

EXERCISE 5.27

a $$V = \frac{nRT}{p} = \frac{0.250 \text{ mol} \times 0.0821 \text{ atm dm}^3 \text{ K}^{-1} \text{ mol}^{-1} \times 290 \text{ K}}{(750/760) \text{ atm}}$$
$= \mathbf{6.03 \text{ dm}^3}$

b $$V = \frac{nRT}{p} = \frac{0.521 \text{ g}}{17.0 \text{ g mol}^{-1}} \times \frac{0.0821 \text{ atm dm}^3 \text{ K}^{-1} \text{ mol}^{-1} \times 323 \text{ K}}{(762/760) \text{ atm}}$$
$= \mathbf{0.811 \text{ dm}^3}$

EXERCISE 5.28

a $n(CaCO_3 \text{ \& } CO_2) = \frac{7.31 \text{ g}}{100 \text{ g mol}^{-1}} = 0.0731$ mol

$$V = \frac{nRT}{p} = \frac{0.0731 \text{ mol} \times 0.0821 \text{ atm dm}^3 \text{ K}^{-1} \text{ mol}^{-1} \times 298 \text{ K}}{(764/760) \text{ atm}}$$
$= \mathbf{1.78 \text{ dm}^3}$

b $n(Zn \text{ \& } H_2) = \frac{1.52 \text{ g}}{65.4 \text{ g mol}^{-1}} = 0.0232$ mol

$$T = \frac{pV}{nR} = \frac{1.01 \text{ atm} \times 0.557 \text{ dm}^3}{0.0232 \text{ mol} \times 0.0821 \text{ atm dm}^3 \text{ K}^{-1} \text{ mol}^{-1}}$$
= **295 K or 22 °C**

c $n(H_2) = \frac{1}{2}n(HCl) = \frac{1}{2} \times 0.050 \text{ dm}^3 \times 0.102 \text{ mol dm}^{-3}$
= 0.00255 mol

$$V = \frac{nRT}{p} = \frac{0.00255 \text{ mol} \times 0.0821 \text{ atm dm}^{-3} \text{ K}^{-1} \text{ mol}^{-1} \times 287 \text{ K}}{(751/760) \text{ atm}}$$

$= \mathbf{0.0608 \text{ dm}^3}$

EXERCISE 5.29

$$p = \frac{nRT}{V} = \frac{5.00 \text{ g}}{18.0 \text{ g mol}^{-1}} \times \frac{0.0821 \text{ atm dm}^3 \text{ K}^{-1} \text{ mol}^{-1} \times 573 \text{ K}}{0.150 \text{ dm}^3} = \mathbf{87.1 \text{ atm}}$$

EXERCISE 5.30

a $n = \frac{m}{M}$

b $pV = nRT = \frac{m}{M} \times RT \therefore \boldsymbol{M = \frac{mRT}{pV}}$

EXERCISE 5.31

$M = \frac{mRT}{pV}$ (Don't try to remember this – when you need it derive it as in Exercise 5.30.)

where $p = \frac{740}{760}$ atm = 0.974 atm

$$\therefore M = \frac{3.72 \text{ g} \times 0.0821 \text{ atm dm}^3 \text{ mol}^{-1} \text{ K}^{-1} \times 373 \text{ K}}{0.974 \text{ atm} \times 2.00 \text{ dm}^3} = \mathbf{58.5 \text{ g mol}^{-1}}$$

EXERCISE 5.32

$$M = \frac{mRT}{pV} = \frac{0.198 \text{ g} \times 0.0821 \text{ atm dm}^3 \text{ K}^{-1} \text{ mol}^{-1} \times 294 \text{ K}}{(759/760) \text{ atm} \times 0.109 \text{ dm}^3} = \mathbf{43.9 \text{ g mol}^{-1}}$$

EXERCISE 5.33

A $M = \dfrac{mRT}{pV} = \dfrac{0.672\ \text{g} \times 0.0821\ \text{atm dm}^3\ \text{K}^{-1}\ \text{mol}^{-1} \times 298\ \text{K}}{1.00\ \text{atm} \times 0.257\ \text{dm}^3}$

$= \mathbf{64.0\ g\ mol^{-1}}$

B $M = \dfrac{mRT}{pV} = \dfrac{0.128\ \text{g} \times 0.0821\ \text{atm dm}^3\ \text{K}^{-1}\ \text{mol}^{-1} \times 290\ \text{K}}{(763/760)\ \text{atm} \times 0.111\ \text{dm}^3}$

$= \mathbf{27.3\ g\ mol^{-1}}$

C $M = \dfrac{mRT}{pV} = \dfrac{1.60\ \text{g} \times 0.0821\ \text{atm dm}^3\ \text{K}^{-1}\ \text{mol}^{-1} \times 294\ \text{K}}{(103/100)\ \text{atm} \times 0.526\ \text{dm}^3}$

$= \mathbf{71.3\ g\ mol^{-1}}$

EXERCISE 5.34

Volume of CO_2 = volume of flask

$= \dfrac{\text{mass of water}}{\text{density of water}} = \dfrac{152.8\ \text{g} - 47.9\ \text{g}}{1.00\ \text{gm cm}^{-3}} = 104.\ 9\ \text{cm}^3$

Mass of empty flask = mass of flask and air − mass of air
= 47.933 g − (volume of air × density of air)
= $47.933\ \text{g} - (104.9\ \text{cm}^3 \times 0.00118\ \text{g cm}^{-3})$
= 47.933 g − 0.124 g = 47.809 g

Mass of CO_2 = mass of flask and CO_2 − mass of flask
= 47.998 g − 47.809 g = 0.189 g

$pV = nRT = \dfrac{mRT}{M}\ \therefore M = \dfrac{mRT}{pV}$

If we use $R = 0.0821\ \text{atm dm}^3\ \text{K}^{-1}\ \text{mol}^{-1}$,

then p must be in atm, i.e. $\dfrac{757\ \text{atm}}{760\ \text{atm}} = 0.996\ \text{atm}$

and V must be in dm^3, i.e. $0.105\ \text{dm}^3$

and T must be in K, i.e. (273 + 26) K = 299 K

$\therefore M = \dfrac{0.189\ \text{g} \times 0.0821\ \text{atm dm}^3\ \text{K}^{-1}\ \text{mol}^{-1} \times 299\ \text{K}}{0.996\ \text{atm} \times 0.105\ \text{dm}^3} = \mathbf{44.4\ g\ mol^{-1}}$

EXERCISE 5.35

$$M = \frac{mRT}{pV} = \frac{0.25\ \text{cm}^3 \times 0.91\ \text{g cm}^{-3} \times 0.0821\ \text{atm dm}^3\ \text{K}^{-1}\ \text{mol}^{-1} \times 433\ \text{K}}{(764/760)\ \text{atm} \times 0.081\ \text{dm}^3} = \mathbf{99\ g\ mol^{-1}}$$

EXERCISE 5.36

D $M = \dfrac{mRT}{pV} = \dfrac{0.184\ \text{g} \times 0.0821\ \text{atm dm}^3\ \text{K}^{-1}\ \text{mol}^{-1} \times 373\ \text{K}}{0.984\ \text{atm} \times 0.0792\ \text{dm}^3}$

$= \mathbf{72.3\ g\ mol^{-1}}$

E $M = \dfrac{mRT}{pV} = \dfrac{0.295\ \text{g} \times 0.0821\ \text{atm dm}^3\ \text{K}^{-1}\ \text{mol}^{-1} \times 413\ \text{K}}{(747/760)\ \text{atm} \times 0.121\ \text{dm}^3}$

$= \mathbf{84.1\ g\ mol^{-1}}$

F $M = \dfrac{mRT}{pV} = \dfrac{0.163\ \text{g} \times 0.0821\ \text{atm dm}^3\ \text{K}^{-1}\ \text{mol}^{-1} \times 374\ \text{K}}{(105/100)\ \text{atm} \times 0.0650\ \text{dm}^3}$

$= \mathbf{73.3\ g\ mol^{-1}}$

EXERCISE 5.37

a Density, $\rho = \dfrac{\text{mass of sample}}{\text{volume of sample}} = \dfrac{M}{V_m}$

Since V_m is constant, $\dfrac{\rho_{SO_2}}{\rho_{H_2}} = \dfrac{M_{SO_2}}{M_{H_2}} = \dfrac{64.0}{2.02} = \dfrac{\mathbf{31.7}}{\mathbf{1}}$ **or 31.7:1**

b $\dfrac{\text{rate of effusion of } H_2}{\text{rate of effusion of } SO_2} = \sqrt{\dfrac{\rho_{SO_2}}{\rho_{H_2}}} = \sqrt{31.7} = \mathbf{5.6}$

i.e. hydrogen escapes 5.6 times as fast as the sulphur dioxide.

EXERCISE 5.38

$\dfrac{\text{rate of effusion of } H_2}{\text{rate of effusion of CO}} = \sqrt{\dfrac{\text{density of CO}}{\text{density of } H_2}} = \sqrt{\dfrac{M_{CO}}{M_{H_2}}}$

Now the rate of effusion = $\dfrac{\text{amount effused}}{\text{time taken}}$

$\therefore$ for a constant amount, rate $\propto \dfrac{1}{\text{time}}$

$\therefore \dfrac{\text{time for effusion of CO}}{\text{time for effusion of } H_2} = \sqrt{\dfrac{M_{CO}}{M_{H_2}}}$

or $\left(\dfrac{t_{CO}}{t_{H_2}}\right)^2 = \dfrac{M_{CO}}{M_{H_2}}$

$\therefore M_{CO} = M_{H_2} \times \left(\dfrac{t_{CO}}{t_{H_2}}\right)^2 = 2.02\ \text{g mol}^{-1} \times \left(\dfrac{93\ \text{s}}{25\ \text{s}}\right)^2 = \mathbf{28.0\ g\ mol^{-1}}$

EXERCISE 5.39

$$\text{Ratio} = \sqrt{\frac{238 + 6(19.0)}{235 + 6(19.0)}} = \sqrt{\frac{352}{349}} = 1.004 \quad \text{or} \quad \mathbf{1.004:1}$$

EXERCISE 5.40

$$\frac{47\text{ s}}{17\text{ s}} = \sqrt{\frac{32\text{ g mol}^{-1}}{M}}$$

$$\therefore\ M = 32\text{ g mol}^{-1} \times \frac{17^2}{47^2} = \mathbf{4.2\ g\ mol^{-1}}$$

$$\frac{47\text{ s}}{t} = \sqrt{\frac{32\text{ g mol}^{-1}}{17\text{ g mol}^{-1}}}$$

$$\therefore\ t = 47\text{ s} \times \sqrt{\frac{17}{32}} = \mathbf{34\ s}$$

EXERCISE 5.41

$$\frac{28\text{ s}}{24\text{ s}} = \sqrt{\frac{44\text{ g mol}^{-1}}{M}}$$

$$\therefore\ M = 44\text{ g mol}^{-1} \times \frac{24^2}{28^2} = \mathbf{32\ g\ mol^{-1}}$$

$$(X_{CO_2} \times 44) + (1 - X_{CO_2})(28) = 32$$

$$\therefore\ X_{CO_2} = \frac{32 - 28}{44 - 28} = \mathbf{0.25}$$

and $X_{CO} = \mathbf{0.75}$

EXERCISE 5.42

After admission of H_2, p_{O_2} is still 0.30 atm
and total pressure = $p_{O_2} + p_{H_2}$ = 0.80 atm
$\therefore\ p_{H_2}$ = 0.80 atm − 0.30 atm = 0.50 atm
After admission of N_2, p_{O_2} is still 0.30 atm and p_{H_2} is still 0.50 atm
and total pressure = $p_{O_2} + p_{H_2} + p_{N_2}$ = 0.90 atm
$\therefore\ p_{N_2}$ = 0.90 atm − 0.80 atm = 0.10 atm
$\mathbf{p_{O_2}}$ **= 0.30 atm,** $\mathbf{p_{H_2}}$ **= 0.50 atm,** $\mathbf{p_{N_2}}$ **= 0.10 atm**

EXERCISE 5.43

$$2NH_3(g) \rightarrow N_2(g) + 3H_2(g)$$
1 mol   3 mol

$$X_{N_2} = \frac{1\text{ mol}}{1\text{ mol} + 3\text{ mol}} = \frac{1}{4} \quad \text{and} \quad X_{H_2} = \frac{3\text{ mol}}{1\text{ mol} + 3\text{ mol}} = \frac{3}{4}$$

$$p_{N_2} = pX_{N_2} = 760\text{ mmHg} \times \frac{1}{4} = \mathbf{190\ mmHg}$$

$$p_{H_2} = pX_{H_2} = 760\text{ mmHg} \times \frac{3}{4} = \mathbf{570\ mmHg}$$

EXERCISE 5.44

$$p_{O_2} = pX_{O_2} = p \times \frac{\text{amount of } O_2}{\text{total amount}}$$

$$\therefore\ \text{total amount} = \text{amount of } O_2 \times \frac{p}{p_{O_2}}$$

$$= 0.25\text{ mol} \times \frac{0.90\text{ atm}}{0.30\text{ atm}} = \mathbf{0.75\ mol}$$

$$p_{H_2} = pX_{H_2} = p \times \frac{\text{amount of } H_2}{\text{total amount}}$$

$$\therefore\ \text{amount of } H_2 = \text{total amount} \times \frac{p_{H_2}}{p}$$

$$= 0.75\text{ mol} \times \frac{0.50\text{ atm}}{0.90\text{ atm}} = \mathbf{0.42\ mol}$$

$$p_{N_2} = pX_{N_2} = p \times \frac{\text{amount of } N_2}{\text{total amount}}$$

$$\therefore\ \text{amount of } N_2 = \text{total amount} \times \frac{p_{N_2}}{p}$$

$$= 0.75\text{ mol} \times \frac{0.10\text{ atm}}{0.90\text{ atm}} = \mathbf{0.08\ mol}$$

(Check: 0.25 mol of O_2 + 0.42 mol of H_2 + 0.08 mol of N_2 = 0.75 mol)

EXERCISE 5.45

$$X_{H_2O} = \frac{\text{amount of } H_2O}{\text{total amount}} = \frac{\text{volume of } H_2O}{\text{total volume}} \quad \text{(applying Avogadro's theory)}$$

$$= \frac{85.0\text{ cm}^3 - 82.0\text{ cm}^3}{85.0\text{ cm}^3} = 0.0353$$

$p_{H_2O} = pX_{H_2O}$ = 1.00 atm × 0.0353 = **0.0353 atm**

EXERCISE 5.46

$p_{O_2} = pX_{O_2}$

$\therefore \quad X_{O_2} = \dfrac{p_{O_2}}{p}$

$= \dfrac{2.0 \times 10^4 \text{ N m}^{-2}}{5.0 \times 10^5 \text{ N m}^{-2}} = 4.0 \times 10^{-2}$

By Avogadro's theory, mole fraction = volume fraction.
Also, volume percentage = volume fraction × 100
$= 4.0 \times 10^{-2} \times 100 =$ **4.0%**

EXERCISE 5.47

$p_{H_2} = 2.4 \text{ atm} \times \dfrac{75}{180} = \mathbf{1.0\ atm}$

$p_{N_2} = 2.4 \text{ atm} \times \dfrac{15}{180} = \mathbf{0.2\ atm}$

$p_{CO_2} = 2.4 \text{ atm} \times \dfrac{90}{180} = \mathbf{1.2\ atm.}$

EXERCISE 5.48

$p_{O_2} = 1.00 \text{ atm}$

$p_{N_2} = p_{O_2} \times \dfrac{300}{500} = \mathbf{0.600\ atm}$

$p_{CO_2} = 3.10 \text{ atm} - 0.60 \text{ atm} - 1.00 \text{ atm} = \mathbf{1.50\ atm}$

$V = 500 \text{ cm}^3 \times \dfrac{1.50 \text{ atm}}{1.00 \text{ atm}} = \mathbf{750\ cm^3}$

EXERCISE 5.49

$p_{O_2} = 1.00 \text{ atm} \times \dfrac{200 \text{ cm}^3}{700 \text{ cm}^3} = \mathbf{0.286\ atm}$

$p_{He} = 2.00 \text{ atm} \times \dfrac{500 \text{ cm}^3}{700 \text{ cm}^3} = \mathbf{1.43\ atm}$

$p = (1.43 + 0.286) \text{ atm} = \mathbf{1.72\ atm}$

EXERCISE 5.50

$n = \dfrac{0.0232 \text{ g}}{28.0 \text{ g mol}^{-1}} + \dfrac{0.0417 \text{ g}}{28.0 \text{ g mol}^{-1}} = 0.00232 \text{ mol}$

$p = \dfrac{nRT}{V}$

$= \dfrac{0.00232 \text{ mol} \times 0.0821 \text{ atm dm}^3 \text{ K}^{-1} \text{ mol}^{-1} \times 296 \text{ K}}{0.600 \text{ dm}^3} = \mathbf{0.0940\ atm}$

$p_{N_2} = 0.0940 \text{ atm} \times \dfrac{0.000829 \text{ mol}}{0.00232 \text{ mol}} = \mathbf{0.0336\ atm}$

$p_{CO} = 0.0940 \text{ atm} \times \dfrac{0.00149 \text{ mol}}{0.00232 \text{ mol}} = \mathbf{0.0604\ atm}$

EXERCISE 5.51

$n = \dfrac{82.2 \text{ g}}{4.00 \text{ g mol}^{-1}} + \dfrac{27.4 \text{ g}}{32.0 \text{ g mol}^{-1}} = 20.55 \text{ mol} + 0.86 \text{ mol} = 21.4 \text{ mol}$

$V = \dfrac{nRT}{p} = \dfrac{21.4 \text{ mol} \times 0.0821 \text{ atm dm}^3 \text{ K}^{-1} \text{ mol}^{-1} \times 284 \text{ K}}{25.0 \text{ atm}} = \mathbf{20.0\ dm^3}$

$p_{He} = 25.0 \text{ atm} \times \dfrac{20.6 \text{ mol}}{21.4 \text{ mol}} = \mathbf{24.1\ atm}$

$p_{O_2} = 25.0 \text{ atm} - 24.1 \text{ atm} = \mathbf{0.9\ atm}$

EXERCISE 5.52

$x\text{CO(g)} + y\text{CH}_4\text{(g)} + (\frac{x}{2} + 2y)\text{O}_2\text{(g)} \rightarrow (x + y)\text{CO}_2\text{(g)} + 2y\text{H}_2\text{O(l)}$

Using relative volumes as stoichiometric coefficients:

volume of initial mixture = $(x + y)$ cm^3 = 15 cm^3

and contraction in volume = volume of oxygen used up = $(\frac{x}{2} + 2y)$ cm^3 = 21 cm^3

Hence, $y = 9$, $x = 6$.

$\therefore \; X_{CO} = \dfrac{6}{15}$ and $p_{CO} = 1 \text{ atm} \times \dfrac{6}{15} = \mathbf{0.40\ atm}$

$X_{CH_4} = \dfrac{9}{15}$ and $p_{CO} = 1 \text{ atm} \times \dfrac{9}{15} = \mathbf{0.60\ atm}$

End-of-chapter questions

5.1 E
5.2 A
5.3 **b** Ideal behaviour, i.e. $p \propto \frac{1}{V}$. Straight line graph.
c Boyle's law. Pressure of fixed mass of gas at constant temperature is inversely proportional to volume. Or pV is constant at constant temperature.
d Gradient = 4.50 atm dm^3
= constant value of pV
But $pV = nRT$ ∴ gradient = nRT
e 27.9
f 0.951 g dm^{-3}
5.4 99 g mol^{-1}
5.5 **a** 273 cm^3 (or 0.273 dm^3)
b 44.8 g mol^{-1}
c $M(C_3H_6)$ = 42.0 g mol^{-1}: $M(C_4H_8)$ = 56.0 g mol^{-1}
M(mixture) is close to $M(C_3H_6)$
∴ C_3H_6 predominates
5.6 8.01×10^{16}
5.7 D
5.8 E
5.9 **a** 159 (strictly 1.6×10^2)
b The measurements are on the vapour, and the equation used refers to the gaseous state.
5.10 **a** 16.0 g mol^{-1}
b 64 g mol^{-1}
5.11 0.88 g
5.12 506 cm^3
5.13 240 atm
5.14 29.1 cm^3
5.15 C_2H_4

CHAPTER 6

EXERCISE 6.1

a i) $K_c = \dfrac{[H_2(g)]\,[Br_2(g)]}{[HBr(g)]^2}$ no unit

ii) $K_c = \dfrac{[SO_3(g)]^2}{[SO_2(g)]^2\,[O_2(g)]}$ $dm^3\ mol^{-1}$

iii) $K_c = \dfrac{[Cu^{2+}(aq)]\,[NH_3(aq)]^4}{[Cu(NH_3)_4^{2+}(aq)]}$ $mol^4\ dm^{-12}$

iv) $K_c = \dfrac{[NO_2(g)]^2}{[NO(g)]^2\,[O_2(g)]}$ $dm^3\ mol^{-1}$

v) $K_c = \dfrac{[P_4(g)]\,[F_2(g)]^{10}}{[PF_5(g)]^4}$ $mol^7\ dm^{-21}$

vi) $K_c = \dfrac{[N_2(g)]\,[O_2(g)]}{[NO(g)]^2}$ no unit

vii) $K_c = \dfrac{[CH_3CO_2C_2H_5(l)]\,[H_2O(l)]}{[C_2H_5OH(l)]\,[CH_3CO_2H(l)]}$ no unit

b In equations i), vi) and vii) the same number of moles appears on each side of the equation. This means that the same number of concentration terms appears at the top and bottom of the equilibrium constant expressions, so that the units of concentration cancel.

EXERCISE 6.2

a $K_c = \dfrac{[CO(g)]\,[Cl_2(g)]}{[COCl_2(g)]}$ $K_c' = \dfrac{[COCl_2(g)]}{[CO(g)]\,[Cl_2(g)]}$

b $K_c = \dfrac{1}{K_c'}$ $\left(\text{or } K_c' = \dfrac{1}{K_c}\right)$

EXERCISE 6.3

a $K_c = \dfrac{[NO_2(g)]}{[N_2O_4(g)]^{1/2}}$ $K_c' = \dfrac{[NO_2(g)]^2}{[N_2O_4(g)]}$

b Squaring K_c from **a** gives $K_c^2 = \dfrac{[NO_2(g)]^2}{[N_2O_4(g)]} = K_c'$

∴ at 100 °C, $K_c = \sqrt{(K_c')} = (0.490\ mol\ dm^{-3})^{\frac{1}{2}} =$ **$0.700\ mol^{1/2}\ dm^{-3/2}$**

and at 200 °C, $K_c = \sqrt{(K_c')} = (18.6\ mol\ dm^{-3})^{\frac{1}{2}} =$ **$4.31\ mol^{1/2}\ dm^{-3/2}$**

EXERCISE 6.4

$K_c = \dfrac{[NO_2]^2}{[N_2O_4]} = \dfrac{(0.010\ mol\ dm^{-3})^2}{0.021\ mol\ dm^{-3}} =$ **$4.8 \times 10^{-3}\ mol\ dm^{-3}$**

EXERCISE 6.5

$K_c = \dfrac{[PCl_3(g)]\,[Cl_2(g)]}{[PCl_5(g)]} = \dfrac{(1.50 \times 10^{-2}\ mol\ dm^3) \times (1.50 \times 10^{-2}\ mol\ dm^{-3})}{1.18 \times 10^{-3}\ mol\ dm^{-3}}$

$K_c =$ **$0.19\ mol\ dm^{-3}$**

EXERCISE 6.6

$$K_c = \frac{[SO_3(g)]^2}{[SO_2(g)]^2\,[O_2(g)]} = \frac{(0.92\ \text{mol dm}^{-3})^2}{(0.23\ \text{mol dm}^{-3})^2\,(1.37\ \text{mol dm}^{-3})} = \mathbf{11.7\ dm^3\ mol^{-1}}$$

EXERCISE 6.7

a

$$[PCl_5(g)] = \frac{0.0042\ \text{mol}}{2.0\ \text{dm}^3} = 0.0021\ \text{mol dm}^{-3}$$

$$[Cl_2(g)] = [PCl_3(g)] = \frac{0.040\ \text{mol}}{2.0\ \text{dm}^3} = 0.020\ \text{mol dm}^{-3}$$

$$\therefore\ K_c = \frac{[PCl_3(g)][Cl_2(g)]}{[PCl_5(g)]} = \frac{(0.020\ \text{mol dm}^{-3})^2}{0.0021\ \text{mol dm}^{-3}} = \mathbf{0.19\ mol\ dm^{-3}}$$

b

i) When K_c is large, the concentrations of products are greater than the concentrations of reactants.

ii) When K_c is small, the concentrations of products are smaller than the concentrations of reactants.

EXERCISE 6.8

a $H_2(g) + I_2(g) \rightleftharpoons 2HI(g)$

b

$$K_c = \frac{[HI(g)]^2}{[H_2(g)]\,[I_2(g)]}$$

c Mixture 1.

$$K_c = \frac{(0.1715\ \text{mol}/1.00\ \text{dm}^3)^2}{(0.02265\ \text{mol}/1.00\ \text{dm}^3)\times(0.02840\ \text{mol}/1.00\ \text{dm}^3)}$$

$$= \frac{0.02941\ \text{mol}^2\ \text{dm}^{-6}}{6.433\times10^{-4}\ \text{mol}^2\ \text{dm}^{-6}} = \mathbf{45.72}$$

Mixture 2.

$$K_c = \frac{(0.1779\ \text{mol}/1.00\ \text{dm}^3)^2}{(0.01699\ \text{mol}/1.00\ \text{dm}^3)\times(0.04057\ \text{mol}/1.00\ \text{dm}^3)}$$

$$= \frac{0.03165\ \text{mol}^2\ \text{dm}^{-6}}{6.893\times10^{-4}\ \text{mol}^2\ \text{dm}^{-6}} = \mathbf{45.92}$$

d

$$K_c = \frac{(0.1715\ \text{mol}/2.00\ \text{dm}^3)^2}{(0.02265\ \text{mol}/2.00\ \text{dm}^3)\times(0.02840\ \text{mol}/2.00\ \text{dm}^3)}$$

$$= \frac{7.353\times10^{-3}\ \text{mol}^2\ \text{dm}^{-6}}{1.608\times10^{-4}\ \text{mol}^2\ \text{dm}^{-6}} = \mathbf{45.72}$$

(Did you notice the short cut? 2.00 dm^3 cancels, giving the same expression as in part **c**.)

e

$$K_c = \frac{(0.1779\ \text{mol}/V\ \text{dm}^3)^2}{(0.01699\ \text{mol}/V\ \text{dm}^3)\times(0.04057\ \text{mol}/V\ \text{dm}^3)}$$

$$= \frac{(0.1779\ \text{mol})^2}{0.01699\ \text{mol}\times0.04057\ \text{mol}}\times\frac{V^2\ \text{dm}^6}{V^2\ \text{dm}^6} = \mathbf{45.92}$$

EXERCISE 6.9

Amount of HCl = amount of NaOH = cV = 0.974 mol dm^{-3} × 0.0105 dm^3
= 0.0102 mol

Mass of HCl (pure) = nM = 0.0102 mol × 36.5 g mol^{-1} = 0.372 g

Mass of H_2O in HCl(aq) = 5.17 g − 0.372 g = 4.80 g

Total mass of H_2O at start = 4.80 g + 0.99 g = 5.79 g

Amount of H_2O at start = m/M = 5.79 g/18.0 g mol^{-1} = 0.322 mol

Total eqm amount of acid = amount of NaOH = cV
= 0.974 mol dm^{-3} × 0.0392 dm^3
= 0.0382 mol

Eqm amount of CH_3CO_2H = total amount − amount of HCl
= (0.0382 − 0.0102) mol − 0.0280 mol

Eqm amount of C_2H_5OH = amount of CH_3CO_2H = 0.0280 mol

Initial amount of $CH_3CO_2C_2H_5$ = m/M = 3.64 g/88.1 g mol^{-1} = 0.0413 mol

Eqm amount of $CH_3CO_2C_2H_5$ = initial amount − amount reacted
= (0.0413 − 0.0280)mol = 0.0133 mol

Eqm amount of H_2O = initial amount − amount reacted
= (0.322 − 0.0280) mol = 0.294 mol

$$K_c = \frac{[C_2H_5OH][CH_3CO_2H]}{[CH_3CO_2C_2H_5][H_2O]} = \frac{0.0280\times0.0280}{0.0133\times0.294}\quad\text{(units cancel)}$$

$$= \mathbf{0.201}$$

EXERCISE 6.10

	$N_2O_4 \rightleftharpoons$	$2NO_2$
Equilibrium concn./mol dm^{-3}	x	1.85×10^{-3}

$$K_c = \frac{[NO_2]^2}{[N_2O_4]}$$

$$\therefore\ 1.06\times10^{-5}\ \text{mol dm}^{-3} = \frac{(1.85\times10^{-3}\ \text{mol dm}^{-3})^2}{x\ \text{mol dm}^{-3}}$$

$$\therefore\ x = \frac{(1.85\times10^{-3})^2}{(1.06\times10^{-5})} = 0.323$$

and $[N_2O_4]$ = **0.323 mol dm^{-3}**

EXERCISE 6.11

	$2H_2S(g) \rightleftharpoons$	$2H_2(g)$ +	$S_2(g)$
Equilibrium concn./mol dm^{-3}	4.84×10^{-3}	x	2.33×10^{-3}

$$K_c = \frac{[H_2(g)]^2\,[S_2(g)]}{[H_2S(g)]^2}$$

$$2.25\times10^{-4}\ \text{mol dm}^{-3} = \frac{(x\ \text{mol dm}^{-3})^2\times(2.33\times10^{-3}\ \text{mol dm}^{-3})}{(4.84\times10^{-3}\ \text{mol dm}^{-3})^2}$$

$$x^2 = \frac{(2.25\times10^{-4})(4.84\times10^{-3})^2}{(2.33\times10^{-3})} = 2.26\times10^{-6}$$

$$x = \sqrt{2.26\times10^{-6}} = 1.50\times10^{-3}$$

$\therefore [H_2(g)]$ = **1.50×10^{-3} mol dm^{-3}**

EXERCISE 6.12

$$K_c = \frac{[HI(g)]^2}{[H_2(g)]\,[I_2(g)]^3} \quad \therefore\ 54.1 = \frac{(3.53\times10^{-3}\ \text{mol dm}^{-3})^2}{(0.48\times10^{-3}\ \text{mol dm}^{-3}\times x)}$$

$x = [I_2(g)] = \mathbf{4.8\times10^{-4}\ mol\ dm^{-3}}$

EXERCISE 6.13

$$K_c = \frac{[CH_3CO_2C_5H_{11}(l)]}{[C_5H_{10}(l)]\,[CH_3CO_2H(l)]}$$

$$540\ \text{dm}^3\ \text{mol}^{-1} = \frac{x}{(5.66\times10^{-3}\ \text{mol dm}^{-3})(2.55\times10^{-3}\ \text{mol dm}^{-3})}$$

$x = [CH_3CO_2C_5H_{11}(l)] = \mathbf{7.79\times10^{-3}\ mol\ dm^{-3}}$

EXERCISE 6.14

$$K_c = \frac{[CH_3CO_2C_2H_5(l)]\,[H_2O(l)]}{[CH_3CO_2H(l)]\,[C_2H_5OH(l)]}$$

Because the same numbers of moles appear on both sides of the equation, it is possible to substitute amounts rather than concentrations.

$$\text{i.e. } 4.0 = \frac{0.66\ \text{mol}\times0.66\ \text{mol}}{0.33\ \text{mol}\times x}$$

x = amount of C_2H_5OH = **0.33 mol**

EXERCISE 6.15

a

	$PCl_5(g) \rightleftharpoons$	$PCl_3(g)$ +	$Cl_2(g)$
Equilibrium amount/mol	x	0.15	0.090

$$K_c = \frac{[PCl_3(g)]\,[Cl_2(g)]}{[PCl_5(g)]}$$

Substituting concentrations gives

$$0.19\ \text{mol dm}^{-3} = \frac{(0.15\ \text{mol}/2.0\ \text{dm}^3)(0.090\ \text{mol}/2.0\ \text{dm}^3)}{(x\ \text{mol}/2.0\ \text{dm}^3)}$$

$$\text{i.e. } 0.19 = \frac{0.075\times0.045}{x/2.0} = \frac{0.075\times0.045\times2.0}{x}$$

$$\therefore\ x = \frac{0.075\times0.045\times2.0}{0.19} = 0.0355$$

$\therefore$ amount of PCl_5 = **0.036 mol**

b mass = amount × molar mass
= 0.0355 mol × 208.5 g mol^{-1} = **7.40 g**

EXERCISE 6.16

Amount of pentene at equilibrium = initial amount – amount reacted
= $(0.020 - 9.0\times10^{-3})$ mol
= 0.011 mol

Amount of ethanoic acid at equilibrium = initial amount – amount reacted
= $(0.010 - 9.0\times10^{-3})$ mol
= 1.0×10^{-3} mol

	CH_3CO_2H +	$C_5H_{10} \rightleftharpoons$	$CH_3CO_2C_5H_{11}$
Initial amount/mol	0.010	0.020	0
Equilibrium amount/mol	1.0×10^{-3}	0.011	9.0×10^{-3}

$$K_c = \frac{[CH_3CO_2C_5H_{11}]}{[CH_3CO_2H][C_5H_{10}]}$$

$$= \frac{(9.0\times10^{-3}\ \text{mol}/0.600\ \text{dm}^3)}{(1.0\times10^{-3}\ \text{mol}/0.600\ \text{dm}^3)\times(0.011\ \text{mol}/0.600\ \text{dm}^3)} = \mathbf{491\ dm^3\ mol^{-1}}$$

EXERCISE 6.17

	$H_2(g)$ +	$I_2(g) \rightleftharpoons$	$2HI(g)$
Initial amount/mol	1.90	1.90	0
Equilibrium amount/mol	(0.40)	(0.40)	(Calculation below)

Amount of H_2 at equilibrium = initial amount – amount reacted.

The equation shows that 2 mol of HI requires 1 mol of H_2 to react.
$\therefore$ to produce 3.00 mol of HI, the amount of H_2 reacted = 1.50 mol

Amount of H_2 at equilibrium = 1.90 mol – 1.50 mol = 0.40 mol

The calculation for I_2 is the same. Since the equation has equal numbers of moles on each side, we can substitute amounts rather than concentrations in the equilibrium law expression. (Volume cancels.)

$$K_c = \frac{[HI(g)]^2}{[H_2(g)][I_2(g)]} = \frac{(3.00\ \text{mol})^2}{(0.40\ \text{mol})\times(0.40\ \text{mol})} = \mathbf{56}$$

EXERCISE 6.18

$$\text{Initial amount of } CH_3CO_2H = \frac{6.0\ \text{g}}{60.1\ \text{g mol}^{-1}} = 0.10\ \text{mol}$$

$$\text{Initial amount of } C_2H_5OH = \frac{6.9\ \text{g}}{46.1\ \text{g mol}^{-1}} = 0.15\ \text{mol}$$

$$\text{Equilibrium amount of } CH_3CO_2C_2H_5 = \frac{7.0\ \text{g}}{88.1\ \text{g mol}^{-1}} = 0.079\ \text{mol}$$

The equation shows that the equilibrium amount of H_2O is also 0.079 mol, and that this is formed from an equal amount of CH_3CO_2H or C_2H_5OH.

Equilibrium amount of CH_3CO_2H = initial amount – amount reacted.
= 0.10 mol – 0.079 mol = 0.021 mol

Equilibrium amount of C_2H_5OH = 0.15 mol − 0.079 mol = 0.071 mol

Initial amount/mol	0.10	0.15	0	0
	CH_3CO_2H +	C_2H_5OH ⇌	$CH_3CO_2C_2H_5$ +	H_2O
Equilibrium amount/mol	0.021	0.071	0.079	0.079

Since the equation has equal numbers of moles on each side, we can substitute amounts rather than concentrations in the equilibrium law expression.

$$K_c = \frac{[CH_3CO_2C_2H_5][H_2O]}{[CH_3CO_2H][C_2H_5OH]} = \frac{(0.079)^2}{0.021 \times 0.071} = \mathbf{4.2}$$

EXERCISE 6.19

Initial amount/mol	20.57	5.22	0
	$H_2(g)$ +	$I_2(g)$ ⇌	$2HI(g)$
Equilibrium amount/mol	15.46	0.11	10.22

Amount of H_2 at equilibrium = initial amount − amount reacted
= (20.57 − ½ × 10.22) mol = 15.46 mol

Amount of I_2 at equilibrium = initial amount − amount reacted
= (5.22 − ½ × 10.22) mol = 0.11 mol

$$K_c = \frac{[HI(g)]^2}{[H_2(g)]\,[I_2(g)]} = \frac{(10.22\text{ mol})^2}{(15.46\text{ mol})\,(0.11\text{ mol})} = \mathbf{61}$$

EXERCISE 6.20

Initial concn./mol dm^{-3}	0.1307	0
	$N_2O_4(l)$ ⇌	$2NO_2(l)$
Equilibrium concn./mol dm^{-3}	0.1300	0.0014

Equilibrium concentration of N_2O_4 = initial concn. − concn. reacted
= (0.1307 − 0.0007) mol dm^{-3}
= 0.1300 mol dm^{-3}

$$K_c = \frac{[NO_2(l)]^2}{[N_2O_4(l)]} = \frac{(0.0014\text{ mol dm}^{-3})^2}{(0.1300\text{ mol dm}^{-3})} = \mathbf{1.51 \times 10^{-5}\ mol\ dm^{-3}}$$

EXERCISE 6.21

Initial amount/mol	2.0	1.0	0	0
	$C_2H_5OH(l)$ +	$CH_3CO_2H(l)$ ⇌	$CH_3CO_2C_2H_5(l)$ +	$H_2O(l)$
Equilibrium amount/mol	1.155	0.155	0.845	0.845

$$K_c = \frac{[CH_3CO_2C_2H_5(l)]\,[H_2O(l)]}{[C_2H_5OH(l)]\,[CH_3CO_2H(l)]} = \frac{(0.845\text{ mol})(0.845\text{ mol})}{(1.155\text{ mol})(0.155\text{ mol})} = \mathbf{3.99}\text{ (volume cancels)}$$

Equilibrium amount of C_2H_5OH = initial amount − amount reacted
= (2.0 − 0.845) mol = 1.155 mol
Equilibrium amount of CH_3CO_2H = initial amount − amount reacted
= (1.0 − 0.845) mol = 0.155 mol

EXERCISE 6.22

Initial amount/mol	3.0	3.0	0	0
	$CO(g)$ +	$H_2O(g)$ ⇌	$CO_2(g)$ +	$H_2(g)$
Equilibrium amount/mol	$3.0 - x$	$3.0 - x$	x	x

$$K_c = \frac{[CO_2(g)][H_2(g)]}{[CO(g)][H_2O(g)]}$$

$$4.00 = \frac{(x\text{ mol})\,(x\text{ mol})}{(3.0-x)\text{ mol}\,(3.0-x)\text{ mol}} = \frac{x^2}{(3.0-x)^2}$$

Taking the square root of both sides,

$$2.00 = \frac{x}{3.0-x}$$

$\therefore\ 6.0 - 2.0x = x$ or $3x = 6.0$

$\therefore\ x = 2.0$ and amount of hydrogen = **2.0 mol**

EXERCISE 6.23

$$\text{Amount of } PCl_5 = \frac{2.085\text{ g}}{208.5\text{ g mol}^{-1}} = 0.0100\text{ mol}$$

Initial amount/mol	0.0100	0	0
	$PCl_5(g)$ ⇌	$PCl_3(g)$ +	$Cl_2(g)$
Equilibrium amount/mol	$0.0100 - x$	x	x

$$K_c = \frac{[PCl_3(g)][Cl_2(g)]}{[PCl_5(g)]}$$

$$0.19\text{ mol dm}^{-3} = \frac{\left(\frac{x}{0.500}\text{ mol dm}^{-3}\right)\left(\frac{x}{0.500}\text{ mol dm}^{-3}\right)}{\left(\frac{0.0100-x}{0.500}\text{ mol dm}^{-3}\right)}$$

$$0.19 = \frac{x^2}{0.500(0.0100-x)} = \frac{x^2}{0.00500 - 0.500x}$$

$$9.5 \times 10^{-4} - 0.095x = x^2 \text{ or } x^2 + 0.095x - (9.5 \times 10^{-4}) = 0$$

$$x = \frac{-b \pm \sqrt{b^2 - 4ac}}{2a} \text{ where } a = 1,\ b = 0.095 \text{ and } c = -(9.5 \times 10^{-4})$$

$$x = \frac{-0.095 \pm \sqrt{(0.095)^2 + (4 \times 9.5 \times 10^{-4})}}{2}$$

$$2x = -0.095 \pm \sqrt{(9.025 \times 10^{-3}) + (3.8 \times 10^{-3})}$$

$$= -0.095 \pm \sqrt{1.2825 \times 10^{-2}} = -0.095 \pm 0.1132$$

$$\therefore x = 0.0091 \text{ or } -0.1041 \text{ (absurd root)}$$

$$[PCl_5(g)] = \frac{(0.0100 - 0.0091) \text{ mol}}{0.500 \text{ dm}^3} = \mathbf{1.8 \times 10^{-3} \text{ mol dm}^{-3}}$$

$$[Cl_2(g)] = [PCl_3(g)] = \frac{0.0091 \text{ mol}}{0.500 \text{ dm}^3} = \mathbf{1.8 \times 10^{-2} \text{ mol dm}^{-3}}$$

EXERCISE 6.24

Initial amount/mol	2.00	2.00	0
	$H_2(g)$ +	$I_2(g)$ $\rightleftharpoons$	$2HI(g)$
Equilibrium amount/mol	$2.00 - x$	$2.00 - x$	$2x$

$$K_c = \frac{[HI(g)]^2}{[H_2(g)]\,[I_2(g)]}$$

$$49.0 = \frac{(2x)^2}{(2.00 - x)\,(2.00 - x)}$$

Taking the square roots, $7.00 = \frac{2x}{2.00 - x}$

$$14.0 - 7.00x = 2x$$

$$9.00x = 14.0 \text{ and } x = 1.56$$

$\therefore$ amount of HI = $2x$ mol = **3.12 mol**

Amount of H_2 = amount of I_2 = $(2.00 - x)$
= **0.44 mol**

EXERCISE 6.25

Initial amount/mol	8.0	6.0	0	0
	$CH_3CO_2H(l)$ +	$C_2H_5OH(l)$ $\rightleftharpoons$	$CH_3CO_2C_2H_5(l)$ +	$H_2O(l)$
Equilibrium amount/mol	$8.0 - x$	$6.0 - x$	x	x

$$K_c = \frac{[CH_3CO_2C_2H_5(l)]\,[H_2O(l)]}{[CH_3CO_2H(l)]\,[C_2H_5OH(l)]}$$

$$4.5 = \frac{x \times x}{(8.0 - x)(6.0 \times x)} = \frac{x^2}{48.0 - 14.0x + x^2}$$

$$3.5x^2 - 63.0x + 216.0 = 0$$

$$x = \frac{-(-63.0) \pm \sqrt{(-63.0)^2 - (4 \times 3.5 \times 216.0)}}{2 \times 3.5}$$

$$= \frac{63.0 \pm \sqrt{945}}{7.0} = 4.61 \text{ or } 13.4 \text{ (absurd root)}$$

$\therefore$ equilibrium amount of water = **4.6 mol**

EXERCISE 6.26

Initial amount/mol	0	2.0	1.0
	$2HI(g)$ $\rightleftharpoons$	$H_2(g)$ +	$I_2(g)$
Equilibrium amount/mol	$2x$	$(2.0 - x)$	$(1.0 - x)$

$$K_c = \frac{[H_2(g)]\,[I_2(g)]}{[HI(g)]^2}$$

$$0.02 = \frac{(2.0 - x)(1.0 - x)}{(2x)^2} = \frac{2.0 - 3.0x + x^2}{4x^2}$$

$$0.92x^2 - 3.0x + 2.0 = 0$$

$$x = \frac{-(-3.0) \pm \sqrt{(-3.0)^2 - (4 \times 0.92 \times 2.0)}}{2 \times 0.92} = \frac{3.0 \pm \sqrt{1.64}}{1.84}$$

$$= 0.93 \text{ or } 2.33 \text{ (absurd root)}$$

amount of HI = $2x = 2 \times 0.93 \text{ mol} = 1.86 \text{ mol}$

$$[HI(g)] = \frac{\text{amount}}{\text{volume}} = \frac{1.86 \text{ mol}}{1.0 \text{ dm}^3} = \mathbf{1.9 \text{ mol dm}^{-3}}$$

$$[H_2(g)] = \frac{(2.0 - 0.93) \text{ mol}}{1.0 \text{ dm}^3} = \mathbf{1.1 \text{ mol dm}^{-3}}$$

$$[I_2(g)] = \frac{(1.0 - 0.93) \text{ mol}}{1.0 \text{ dm}^3} = \mathbf{0.07 \text{ mol dm}^{-3}}$$

EXERCISE 6.27

Initial amount/mol	0.019	0	0
	$PCl_5(g)$ $\rightleftharpoons$	$PCl_3(g)$ +	$Cl_2(g)$
Equilibrium amount/mol	$0.019 - x$	x	x

$$K_c = \frac{[PCl_3(g)]\,[Cl_2(g)]}{[PCl_5(g)]^2}$$

$$0.19 \text{ mol dm}^{-3} = \frac{(x \text{ mol}/0.75 \text{ dm}^3) \times (x \text{ mol}/0.75 \text{ dm}^3)}{(0.019 - x) \text{ mol}/0.75 \text{ dm}^3}$$

$$0.19 = \frac{x^2}{(0.75)(0.019 - x)}$$

$$x^2 + 0.1425x - 2.71 \times 10^{-3} = 0$$

$$x = \frac{-0.1425 \pm \sqrt{(0.1425)^2 - (4)(1)(-2.71 \times 10^{-3})}}{2 \times 1}$$

$$= \frac{-0.1425 \pm \sqrt{0.0311}}{2}$$

$$= 0.017 \text{ or } -0.159 \text{ (absurd root)}$$

amount of PCl_5 at equilibrium = $(0.019 - 0.017)$ mol = **2.0×10^{-3} mol**

EXERCISE 6.28

a $K_p = \dfrac{p_{N_2} \times p_{H_2}^3}{p_{NH_3}^2}$ atm^2

b $K_p = \dfrac{p_{SO_3}^2}{p_{SO_2}^2 \times p_{O_2}}$ atm^{-1}

c $K_p = \dfrac{p_{CO}^2}{p_{CO_2}}$ atm

d $K_p = p_{CO_2}$ atm

e $K_p = p_{NH_3} \times p_{H_2S}$ atm^2

Note that the partial pressure of a solid is taken to be constant (like its concentration) and does not therefore appear in the equilibrium expression.

EXERCISE 6.29

$$K_p = \frac{p_{SO_3}^2}{p_{SO_2}^2 \times p_{O_2}} = \frac{(4.5\text{ atm})^2}{(0.090\text{ atm})^2 \times 0.083\text{ atm}} = \mathbf{3.01 \times 10^4\ atm^{-1}}$$

EXERCISE 6.30

$$K_p = \frac{p_{NO_2}^2}{p_{N_2O_4}} = \frac{(0.67\text{ atm})^2}{0.33\text{ atm}} = \mathbf{1.36\ atm}$$

EXERCISE 6.31

$$K_p = \frac{p_{SO_3}^2}{p_{SO_2}^2 \times p_{O_2}} = \frac{(2.3\text{ atm})^2}{(2.3\text{ atm})^2 \times 4.5\text{ atm}} = \mathbf{0.22\ atm^{-1}}$$

EXERCISE 6.32

$$K_p = \frac{p_{HI}^2}{p_{H_2} \times p_{I_2}} = \frac{(0.40\text{ atm})^2}{(0.25\text{ atm}) \times (0.16\text{ atm})} = \mathbf{4.0}$$

EXERCISE 6.33

a

$$\text{Amount of } H_2 = \frac{12.8\text{ g}}{2.02\text{ g mol}^{-1}} = 6.34\text{ mol}$$

$$\text{Amount of } NH_3 = \frac{25.1\text{ g}}{17.0\text{ g mol}^{-1}} = 1.48\text{ mol}$$

$$\text{Amount of } N_2 = \frac{59.6\text{ g}}{28.0\text{ g mol}^{-1}} = 2.13\text{ mol}$$

$$\text{Total amount} = (6.34 + 1.48 + 2.13)\text{ mol} = 9.95\text{ mol}$$

$$X_{H_2} = \frac{\text{amount of } H_2}{\text{total amount}} = \frac{6.34\text{ mol}}{9.95\text{ mol}} = \mathbf{0.640}$$

$$X_{NH_3} = \frac{1.48\text{ mol}}{9.95\text{ mol}} = \mathbf{0.149}$$

$$X_{N_2} = \frac{2.13\text{ mol}}{9.95\text{ mol}} = \mathbf{0.214}$$

b

Partial pressure = mole fraction × total pressure

$p_{H_2} = 0.640 \times 10.0$ atm = **6.40 atm**

$p_{NH_3} = 0.149 \times 10.0$ atm = **1.49 atm**

$p_{N_2} = 0.214 \times 10.0$ atm = **2.14 atm**

c

$$K_p = \frac{p_{NH_3}^2}{p_{N_2} \times p_{H_2}^3} = \frac{(1.49\text{ atm})^2}{2.14\text{ atm} \times (6.40\text{ atm})^3} = \mathbf{3.96 \times 10^{-3}\ atm^{-2}}$$

EXERCISE 6.34

Total amount of gas = (0.33 + 0.67 + 0.67) mol = 1.67 mol
Partial pressure = mole fraction × total pressure

$$p_{PCl_5} = \frac{0.33\text{ mol}}{1.67\text{ mol}} \times 10.0\text{ atm} = 1.98\text{ atm}$$

$$p_{PCl_3} = p_{Cl_2} = \frac{0.67\text{ mol}}{1.67\text{ mol}} \times 10.0\text{ atm} = 4.01\text{ atm}$$

$$K_p = \frac{p_{PCl_3} \times p_{Cl_2}}{p_{PCl_5}} = \frac{4.01\text{ atm} \times 4.01\text{ atm}}{1.98\text{ atm}} = \mathbf{8.12\ atm}$$

EXERCISE 6.35

Total amount of gas = (0.224 + 0.142) mol = 0.366 mol
Partial pressure = mole fraction × total pressure

$$p_{CO_2} = \frac{0.224}{0.366} \times 1.83\text{ atm} = \mathbf{1.12\ atm}$$

$$p_{NH_3} = \frac{0.142}{0.366} \times 1.83\text{ atm} = \mathbf{0.710\ atm}$$

$$K_p = p_{NH_3}^2 \times p_{CO_2} = (0.710\text{ atm})^2 \times 1.12\text{ atm} = \mathbf{0.565\ atm^3}$$

EXERCISE 6.36

Total amount of gas = $(6.2 \times 10^{-3} + 6.2 \times 10^{-3} + 0.994 + 0.994)$ mol = 2.00 mol

Partial pressure = mole fraction × total pressure

$$p_{CO} = \frac{6.2 \times 10^{-3}}{2.00} \times 2.0\text{ atm} = 6.2 \times 10^{-3}\text{ atm}$$

$$p_{H_2O} = \frac{6.2 \times 10^{-3}}{2.00} \times 2.0\text{ atm} = 6.2 \times 10^{-3}\text{ atm}$$

$$p_{H_2} = \frac{0.994}{2.00} \times 2.0\text{ atm} = 0.994\text{ atm}$$

$$p_{CO_2} = \frac{0.994}{2.00} \times 2.0\text{ atm} = 0.994\text{ atm}$$

$$K_p = \frac{p_{CO} \times p_{H_2O}}{p_{CO_2} \times p_{H_2}} = \frac{(6.2 \times 10^{-3}\text{ atm}) \times (6.2 \times 10^{-3}\text{ atm})}{0.994\text{ atm} \times 0.994\text{ atm}} = \mathbf{3.89 \times 10^{-5}}$$

EXERCISE 6.37

Total amount of gas = (0.4 + 0.6) mol = 1.0 mol

$$p_I = \frac{0.40}{1.00} \times 1.0\text{ atm} = 0.40\text{ atm} \qquad p_{I_2} = \frac{0.60}{1.00} \times 1.0\text{ atm} = 0.60\text{ atm}$$

$$K_p = \frac{p_I^2}{p_{I_2}} = \frac{(0.40\text{ atm})^2}{0.60\text{ atm}} = \mathbf{0.27\text{ atm}}$$

EXERCISE 6.38

Total amount of gas = (0.560 + 0.060 + 1.27) mol = 1.89 mol

$$p_{H_2} = \frac{0.560}{1.89} \times 2.00\text{ atm} = 0.593\text{ atm}$$

$$p_{I_2} = \frac{0.060}{1.89} \times 2.00\text{ atm} = 0.0635\text{ atm}$$

$$p_{HI} = \frac{1.27}{1.89} \times 2.00\text{ atm} = 1.34\text{ atm}$$

$$K_p = \frac{p_{HI}^2}{p_{H_2} \times p_{I_2}} = \frac{(1.34\text{ atm})^2}{0.593\text{ atm} \times 0.0635\text{ atm}} = \mathbf{47.7}$$

EXERCISE 6.39

Rearranging the expression $K_p = K_c (RT)^{\Delta n}$ and putting $\Delta n = 1$

$$K_c = \frac{K_p}{(RT)^{\Delta n}} = \frac{0.811\text{ atm}}{0.0821\text{ atm dm}^3\text{ K}^{-1}\text{ mol}^{-1} \times 523\text{ K}} = \mathbf{1.89 \times 10^{-2}\text{ mol dm}^{-3}}$$

EXERCISE 6.40

$$K_p = K_c (RT)^{\Delta n} = 2.0\text{ dm}^6\text{ mol}^{-2} \times (0.0821\text{ atm dm}^3\text{ K}^{-1}\text{ mol}^{-1} \times 620\text{ K})^{-2} = \mathbf{7.72 \times 10^{-4}\text{ atm}^{-2}}$$

EXERCISE 6.41

For the equation, $\Delta n = 0$

$\therefore (RT)^{\Delta n} = 1$ and $K_p = K_c$

EXERCISE 6.42

$$\text{Amount of } SO_3 \text{ at equilibrium} = 2.00\text{ mol} \times \frac{20}{100} = 0.40\text{ mol}$$

	$2SO_2(g)$	$+ O_2(g)$	$\rightleftharpoons 2SO_3(g)$
Initial amount/mol	2.0	2.0	0
Equilibrium amount/mol	1.60	1.80	0.40

Total amount of gas = (1.60 + 1.80 + 0.40) mol = 3.80 mol

$$p_{SO_3} = X_{SO_3} p_T = \frac{0.40}{3.80} \times p_T$$

$$p_{SO_2} = \frac{1.60}{3.80} \times p_T \qquad p_{O_2} = \frac{1.80}{3.80} \times p_T$$

$$K_p = \frac{p_{SO_3}^2}{p_{SO_2}^2 \times p_{O_2}}$$

$$0.13\text{ atm}^{-1} = \frac{(0.40/3.80)^2 p_T^2}{(1.60/3.80)^2 p_T^2 (1.80/3.80) p_T} = \frac{0.0111}{0.177 \times 0.474 p_T}$$

$$p_T = \frac{0.0111}{0.13\text{ atm}^{-1} \times 0.177 \times 0.474} = \mathbf{1.02\text{ atm}}$$

EXERCISE 6.43

	$2SO_2(g)$	$+ O_2(g)$	$\rightleftharpoons 2SO_3(g)$
Initial amount/mol	2	1	0
Equilibrium amount/mol	4/3	2/3	2/3

Total amount of gas = 4/3 + 2/3 + 2/3 = 8/3 mol

Partial pressure = mole fraction × total pressure

$$p_{SO_3} = \frac{(2/3)}{(8/3)} \times 9\text{ atm} = 9/4\text{ atm}$$

$$p_{SO_2} = \frac{(4/3)}{(8/3)} \times 9\text{ atm} = 9/2\text{ atm}$$

$$p_{O_2} = \frac{(2/3)}{(8/3)} \times 9\text{ atm} = 9/4\text{ atm}$$

$$K_p = \frac{p_{SO_3}^2}{p_{SO_2}^2 \times p_{O_2}} = \frac{(9/4)^2\text{ atm}^2}{(9/2)^2\text{ atm}^2 \times 9/4\text{ atm}} = \frac{1}{9}\text{ atm}^{-1} = \mathbf{0.11\text{ atm}^{-1}}$$

EXERCISE 6.44

Total amount of gas = (0.96 + 0.04 + 0.02) mol = 1.02 mol

$$p_{NO_2} = \frac{0.96}{1.02} \times p_T \qquad p_{NO} = \frac{0.04}{1.02} \times p_T \qquad p_{O_2} = \frac{0.02}{1.02} \times p_T$$

$$K_p = \frac{p_{NO}^2 \times p_{O_2}}{p_{NO_2}^2}$$

$$6.8 \times 10^{-6} \text{ atm} = \frac{(0.04/1.02)^2 p_T^2 (0.02/1.02) p_T}{(0.96/1.02)^2 p_T^2}$$

$$p_T = \frac{6.8 \times 10^{-6} \text{ atm} \times 0.886}{0.00154 \times 0.0196} = \mathbf{0.20\ atm}$$

EXERCISE 6.45

Total amount of gas = (2.0 + 1.0 + 4.0) mol = 7.0 mol

$$K_p = \frac{p_{CO} \times p_{H_2}}{p_{H_2O}}$$

$$3.72 \text{ atm} = \frac{\frac{2.0}{7.0} \times p_T \times \frac{1.0}{7.0} \times p_T}{\frac{4.0}{7.0} \times p_T}$$

$$p_T = \frac{3.72 \text{ atm} \times (4.0/7.0)}{(2.0/7.0) \times (1.0/7.0)} = \mathbf{52\ atm}$$

EXERCISE 6.46

Amount of N_2O_4 at equilibrium = initial amount – amount reacted
= (1.00 – 0.66) mol = 0.34 mol

Amount of NO_2 formed = 2 × amount N_2O_4 reacted
= 2 × 0.66 mol = 1.32 mol

	$N_2O_4(g) \rightleftharpoons$	$2NO_2(g)$
Initial amount/mol	1.00	0
Equilibrium amount/mol	0.34	1.32

Total amount of gas = (0.34 + 1.32) mol = 1.66 mol
Partial pressure = mole fraction × total pressure

$$p_{N_2O_4} = \frac{0.34 \text{ mol}}{1.66 \text{ mol}} \times 98.3 \text{ kPa} = 20.1 \text{ kPa}$$

$$p_{NO_2} = \frac{1.32 \text{ mol}}{1.66 \text{ mol}} \times 98.3 \text{ kPa} = 78.2 \text{ kPa}$$

$$K_p = \frac{p_{NO_2}^2}{p_{N_2O_4}} = \frac{(78.2 \text{ kPa})^2}{20.1 \text{ kPa}} = \mathbf{304\ kPa}$$

Since 100 kPa = 760 mmHg = 1.00 atm

$$K_p = 304 \text{ kPa} \times \frac{760 \text{ mmHg}}{100 \text{ kPa}} = \mathbf{2310\ mmHg}$$

and

$$K_p = 304 \text{ kPa} \times \frac{1.00 \text{ atm}}{100 \text{ kPa}} = \mathbf{3.04\ atm}$$

EXERCISE 6.47

a

	$2NO_2(g) \rightleftharpoons$	$2NO(g) +$	$O_2(g)$
Equilibrium amount/mol	0.96	0.040	0.020

Total amount of gas = (0.96 + 0.04 + 0.02) mol = 1.02 mol

$$p_{NO_2} = \frac{0.96}{1.02} \times 0.20 \text{ atm} = 0.19 \text{ atm}$$

$$p_{NO} = \frac{0.040}{1.02} \times 0.20 \text{ atm} = 0.0078 \text{ atm}$$

$$p_{O_2} = \frac{0.02}{1.02} \times 0.20 \text{ atm} = 0.0039 \text{ atm}$$

$$K_p = \frac{p_{NO}^2 \times p_{O_2}}{p_{NO_2}^2} = \frac{0.0078^2 \text{ atm}^2 \times 0.0039 \text{ atm}}{0.19^2 \text{ atm}^2} = \mathbf{6.6 \times 10^{-6}\ atm}$$

b Average molar mass = $X_{NO_2} M_{NO_2} + X_{NO} M_{NO} + X_{O_2} M_{O_2}$

$$\left(\frac{0.96}{1.02} \times 46.0 \text{ g mol}^{-1}\right) + \left(\frac{0.040}{1.02} \times 30.0 \text{ g mol}^{-1}\right) + \left(\frac{0.020}{1.02} \times 32.0 \text{ g mol}^{-1}\right)$$

$$= (43.3 + 1.18 + 0.63) \text{ g mol}^{-1} = \mathbf{45.1\ g\ mol^{-1}}$$

EXERCISE 6.48

a $M_{avg} = X_{CO}M_{CO} + X_{CO_2} M_{CO_2}$
Since $X_{CO} + X_{CO_2} = 1$, $X_{CO_2} = 1 - X_{CO}$
$36 \text{ g mol}^{-1} = (X_{CO} \times 28 \text{ g mol}^{-1}) + (1 - X_{CO}) \times 44 \text{ g mol}^{-1}$
$36 = 28X_{CO} + 44 - 44X_{CO}$

$$X_{CO} = \frac{44 - 36}{44 - 28} = \frac{8}{16} = \mathbf{0.5}$$

b $p_{CO} = X_{CO}p_T = 0.5 \times 12 \text{ atm} = 6 \text{ atm}$
$p_{CO_2} = X_{CO_2}p_T = 0.5 \times 12 \text{ atm} = 6 \text{ atm}$

$$K_p = \frac{p_{CO}^2}{p_{CO_2}} = \frac{(6 \text{ atm})^2}{6 \text{ atm}} = \mathbf{6\ atm}$$

c Let $x = X_{CO}$ and $1 - x = X_{CO_2}$

$$K_p = \frac{(X_{CO} \times p_T)^2}{X_{CO_2} \times p_T}$$

$$\therefore\ 6 \text{ atm} = \frac{(x \times 2 \text{ atm})^2}{(1 - x) \times 2 \text{ atm}} \text{ or } 6 = \frac{2x^2}{1 - x}$$

$$\therefore\ 2x^2 = 6 - 6x$$

$$x^2 + 3x - 3 = 0$$

$$x = \frac{-3 \pm \sqrt{9 + 12}}{2} = 0.8 \text{ or } -3.8 \text{ (absurd root)}$$

$$\therefore\ X_{CO} = \mathbf{0.8}$$

EXERCISE 6.49

a Total amount of gas = (1.0 + 3.6 + 13.5) mol = 18.1 mol

$$p_{NH_3} = \frac{1.0}{18.1} \times 2.0 \text{ atm} = 0.11 \text{ atm}$$

$$p_{H_2} = \frac{3.6}{18.1} \times 2.0 \text{ atm} = 0.40 \text{ atm}$$

$$p_{N_2} = \frac{13.5}{18.1} \times 2.0 \text{ atm} = 1.49 \text{ atm}$$

$$K_p = \frac{p^2_{NH_3}}{p^3_{H_2} \times p_{N_2}} = \frac{(0.11 \text{ atm})^2}{(0.40 \text{ atm})^3 \times 1.49 \text{ atm}} = \mathbf{0.13\ atm^{-2}}$$

b

$$M_{avg} = X_{NH_3}M_{NH_3} + X_{H_2}M_{H_2} + X_{N_2}M_{N_2}$$

$$= \frac{1.0}{18.1} \times 17.0 \text{ g mol}^{-1} + \frac{3.6}{18.1} \times 2.0 \text{ g mol}^{-1} + \frac{13.5}{18.1} \times 28.0 \text{ g mol}^{-1}$$

$$= (0.94 + 0.40 + 20.9) \text{ g mol}^{-1} = \mathbf{22.2\ g\ mol^{-1}}$$

EXERCISE 6.50

a Total amount of gas = (0.20 + 0.010 + 3.8) mol = 4.01 mol

$$p_{PCl_5} = \frac{0.20}{4.01} \times 3.0 \text{ atm} = 0.15 \text{ atm}$$

$$p_{PCl_3} = \frac{0.010}{4.01} \times 3.0 \text{ atm} = 7.48 \times 10^{-3} \text{ atm}$$

$$p_{Cl_2} = \frac{3.8}{4.01} \times 3.0 \text{ atm} = 2.84 \text{ atm}$$

$$K_p = \frac{p_{PCl_3} \times p_{Cl_2}}{p_{PCl_5}} = \frac{7.48 \times 10^{-3} \text{ atm} \times 2.84 \text{ atm}}{0.15 \text{ atm}} = \mathbf{0.14\ atm}$$

b

$$M_{avg} = X_{PCl_5}M_{PCl_5} + X_{PCl_3}M_{PCl_3} + X_{Cl_2}M_{Cl_2}$$

$$= \frac{0.20}{4.01} \times 208.5 \text{ g mol}^{-1} + \frac{0.010}{4.01} \times 137.5 \text{ g mol}^{-1} + \frac{3.8}{4.01} \times 71.0 \text{ g mol}^{-1}$$

$$= (10.40 + 0.34 + 67.28) \text{ g mol}^{-1} = \mathbf{78\ g\ mol^{-1}}$$

EXERCISE 6.51

a

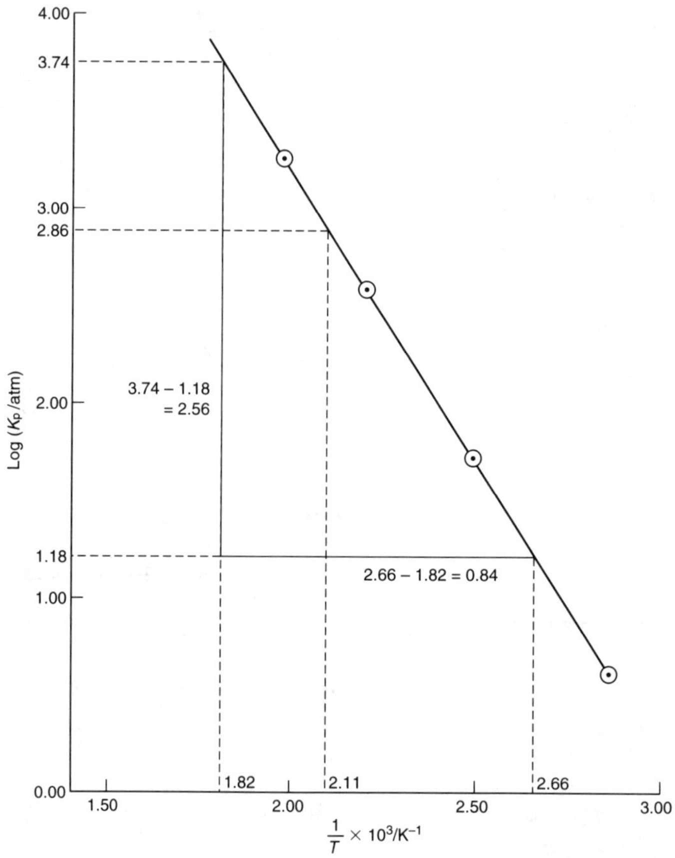

b The slope of the graph = -3.05×10^3 K (note the minus sign)

Since $-\frac{\Delta H^\circ}{2.30R}$ = slope

$\Delta H^\circ = 2.30 \times 8.31 \text{ J mol}^{-1} \text{ K}^{-1} \times 3.05 \times 10^3 \text{ K} = 58\,295 \text{ J mol}^{-1}$
$= \mathbf{58.3\ kJ\ mol^{-1}}$

c i) When T = 375 K, $1/T = 2.66 \times 10^{-3}$ K^{-1}
Interpolation of graph gives log K_p = **1.18**

ii) When T = 475 K, $1/T = 2.11 \times 10^{-3}$ K^{-1}
Interpolation of graph gives log K_p = **2.86**

iii) When T = 550 K, $1/T = 1.82 \times 10^{-3}$ K^{-1}
Extrapolation of graph gives log K_p = **3.74**

The assumption you made was that there is a linear relationship between log K_p and $1/T$ (i.e. $\Delta H^{\ominus}$ is constant).

d An increase in temperature will shift the position of equilibrium to the right resulting in the production of more NO_2. This was deduced from the graph because an increase in temperature (a decrease in the value of $1/T$) increases the value of K_p.

EXERCISE 6.52

a The slope of the graph = $+4.92 \times 10^3$ K

$$-\frac{\Delta H^{\ominus}}{2.30R} = \text{slope}$$

$$\therefore \Delta H^{\ominus} = -2.30 \times 8.31\ \text{J mol}^{-1}\ \text{K}^{-1} \times 4.92 \times 10^3\ \text{K} = -94\,036\ \text{J mol}^{-1} = \mathbf{-94.0\ kJ\ mol^{-1}}$$

b An increase in temperature will shift the position of equilibrium to the left, resulting in the production of more N_2 and H_2. This was deduced from the graph because an increase in temperature (a decrease in the value of $1/T$) results in a decrease in the value of K_p.

c i) The gradient for Exercise 6.51 is negative and the gradient for this Exercise is positive.

ii) The sign of $\Delta H^{\ominus}$ determines the sign of the slope. For an exothermic reaction ($\Delta H^{\ominus}$ is negative), the value of $-\Delta H^{\ominus}/2.30\ R$ is positive, giving a positive gradient. For an endothermic reaction where $\Delta H^{\ominus}$ is positive, the value of $-\Delta H^{\ominus}/2.30R$ is negative, giving a negative gradient.

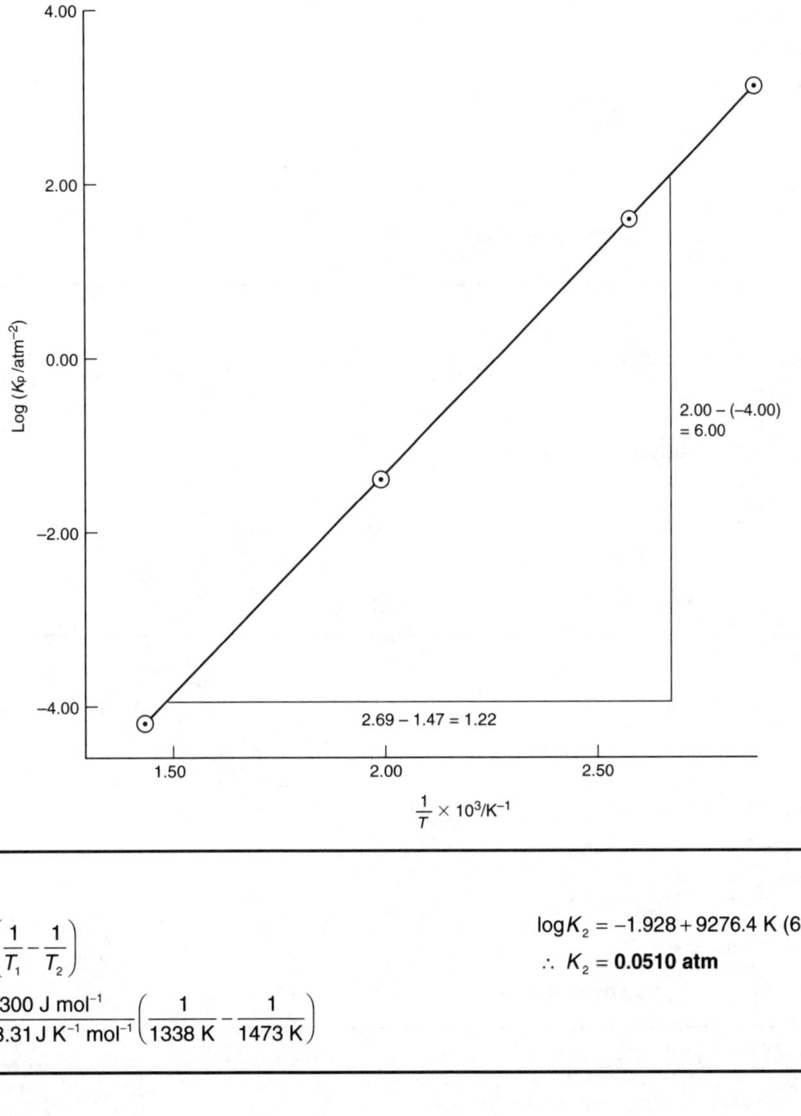

EXERCISE 6.53

$$\log\frac{K_2}{K_1} = \frac{\Delta H^{\ominus}}{2.30R}\left(\frac{1}{T_1} - \frac{1}{T_2}\right)$$

$$\log K_2 - \log 0.0118 = \frac{177\,300\ \text{J mol}^{-1}}{2.30 \times 8.31\ \text{J K}^{-1}\ \text{mol}^{-1}}\left(\frac{1}{1338\ \text{K}} - \frac{1}{1473\ \text{K}}\right)$$

$$\log K_2 = -1.928 + 9276.4\ \text{K}\ (6.85 \times 10^{-5}\ \text{K}^{-1}) = -1.293$$

$$\therefore K_2 = \mathbf{0.0510\ atm}$$

EXERCISE 6.54

$$\log\frac{K_2}{K_1} = \frac{\Delta H^{\ominus}}{2.30R}\left(\frac{1}{T_1} - \frac{1}{T_2}\right)$$

$$\therefore \Delta H^{\ominus} = \frac{2.30R \times \log\frac{K_2}{K_1}}{\left(\frac{1}{T_1} - \frac{1}{T_2}\right)} = \frac{2.30 \times 8.31\ \text{J K}^{-1}\ \text{mol}^{-1} \times \log\left(\frac{3.74}{2.44}\right)}{\left(\frac{1}{1000\ \text{K}} - \frac{1}{1200\ \text{K}}\right)}$$

$$= \frac{3.545\ \text{J K}^{-1}\ \text{mol}^{-1}}{1.667 \times 10^{-4}\ \text{K}^{-1}} = 2.13 \times 10^4\ \text{J mol}^{-1} = \mathbf{21.3\ kJ\ mol^{-1}}$$

EXERCISE 6.55

a $K_p = \dfrac{p^2_{NO}}{p_{N_2} \times p_{O_2}}$

b From the expression for K_p above, $p^2_{NO} = K_p \times p_{N_2} \times p_{O_2}$
The values of K_p and p_{N_2} remain constant, so that $p_{NO} \propto \sqrt{p_{O_2}}$
p_{O_2} has the values 0.05 atm in the fourth column and 0.2 atm in the third.
p_{NO} is proportional to the square roots of these values.
Since $0.05 = \frac{1}{4} \times 0.2$, $\sqrt{0.05} = \frac{1}{2} \times \sqrt{0.2}$
and the value of p_{NO} in the fourth column is half that in the third.

c The table shows that the value of K_p increases with temperature. This indicates a higher yield of NO with increased temperature. Since there are the same number of molecules either side of the equation, changing the pressure will have no effect on the yield of NO.

d The reaction is endothermic, because K_p increases with temperature. Le Chatelier's principle predicts that an increase in temperature will favour an endothermic reaction.

EXERCISE 6.56

$$\log\frac{K_2}{K_1} = \frac{\Delta H^{\ominus}}{2.30R}\left(\frac{1}{T_1} - \frac{1}{T_2}\right)$$

Taking the values of K_p at 1800 K and 2000 K

$$\therefore \quad \Delta H^{\ominus} = \frac{2.30R \times \log\dfrac{K_2}{K_1}}{\left(\dfrac{1}{T_1} - \dfrac{1}{T_2}\right)} = \frac{2.30 \times 8.31\ \text{J K}^{-1}\ \text{mol}^{-1} \times \log\left(\dfrac{4.08}{1.21}\right)}{\left(\dfrac{1}{1800\ \text{K}} - \dfrac{1}{2000\ \text{K}}\right)}$$

$$= \frac{10.089\ \text{J K}^{-1}\ \text{mol}^{-1}}{5.56 \times 10^{-5}\ \text{K}^{-1}} = 1.81 \times 10^5\ \text{J mol}^{-1} = \mathbf{181\ kJ\ mol^{-1}}$$

EXERCISE 6.57

a $CdCO_3(s) \rightleftharpoons Cd^{2+}(aq) + CO_3^{2-}(aq)$

Equilibrium concn./mol dm^{-3}: 1.58×10^{-7} $\quad 1.58 \times 10^{-7}$

$K_s = [Cd^{2+}(aq)]\,[CO_3^{2-}(aq)] = 1.58 \times 10^{-7}$ mol dm^{-3} $\times\ 1.58 \times 10^{-7}$ mol dm^{-3}

$= \mathbf{2.50 \times 10^{-14}\ mol^2\ dm^{-6}}$

b $CaF_2(s) \rightleftharpoons Ca^{2+}(aq) + 2F^-(aq)$

Equilibrium concn./mol dm^{-3}: 2.15×10^{-4} $\quad 2 \times 2.15 \times 10^{-4}$

$K_s = [Ca^{2+}(aq)]\,[F^-(aq)]^2 = 2.15 \times 10^{-4}$ mol dm^{-3} $\times\ (4.30 \times 10^{-4}$ mol dm$^{-3})^2$
$= \mathbf{3.98 \times 10^{-11}\ mol^3\ dm^{-9}}$

c $Cr(OH)_3(s) \rightleftharpoons Cr^{3+}(aq) + 3OH^-(aq)$

Equilibrium concn./mol dm^{-3}: 1.39×10^{-8} $\quad 3 \times 1.39 \times 10^{-8}$

$K_s = [Cr^{3+}(aq)]\,[OH^-(aq)]^3$
$= 1.39 \times 10^{-8}$ mol dm^{-3} $\times\ (3 \times 1.39 \times 10^{-8}$ mol dm$^{-3})^3$
$= \mathbf{1.01 \times 10^{-30}\ mol^4\ dm^{-12}}$

EXERCISE 6.58

Let A refer to $H^+(aq)$ and B refer to $OH^-(aq)$

$$\frac{c_A V_A}{c_B V_B} = \frac{1}{1}$$

$$\frac{0.100\ \text{mol dm}^{-3} \times 11.45\ \text{cm}^3}{c_B \times 25.0\ \text{cm}^3} = 1$$

$$\therefore c_B = \frac{0.100\ \text{mol dm}^{-3} \times 11.45\ \text{cm}^3}{25.0\ \text{cm}^3}$$

$= \mathbf{0.0458\ mol\ dm^{-3}}$

$[Ca^{2+}(aq)] = \frac{1}{2}[OH^-(aq)] = \mathbf{0.0229\ mol\ dm^{-3}}$

Solubility of $Ca(OH)_2$ = $[Ca^{2+}(aq)]$ = 0.0229 mol dm^{-3} (at 15 °C)
$\therefore K_s = [Ca^{2+}(aq)][OH^-(aq)]^2$
$= 0.0229$ mol dm^{-3} $\times\ (0.0458$ mol dm$^{-3})^2$
$= \mathbf{4.80 \times 10^{-5}\ mol^3\ dm^{-9}}$

EXERCISE 6.59

a $CuS(s) \rightleftharpoons Cu^{2+}(aq) + S^{2-}(aq)$

Equilibrium concn./mol dm^{-3}: x $\quad x$

$K_s = [Cu^{2+}(aq)]\,[S^{2-}(aq)]$

6.3×10^{-36} mol^2 dm^{-6} = x mol dm^{-3} $\times$ x mol dm^{-3}

$x^2 = 6.3 \times 10^{-36}$ $\quad \therefore x = 2.5 \times 10^{-18}$

and solubility = $\mathbf{2.5 \times 10^{-18}\ mol\ dm^{-3}}$

b $Fe(OH)_2(s) \rightleftharpoons Fe^{2+}(aq) + 2OH^-(aq)$

Equilibrium concn./mol dm^{-3}: x $\quad 2x$

$K_s = [Fe^{2+}(aq)]\,[OH^-(aq)]^2$

6.0×10^{-15} mol^3 dm^{-9} = x mol dm^{-3} $\times$ $(2x$ mol dm$^{-3})^2$
$\therefore 4x^3 = 6.0 \times 10^{-15}$
$x = (1.5 \times 10^{-15})^{1/3} = 1.1 \times 10^{-5}$
and solubility = $\mathbf{1.1 \times 10^{-5}\ mol\ dm^{-3}}$

c $Ag_3PO_4(s) \rightleftharpoons 3Ag^+(aq) + PO_4^{3-}(aq)$

Equilibrium concn./mol dm^{-3}: $3x$, x

$K_s = [Ag^+(aq)]^3\,[PO_4^{3-}(aq)]$

1.25×10^{-20} mol^4 dm^{-12} = $(3x$ mol dm$^{-3})^3 \times x$ mol dm^{-3}

$\therefore 27x^4 = 1.25 \times 10^{-20}$

$x = (4.63 \times 10^{-22})^{1/4} = 4.64 \times 10^{-6}$

and solubility = **4.64×10^{-6} mol dm^{-3}**

EXERCISE 6.60

a Initial concn./mol dm^{-3}: 0, 0.10

$SrSO_4(s) \rightleftharpoons Sr^{2+}(aq) + SO_4^{2-}(aq)$

Equilibrium concn./mol dm^{-3}: x, $(0.10 + x)$

$K_s = [Sr^{2+}(aq)][SO_4^{2-}(aq)]$

4.0×10^{-7} mol^2 dm^{-6} = $(x$ mol dm$^{-3})$ $(0.10 + x)$ mol dm^{-3}

Assume that $(0.10 + x) = 0.10$

$\therefore 4.0 \times 10^{-7} = x \times 0.10$

$\therefore x = 4.0 \times 10^{-6}$

and solubility = **4.0×10^{-6} mol dm^{-3}**

b Initial concn./mol dm^{-3}: 0, 0.20

$MgF_2(s) \rightleftharpoons Mg^{2+}(aq) + 2F^-(aq)$

Equilibrium concn./mol dm^{-3}: x, $(2x + 0.20)$

$K_s = [Mg^{2+}(aq)][F^-(aq)]^2$

7.2×10^{-9} mol^3 dm^{-9} = $(x$ mol dm$^{-3})$ $(2x + 0.20)^2$ mol^2 dm^{-6}

Assume that $(2x + 0.20) \approx 0.20$

$7.2 \times 10^{-9} = x \times (0.20)^2$

$\therefore x = 1.8 \times 10^{-7}$

and solubility = **1.8×10^{-7} mol dm^{-3}**

EXERCISE 6.61

$K_s\,(SrCO_3) = 1.1 \times 10^{-10}$ mol^2 dm^{-6}

Since the volume is doubled

Initial concn./mol dm^{-3}: 0.50

$SrCO_3(s) \rightleftharpoons Sr^{2+}(aq) + CO_3^{2-}(aq)$

Equilibrium concn./mol dm^{-3}: x, $(x + 0.50)$

$K_s = [Sr^{2+}(aq)][CO_3^{2-}(aq)]$

1.1×10^{-10} mol^2 dm^{-6} = x mol dm$^{-3} \times (x + 0.50)$ mol dm^{-3}

Assume that $x \ll 0.5$ i.e. $x + 0.50 \approx 0.50$

$$\therefore x = \frac{1.1 \times 10^{-10}}{0.50} = 2.2 \times 10^{-10}$$

and $[Sr^{2+}(aq)]$ = **2.2×10^{-10} mol dm^{-3}**

EXERCISE 6.62

Before mixing, $[Ba^{2+}(aq)] = 0.10$ mol dm^{-3}.
Mixing changes volume from 150 cm^3 to 200 cm^3.
Then $[Ba^{2+}(aq)] = 0.10$ mol dm$^{-3} \times (150/200) = 0.075$ mol dm^{-3}
Before mixing, $[F^-(aq)] = 0.050$ mol dm^{-3}.
Mixing changes volume from 50 cm^3 to 200 cm^3.

Then, $[F^-(aq)] = 0.050$ mol dm$^{-3} \times (50/200) = 0.0125$ mol dm^{-3}
Ion product = $[Ba^{2+}(aq)][F^-(aq)]^2$
= 0.075 mol dm$^{-3} \times (0.0125$ mol dm$^{-3})^2$
= 1.2×10^{-5} mol^3 dm^{-9}
Ion product is greater than K_s. Therefore **precipitation occurs**.

EXERCISE 6.63

a Ag^+Cl^- : 2 concentration terms, $\therefore$ unit = (mol dm$^{-3})^2$ = mol^2 dm^{-6}

$Pb^{2+}2Br^-$: 3 concentration terms, $\therefore$ unit = (mol dm$^{-3})^3$ = mol^3 dm^{-9}

$Ag^+BrO_3^-$: 2 concentration terms, $\therefore$ unit = (mol dm$^{-3})^2$ = mol^2 dm^{-6}

$Mg^{2+}2OH^-$: 3 concentration terms, $\therefore$ unit = (mol dm$^{-3})^3$ = mol^3 dm^{-9}

b When equal volumes of the solution are mixed, their concentrations, before reaction, are halved: i.e. concentrations of all ions required for ion product calculations = 5.0×10^{-4} mol dm^{-3}

i) Ion product = $[Ag^+(aq)]\,[Cl^-(aq)]$ = $(5.0 \times 10^{-4}$ mol dm$^{-3})^2$ = 2.5×10^{-7} mol^2 dm^{-6}

$> K_s \;\therefore$ **precipitation occurs**

ii) Ion product = $[Pb^{2+}(aq)][Br^-(aq)]^2$ = $(5.0 \times 10^{-4}$ mol dm$^{-3})^3$ = 1.25×10^{-10} mol^3 dm^{-9}

$< K_s \;\therefore$ **no precipitation**

iii) Ion product = $[Ag^+(aq)][BrO_3^-(aq))]$ = $(5.0 \times 10^{-4}$ mol dm$^{-3})^2$ = 2.5×10^{-7} mol^2 dm^{-6}

$< K_s \;\therefore$ **no precipitation**

iv) Ion product = $[Mg^{2+}(aq)][OH^-(aq)]^2$ = $(5.0 \times 10^{-4}$ mol dm$^{-3})^3$ = 1.25×10^{-10} mol^3 dm^{-9}

$> K_s \;\therefore$ **precipitation occurs**

EXERCISE 6.64

$$K_d = \frac{[\text{acid(aq)}]}{[\text{acid(ether)}]}$$

$$K_d = \frac{0.0759 \text{ mol dm}^{-3}}{0.0114 \text{ mol dm}^{-3}} = 6.66$$

$$K_d = \frac{0.108 \text{ mol dm}^{-3}}{0.0162 \text{ mol dm}^{-3}} = 6.67$$

$$K_d = \frac{0.158 \text{ mol dm}^{-3}}{0.0237 \text{ mol dm}^{-3}} = 6.67$$

$$K_d = \frac{0.300 \text{ mol dm}^{-3}}{0.0451 \text{ mol dm}^{-3}} = 6.65$$

Average K_d = **6.66**

Or, from the graph (see next page)

$$\text{slope} = \frac{\Delta[\text{acid(aq)}]}{\Delta[\text{acid(ether)}]}$$

$$K_d = \frac{(0.268 - 0.050)}{(0.0400 - 0.0072)}$$

$$= \frac{0.218}{0.0328} = \mathbf{6.65}$$

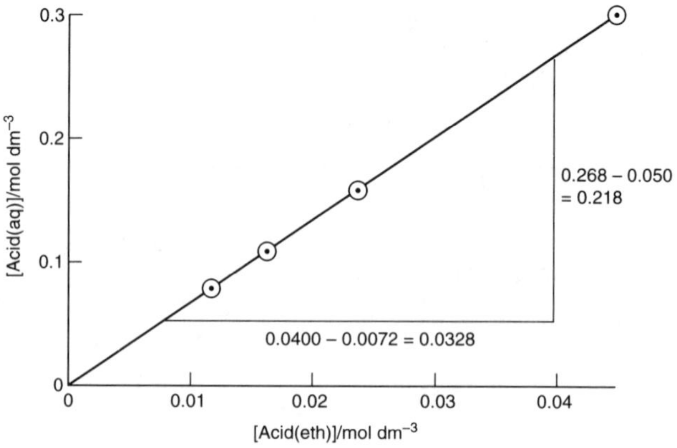

EXERCISE 6.65

a

Initial amount/mol	0.010	0
	NH_3(tce) $\rightleftharpoons$	NH_3(aq)
Equilibrium amount/mol	x	$0.010 - x$

$$K_d = \frac{[NH_3(aq)]}{[NH_3(tce)]}$$

$$290 = \frac{(0.010 - x)\ \text{mol}/0.10\ \text{dm}^3}{x\ \text{mol}/0.10\ \text{dm}^3} = \frac{(0.010 - x)}{x}$$

$$290x = 0.010 - x$$

$$x = \frac{0.010}{291} = 3.4 \times 10^{-5}$$

∴ the amount of ammonia remaining = x mol = **3.4 × 10^{-5} mol**

b After the first addition of water, the remaining amount of ammonia, x mol, is given by the same method as in **a**:

$$290 = \frac{(0.010 - x_1)\ \text{mol}/0.025\ \text{dm}^3}{x_1\ \text{mol}/0.10\ \text{dm}^3} = \frac{4(0.010 - x_1)}{x_1}$$

$$290x_1 = 0.040 - 4x_1$$

$$x_1 = \frac{0.040}{294} = 1.4 \times 10^{-4}$$

Note that this is 0.010 × (4/294)
or

$$\text{remaining amount} = \text{original amount} \times \frac{V(\text{tce})/V(\text{aq})}{K_d + V(\text{tce})/V(\text{aq})}$$

Thus, after the second addition

$$x_2 = \left(0.010 \times \frac{4}{294}\right) \times \frac{4}{294} = 0.010 \times \left(\frac{4}{294}\right)^2$$

and after the fourth addition

$$x_4 = 0.010 \times \left(\frac{4}{294}\right)^4 = 3.4 \times 10^{-10}$$

The amount finally remaining = x_4 mol = **3.4 × 10^{-10} mol**

Note that the amount remaining is reduced by a factor of 10^{-5}: clearly, the extraction is made far more efficient by dividing the extracting liquid into several portions.

EXERCISE 6.66

If the molecular formula in trichloromethane is $(CH_3CO_2H)_n$, then

$$K_d = \frac{[CH_3CO_2(aq)]^n}{[CH_3CO_2H(tcm)]} \quad (\text{tcm} = \text{trichloromethane})$$

Taking logarithms,

$$\log K_d = n \log [CH_3CO_2H(aq)] - \log [CH_3CO_2H(tcm)]$$

Rearranging this equation into the form $y = mx + c$

$$\log[CH_3CO_2H(tcm)] = n \log[CH_3CO_2H\ (aq)] - \log K_d$$

∴ a plot of log $[CH_3CO_2H(tcm)]$ against log $[CH_3CO_2H(aq)]$ should be a straight line with slope n. (Here we may use concentrations in g dm^{-3} because they are proportional to concentrations in mol dm^{-3}.)

log ($[CH_3CO_2H(tcm)]$/g dm^{-3})	1.24	1.64	1.93
log ($[CH_3CO_2H(aq)]$/g dm^{-3})	2.47	2.68	2.81

From the graph, the slope, $n = \frac{1.90 - 1.40}{2.80 - 2.55} = \frac{0.50}{0.25} = 2.0$

∴ the formula of ethanoic acid in trichloromethane is $\mathbf{(CH_3CO_2H)_2}$

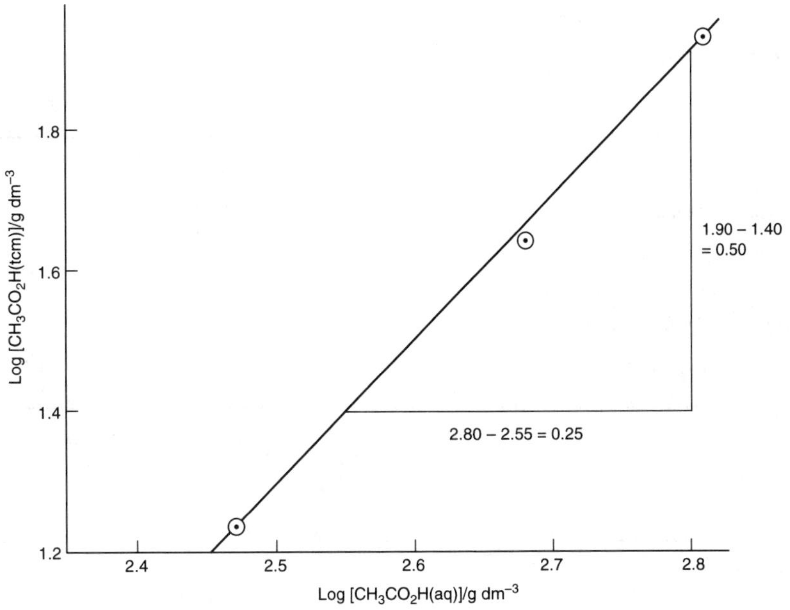

End-of-chapter questions

6.1 0.17 $dm^6\ mol^{-2}$
6.2 ⅔ mol
6.3 59.6 g
6.4 1.6 atm
6.5 **a** 0.426
b 0.316 atm
c 0.205
6.6 D
6.7 3
6.8 4.0
6.9 **a** 1.64×10^6 N m^{-2}
b Yes.

6.10 **a** $K_s = [Pb^{2+}(aq)]\ [Cl^-(aq)]^2$
b i) 1.6×10^{-2} mol dm^{-3}
ii) 4.0×10^{-3} mol
6.11 **a** $K_c = [SO_4^{2-}(aq)]/[I^-(aq)]^2$
b The concentration of a solid is constant; therefore, changing the amount of
solid present has no effect on the equilibrium position.
c i) 0.100 mol dm^{-3}
ii) 0.0620 mol dm^{-3}
iii) 0.038 mol dm^{-3}
iv) 0.019 mol dm^{-3}
v) 4.94 $dm^3\ mol^{-1}$

CHAPTER 7

EXERCISE 7.1

$H_2O(l) \rightleftharpoons H^+(aq) + OH^-(aq)$

$$K_c = \frac{[H^+(aq)][OH^-(aq)]}{[H_2O(l)]}$$

$$1.80 \times 10^{-16}\ \text{mol dm}^{-3} = \frac{[H^+(aq)][OH^-(aq)]}{55.6\ \text{mol dm}^{-3}}$$

$[H^+(aq)][OH^-(aq)] = 1.80 \times 10^{-16}\ \text{mol dm}^{-3} \times 55.6\ \text{mol dm}^{-3}$
$= 1.00 \times 10^{-14}\ \text{mol}^2\ \text{dm}^{-6}$

From the equation, one mole of OH^- and one mol of H^+ are produced for every mole of H_2O ionised.

$$[H^+(aq)] = [OH^-(aq)] = \sqrt{1.00 \times 10^{-14}\ \text{mol}^2\ \text{dm}^{-6}} = 1.00 \times 10^{-7}\ \text{mol dm}^{-3}$$

The fraction ionised is given by:

$$\frac{\text{amount ionised}}{\text{initial amount}} = \frac{1.00 \times 10^{-7}\ \text{mol}}{55.6\ \text{mol}} = \mathbf{1.80 \times 10^{-9}}$$

(Approximately 18 molecules ionised in 10 000 000 000.)

EXERCISE 7.2

a In 0.010 M HCl, [H+(aq)] = 0.010 mol dm^{-3}

$$K_w = [H^+(aq)][OH^-(aq)]$$

$$\therefore [OH^-(aq)] = \frac{K_w}{[H^+(aq)]} = \frac{1.0 \times 10^{-14}\ \text{mol}^2\ \text{dm}^{-6}}{0.010\ \text{mol dm}^{-3}}$$

$$= \mathbf{1.0 \times 10^{-12}\ mol\ dm^{-3}}$$

b In 0.10 M H_2SO_4, $[H^+(aq)] = 2 \times 0.10\ mol\ dm^{-3} = 0.20\ mol\ dm^{-3}$

$$K_w = [H^+(aq)][OH^-(aq)]$$

$$\therefore [OH^-(aq)] = \frac{K_w}{[H^+(aq)]} = \frac{1.0 \times 10^{-14}\ mol^2\ dm^{-6}}{0.20\ mol\ dm^{-3}}$$

$$= \mathbf{5.0 \times 10^{-14}\ mol\ dm^{-3}}$$

EXERCISE 7.3

a In 0.010 M KOH, $[OH^-(aq)] = 0.010\ mol\ dm^{-3}$

$$K_w = [H^+(aq)][OH^-(aq)]$$

$$\therefore [H^+(aq)] = \frac{K_w}{[OH^-(aq)]} = \frac{1.0 \times 10^{-14}\ mol^2\ dm^{-6}}{0.010\ mol\ dm^{-3}}$$

$$= \mathbf{1.0 \times 10^{-12}\ mol\ dm^{-3}}$$

b In 0.050 M $Ba(OH)_2$,
$[OH^-(aq)] = 2 \times 0.050\ mol\ dm^{-3} = 0.10\ mol\ dm^{-3}$

$$\therefore [H^+(aq)] = \frac{K_w}{[OH^-(aq)]} = \frac{1.0 \times 10^{-14}\ mol^2\ dm^{-6}}{0.10\ mol\ dm^{-3}}$$

$$= \mathbf{1.0 \times 10^{-13}\ mol\ dm^{-3}}$$

EXERCISE 7.4

$[H^+(aq)] = 10\ mol\ dm^{-3} = 1.0 \times 10^1\ mol\ dm^{-3}$
$pH = -\log [H^+(aq)] = -\log (1 \times 10^1) = \mathbf{-1}$

EXERCISE 7.5

A: pH = **3**
B: pH = **6**
C: pH = **9**

EXERCISE 7.6

$$[H^+(aq)] = \frac{K_w}{[OH^-(aq)]} = \frac{1.0 \times 10^{-14}\ mol^2\ dm^{-6}}{10\ mol\ dm^{-3}} = 1.0 \times 10^{-15}\ mol\ dm^{-3}$$

$$pH = -\log [H^+(aq)] = -\log (1.0 \times 10^{-15}) = \mathbf{15}$$

EXERCISE 7.7

a $K_w = [H^+(aq)][OH^-(aq)]$

$$[H^+(aq)] = \frac{K_w}{[OH^-(aq)]} = \frac{1.0 \times 10^{-14}\ mol^2\ dm^{-6}}{1.0 \times 10^{-8}\ mol\ dm^{-3}}$$

$$= 1.0 \times 10^{-6}\ mol\ dm^{-3}$$

$$pH = -\log [H^+(aq)] = -\log (1.0 \times 10^{-6}) = \mathbf{6.0}$$

b
$$[H^+(aq)] = \frac{K_w}{[OH^-(aq)]} = \frac{1.0 \times 10^{-14}\ mol^2\ dm^{-6}}{0.10\ mol\ dm^{-3}}$$

$$= 1.0 \times 10^{-13}\ mol\ dm^{-3}$$

$$pH = -\log [H^+(aq)] = -\log (1.0 \times 10^{-13}) = \mathbf{13.0}$$

c
$$[H^+(aq)] = \frac{K_w}{[OH^-(aq)]} = \frac{1.0 \times 10^{-14}\ mol^2\ dm^{-6}}{1.0 \times 10^{-4}\ mol\ dm^{-3}}$$

$$= 1.0 \times 10^{-10}\ mol\ dm^{-3}$$

$$pH = -\log [H^+(aq)] = -\log (1.0 \times 10^{-10}) = \mathbf{10.0}$$

EXERCISE 7.8

$pH = -\log [H^+(aq)]$
$= -\log (7.0 \times 10^{-7}) = -(-6.1549) = \mathbf{6.2}$

EXERCISE 7.9

$pH = -\log [H^+(aq)] = -\log (3.2 \times 10^{-9}) = -(-8.4948) = \mathbf{8.5}$

EXERCISE 7.10

$K_w = [H^+(aq)][OH^-(aq)]$
In pure water:

$$[H^+(aq)] = [OH^-(aq)] \quad \text{so that} \quad K_w = [H^+(aq)]^2$$

$$[H^+(aq)] = \sqrt{5.47 \times 10^{-14}\ mol^2\ dm^{-6}} = 2.34 \times 10^{-7}\ mol\ dm^{-3}$$

$$pH = -\log [H^+(aq)] = -\log (2.34 \times 10^{-7}) = -(-6.6308) = \mathbf{6.63}$$

EXERCISE 7.11

a $pH = -\log [H^+(aq)] = 9.0$
$\log [H^+(aq)] = -9.0$
$\therefore [H^+(aq)] = 10^{-9.0} = \mathbf{1.0 \times 10^{-9}\ mol\ dm^{-3}}$

b $pH = -\log [H^+(aq)] = 2.0$
$\log [H^+(aq)] = -2.0$
$\therefore [H^+(aq)] = 10^{-2.0} = \mathbf{1.0 \times 10^{-2}\ mol\ dm^{-3}}$

EXERCISE 7.12

a $pH = -\log [H^+(aq)] = 9.20$
$\log [H^+(aq)] = -9.20$
$\therefore [H^+(aq)] = 10^{-9.2} = \mathbf{6.31 \times 10^{-10}\ mol\ dm^{-3}}$

b $pH = -\log [H^+(aq)] = 2.63$
$\log [H^+(aq)] = -2.63$
$\therefore [H^+(aq)] = 10^{-2.63} = \mathbf{2.34 \times 10^{-3}\ mol\ dm^{-3}}$

EXERCISE 7.13

First, calculate $[H^+(aq)]$ from pH:

$$pH = -\log [H^+(aq)] = 4.17$$
$$\log [H^+(aq)] = -4.17$$
$$\therefore [H^+(aq)] = 10^{-4.17} = 6.76 \times 10^{-5}\ mol\ dm^{-3}$$

At equilibrium, the concentrations of hydrogen ion and hydrogencarbonate ion are the same ($6.76 \times 10^{-5}\ mol\ dm^{-3}$):

Initial concn./mol dm^{-3}	0.010		0		0
	$H_2CO_3(aq)$	$\rightleftharpoons$	$H^+(aq)$	+	$HCO_3^-(aq)$
Equilibrium concn./mol dm^{-3}	$(0.010 - 6.76 \times 10^{-5})$		6.76×10^{-5}		6.76×10^{-5}

$$K_a = \frac{[H^+(aq)][HCO_3^-(aq)]}{[H_2CO_3(aq)]} = \frac{(6.76 \times 10^{-5}\ mol\ dm^{-3})^2}{(0.010 - 6.76 \times 10^{-5}\ mol\ dm^{-3})}$$

Since $6.76 \times 10^{-5} \ll 0.010$, then $0.010 - 6.76 \times 10^{-5} \simeq 0.010$

$$\therefore K_a = \frac{(6.76 \times 10^{-5}\ mol\ dm^{-3})^2}{(0.010\ mol\ dm^{-3})} = \mathbf{4.57 \times 10^{-7}\ mol\ dm^{-3}}$$

($4.60 \times 10^{-7}\ mol\ dm^{-3}$ without approximation)

EXERCISE 7.14

First, calculate $[H^+(aq)]$ from pH:

$$pH = -\log [H^+(aq)] = 3.28$$
$$\log [H^+(aq)] = -3.28$$
$$\therefore [H^+(aq)] = 10^{-3.28} = 5.25 \times 10^{-4}\ mol\ dm^{-3}$$

At equilibrium, the concentration of hydrogen ion is equal to the concentration of benzoate ion, i.e. $[C_6H_5CO_2^-(aq)] = 5.25 \times 10^{-4}\ mol\ dm^{-3}$:

Initial concn./mol dm^{-3}	0.0050		0		0
	$C_6H_5CO_2H(aq)$	$\rightleftharpoons$	$H^+(aq)$	+	$C_6H_5CO_2^-(aq)$
Equilibrium concn./mol dm^{-3}	$(0.0050 - 5.25 \times 10^{-4})$		5.25×10^{-4}		5.25×10^{-4}

$$K_a = \frac{[H^+(aq)][C_6H_5CO_2^-(aq)]}{[C_6H_5CO_2H(aq)]}$$

$$= \frac{(5.25 \times 10^{-4}\ mol\ dm^{-3})(5.25 \times 10^{-4}\ mol\ dm^{-3})}{(0.004475\ mol\ dm^{-3})}$$

$$= \mathbf{6.16 \times 10^{-5}\ mol\ dm^{-3}}$$

(5.51×10^{-5} using approximation – clearly not valid.)

EXERCISE 7.15

a $[H^+(aq)] = \dfrac{K_w}{[OH^-(aq)]} = \dfrac{1.00 \times 10^{-14}\ mol^2\ dm^{-6}}{0.0500\ mol\ dm^{-3}}$

$= 2.00 \times 10^{-13}\ mol\ dm^{-3}$

$pH = -\log [H^+(aq)] = -\log (2.00 \times 10^{-13}) = \mathbf{12.7}$

b

Initial concn./mol dm^{-3}	0.100		0		0
	$CH_3CO_2H(aq)$	$\rightleftharpoons$	$H^+(aq)$	+	$CH_3CO_2^-(aq)$
Equilibrium concn./mol dm^{-3}	$0.100 - x$		x		x

$$K_a = \frac{[H^+(aq)][CH_3CO_2^-(aq)]}{[CH_3CO_2H(aq)]}$$

$$1.69 \times 10^{-5}\ mol\ dm^{-3} = \frac{(x\ mol\ dm^{-3})(x\ mol\ dm^{-3})}{(0.100 - x)\ mol\ dm^{-3}}$$

Assuming that $x \ll 0.100$, $\left(\dfrac{c_0}{K_a} = \dfrac{0.100}{1.69 \times 10^{-5}} = >1000\right)$

$(0.100 - x) \simeq 0.100$

$$\therefore 1.69 \times 10^{-5} = \frac{x^2}{0.100}$$

$x = \sqrt{1.69 \times 10^{-6}} = 1.30 \times 10^{-3}$
$\therefore [H^+(aq)] = 1.30 \times 10^{-3}\ mol\ dm^{-3}$
$pH = -\log [H^+(aq)] = -\log (1.30 \times 10^{-3}) = -(-2.9) = \mathbf{2.9}$

EXERCISE 7.16

Initial concn./mol dm^{-3}	0.100		0		0
	$HClO_2(aq)$	$\rightleftharpoons$	$H^+(aq)$	+	$ClO_2^-(aq)$
Equilibrium concn./mol dm^{-3}	$0.100 - x$		x		x

$$K_a = \frac{[H^+(aq)][ClO_2^-(aq)]}{[HClO_2(aq)]}$$

$$1.0 \times 10^{-2}\ mol\ dm^{-3} = \frac{(x\ mol\ dm^{-3})^2}{(0.100 - x)\ mol\ dm^{-3}}$$

Put $(0.100 - x) = 0.100$

$$1.0 \times 10^{-2} = \frac{x^2}{0.100}$$

$$x = \sqrt{1.0 \times 10^{-3}} = 0.0316$$

Now x is obviously large when compared with the initial concentration and so must be recalculated.

Substituting this value of x in the denominator and recalculating:

$$x = \sqrt{1.0 \times 10^{-2}(0.100 - 0.0316)} = 0.026$$

$\therefore$ $[H^+(aq)] = 0.026$ mol dm^{-3} and pH = $-\log(0.026)$ = **1.6**

(Alternatively, solving the quadratic equation by the usual method produces $x = 0.027$ which again gives pH = 1.6.)

EXERCISE 7.17

The method is the same as in the Worked Example on page 149.
The pH values, and intermediate answers, are shown below in tabular form.

Volume of alkali /cm^3	**Amount of alkali /mol**	**Amount of acid /mol**	**Excess /mol**	**[OH$^-$(aq)] /mol dm^{-3}**	**[H$^+$(aq)] /mol dm^{-3}**	**pH**
0.0	–	0.002 50	0.002 50	–	0.100	1.00
5.0	0.000 50	0.002 50	0.002 00	–	0.0667	1.18
10.0	0.001 00	0.002 50	0.001 50	–	0.0429	1.37
20.0	0.002 00	0.002 50	0.000 50	–	0.0111	1.95
24.0	0.002 40	0.002 50	0.000 10	–	0.002 04	2.69
25.0	0.002 50	0.002 50	–	1.00×10^{-7}	1.00×10^{-7}	7.00
26.0	0.002 60	0.002 50	0.000 10	0.001 96	5.10×10^{-12}	11.3
30.0	0.003 00	0.002 50	0.000 50	0.009 09	1.10×10^{-12}	12.0

EXERCISE 7.18

a Initial amount of CH_3CO_2H = concentration × volume
= 0.10 mol dm^{-3} × 0.025 dm^3
= 2.5×10^{-3} mol

Amount of CH_3CO_2H at half neutralisation point = 1.25×10^{-3} mol

$$\text{Volume of NaOH required for half neutralisation} = \frac{\text{amount}}{\text{concentration}}$$

$$= \frac{1.25 \times 10^{-3}\ \text{mol}}{0.10\ \text{mol dm}^{-3}}$$

$$= 1.25 \times 10^{-2}\ \text{dm}^3$$

Total volume at half neutralisation point = volume of CH_3CO_2H solution + volume of NaOH solution added = 0.0375 dm^3

$$[CH_3CO_2H(aq)] = \frac{\text{amount}}{\text{volume}} = \frac{1.25 \times 10^{-3}\ \text{mol}}{0.0375\ \text{dm}^3} = \mathbf{0.033\ mol\ dm^{-3}}$$

Since half the original acid has been neutralised and converted to ethanoate ions,
$[CH_3CO_2^-(aq)] = [CH_3CO_2H(aq)]$ = **0.033 mol dm^{-3}**

b $$K_a = \frac{[H^+(aq)][CH_3CO_2^-(aq)]}{[CH_3CO_2H(aq)]}$$

$$\therefore [H^+(aq)] = K_a \frac{[CH_3CO_2H(aq)]}{[CH_3CO_2^-(aq)]} = K_a \times 1 = 1.7 \times 10^{-5}\ \text{mol dm}^{-3}$$

pH = $-\log[H^+(aq)] = -\log(1.7 \times 10^{-5})$ = **4.8**

EXERCISE 7.19

$HA(aq) \rightleftharpoons H^+(aq) + A^-(aq)$

$$K_a = \frac{[H^+(aq)][A^-(aq)]}{[HA(aq)]}$$

$$[H^+(aq)] = K_a \times \frac{[HA(aq)]}{[A^-(aq)]}$$

$$\log[H^+(aq)] = \log K_a + \log\frac{[HA(aq)]}{[A^-(aq)]}$$

Multiplying through by −1:

$$-\log[H^+(aq)] = -\log K_a - \log\frac{[HA(aq)]}{[A^-(aq)]}$$

since $-\log[H^+(aq)]$ = pH and $-\log K_a = pK_a$

$$pH = pK_a - \log\frac{[HA(aq)]}{[A^-(aq)]}$$

EXERCISE 7.20

$C_2H_5CO_2H(aq) \rightleftharpoons C_2H_5CO_2^-(aq) + H^+(aq)$

$$K_a = \frac{[H^+(aq)][C_2H_5CO_2^-(aq)]}{[C_2H_5CO_2H(aq)]}$$

$$\therefore [H^+(aq)] = K_a \frac{[C_2H_5CO_2H(aq)]}{[C_2H_5CO_2^-(aq)]}$$

$$= 1.3 \times 10^{-5}\ \text{mol dm}^{-3} \times \frac{0.10\ \text{mol dm}^{-3}}{0.050\ \text{mol dm}^{-3}}$$

$$= 2.6 \times 10^{-5}\ \text{mol dm}^{-3}$$

pH = $-\log[H^+] = -\log(2.6 \times 10^{-5}) = -(-4.6)$ = **4.6**

Note that in this type of calculation it is usual to assume that the concentrations of free acid molecules and anions are equal to the concentrations obtained from the amounts used to make the solution; i.e. we assume that any tendency for the dissolved acid molecules to dissociate is balanced by the tendency for the dissolved anions to combine with hydrogen ions.

EXERCISE 7.21

a $C_3H_7CO_2H(aq) \rightleftharpoons C_3H_7CO_2^-(aq) + H^+(aq)$

$$K_a = \frac{[C_3H_7CO_2^-(aq)][H^+(aq)]}{[C_3H_7CO_2H(aq)]}$$

$$\therefore [H^+(aq)] = K_a \frac{[C_3H_7CO_2H(aq)]}{[C_3H_7CO_2^-(aq)]}$$

$$= 1.5 \times 10^{-5}\ \text{mol dm}^{-3} \times \frac{1.0\ \text{mol dm}^{-3}}{1.0\ \text{mol dm}^{-3}}$$

$$= 1.5 \times 10^{-5}\ \text{mol dm}^{-3}$$

$pH = -\log [H^+(aq)] = -\log (1.5 \times 10^{-5}) = -(-4.8) = \mathbf{4.8}$

b $$[H^+(aq)] = 1.5 \times 10^{-5}\ \text{mol dm}^{-3} \times \frac{2.0\ \text{mol dm}^{-3}}{1.0\ \text{mol dm}^{-3}}$$

$$= 3.0 \times 10^{-5}\ \text{mol dm}^{-3}$$

$pH = -\log [H^+(aq)] = -\log (3.0 \times 10^{-5}) = -(-4.5) = \mathbf{4.5}$

c $$[H^+(aq)] = 1.5 \times 10^{-5}\ \text{mol dm}^{-3} \times \frac{2.0\ \text{mol dm}^{-3}}{0.50\ \text{mol dm}^{-3}}$$

$$= 6.0 \times 10^{-5}\ \text{mol dm}^{-3}$$

$pH = -\log [H^+(aq)] = -\log (6.0 \times 10^{-5}) = -(-4.2) = \mathbf{4.2}$

d $$[H^+(aq)] = 1.5 \times 10^{-5}\ \text{mol dm}^{-3} \times \frac{0.50\ \text{mol dm}^{-3}}{2.0\ \text{mol dm}^{-3}}$$

$$= 3.75 \times 10^{-6}\ \text{mol dm}^{-3}$$

$pH = -\log [H^+(aq)] = -\log (3.75 \times 10^{-6}) = -(-5.4) = \mathbf{5.4}$

EXERCISE 7.22

a $$pH = pK_a - \log \frac{[\text{acid}]}{[\text{conjugate base}]}$$

$$\therefore \log \frac{[CH_3CO_2H(aq)]}{[CH_3CO_2^-(aq)]} = pK_a - pH = 4.74 - 4.70 = 0.04$$

$$\therefore \frac{[CH_3CO_2H(aq)]}{[CH_3CO_2^-(aq)]} = 10^{0.04} = \mathbf{1.1}$$

i.e. the solutions should be mixed in the ratio 1.1 volumes of ethanoic acid to one volume of sodium ethanoate.

b $$\log \frac{[CH_3CO_2H(aq)]}{[CH_3CO_2^-(aq)]} = pK_a - pH = 4.74 - 4.40 = 0.34$$

$$\therefore \frac{[CH_3CO_2H(aq)]}{[CH_3CO_2^-(aq)]} = 10^{0.34} = \mathbf{2.2}$$

i.e. the solutions should be mixed in the ratio 2.2 volumes of ethanoic acid to one volume of sodium ethanoate.

Note that we again assume that, due to the presence of ethanoate ions, the ionisation of ethanoic acid is suppressed. Therefore, the concentration of hydrogen ions is small compared to the concentration of either acid or base, so that we can take initial concentrations as being equal to equilibrium concentrations.

EXERCISE 7.23

a $pK_a = -\log K_a = -\log (1.58 \times 10^{-4}) = 3.8$

$$pH = pK_a - \log \frac{[HCO_2H(aq)]}{[HCO_2^-(aq)]}$$

Since $pH = pK_a$, $$\log \frac{[HCO_2H(aq)]}{[HCO_2^-(aq)]} = 0$$

i.e. $[HCO_2^-(aq)] = [HCO_2H(aq)] = \mathbf{0.50\ mol\ dm^{-3}}$

b $$pH = pK_a - \log \frac{[HCO_2H(aq)]}{[HCO_2^-(aq)]}$$

$$4.1 = 3.8 - \log \frac{[HCO_2H(aq)]}{[HCO_2^-(aq)]}$$

$$\therefore \log \frac{[HCO_2H(aq)]}{[HCO_2^-(aq)]} = 3.8 - 4.1 = -0.3$$

$$\therefore \frac{[HCO_2H(aq)]}{[HCO_2^-(aq)]} = 10^{-0.3} = 5 \times 10^{-1} \text{ or } 0.5$$

$\therefore [HCO_2^-(aq)] = 2[HCO_2H(aq)] = \mathbf{1.0\ mol\ dm^{-3}}$

EXERCISE 7.24

$NH_4^+(aq) \rightleftharpoons NH_3(aq) + H^+(aq)$

$$K_a = \frac{[NH_3(aq)][H^+(aq)]}{[NH_4^+(aq)]}$$

Amount of NH_4^+ added = cV = 0.20 mol dm^{-3} × 0.75 dm^3 = 0.15 mol
Amount of NH_3 added = cV = 0.10 mol dm^{-3} × 0.75 dm^3 = 0.075 mol

$$[H^+(aq)] = K_a \frac{[NH_4^+(aq)]}{[NH_3(aq)]}$$

$$= 6.00 \times 10^{-10}\ \text{mol dm}^{-3} \times \frac{0.15\ \text{mol}/1.5\ \text{dm}^3}{0.075\ \text{mol}/1.5\ \text{dm}^3}$$

$$= 1.2 \times 10^{-9}\ \text{mol dm}^{-3}$$

$pH = -\log [H^+(aq)] = -\log (1.2 \times 10^{-9}) = -(-8.9) = \mathbf{8.9}$

EXERCISE 7.25

a Phenolphthalein has a pH range of 8–10. It would, therefore, remain in its acid form (colourless) beyond the equivalence point at pH 5. The pink colour would not appear until about pH 8 which would require an excess of ammonia.

b Methyl red. $pK_{In} = 5.1$. This indicator would show an intermediate colour between pH 4.2 and pH 6.3 which, of course, includes pH 5, the equivalence point.

EXERCISE 7.26

a Yellow.

b $pH = pK_{In} - \log \dfrac{[HIn(aq)]}{[In^-(aq)]}$

At the beginning of the colour change, $[HIn(aq)]/[In^-(aq)] = 10$

$\therefore pH = pK_{In} - \log 10 = 5.1 - 1 = 4.1$

At the end of the colour change:

$[HIn(aq)]/[In^-(aq)] = 0.1$
$\therefore pH = pK_{In} - \log 0.1 = 5.1 + 1 = 6.1$
pH range = **4.1 to 6.1**

c 0.010 M HCl is diluted from 1 cm^3 to 500 cm^3

$\therefore$ new concentration $= 0.010\ mol\ dm^{-3} \times 1/500$
$= 2.0 \times 10^{-5}\ mol\ dm^{-3}$

Since HCl is completely ionised:

$[H^+(aq)] = [HCl(aq)]$
$= 2.0 \times 10^{-5}\ mol\ dm^{-3}$
$pH = -\log [H^+(aq)] = -\log (2.0 \times 10^{-5}) = -(-4.7) =$ **4.7**

d $pH = pK_{In} - \log \dfrac{[HIn(aq)]}{[In^-(aq)]}$

Substituting the appropriate values:

$$4.7 = 5.1 - \log \frac{[red]}{[yellow]}$$

$$\log \frac{[red]}{[yellow]} = 5.1 - 4.7 = 0.4$$

$$\therefore \frac{[red]}{[yellow]} = 10^{0.4} = \mathbf{2.5}$$

End-of-chapter questions

7.1 B
7.2 D
7.3 B
7.4 D
7.5 $pH = -lg([H^+(aq)]/mol\ dm^{-3}$
a 3.30
b 10.70
7.6 **a** HCl is a strong acid.
pH = 2.00
Ethanoic acid is a weak acid.
pH = 2.38
For pure water at 298 K, pH = 7.00
For pure water at 363 K, pH = 6.63
7.7 **a** $4.0 \times 10^{-4}\ mol\ dm^{-3}$
b $4.0 \times 10^{-4}\ mol\ dm^{-3}$
c $1.6 \times 10^{-5}\ mol\ dm^{-3}$
7.8. 4.75
7.9. **b** $8.2 \times 10^{-2}\ mol\ dm^{-3}$
c $2.0 \times 10^{-5}\ mol\ dm^{-3}$
7.10 **a** $K_a = \dfrac{[H^+(aq)][Cy(aq)]}{[CyH^+(aq)]}$
b $5.00 \times 10^{-5}\ mol\ dm^{-3}$
c 20 : 1 Red
7.11 **a** −6.10
b 20 : 1
c $[base] = 2.4 \times 10^{-2}\ mol\ dm^{-3}$
$[acid] = 1.2 \times 10^{-3}\ mol\ dm^{-3}$

CHAPTER 8

EXERCISE 8.1

a
$Pt(s), H_2(g) \mid H^+(aq, 1.0\ M) \;\vdots\vdots\; Zn^{2+}(aq, 1.0\ M) \mid Zn(s)$ $\Delta E° = -0.76\ V$
$Pt(s), H_2(g) \mid H^+(aq, 1.0\ M) \;\vdots\vdots\; Cu^{2+}(aq, 1.0\ M) \mid Cu(s)$ $\Delta E° = +0.35\ V$
$Pt(s), H_2(g) \mid H^+(aq, 1.0\ M) \;\vdots\vdots\; Ag^+(aq, 1.0\ M) \mid Ag(s)$ $\Delta E° = +0.81\ V$

b
$Zn^{2+}(aq) + 2e^- \rightleftharpoons Zn(s)$ $E^{\ominus} = -0.76\ V$
$Cu^{2+}(aq) + 2e^- \rightleftharpoons Cu(s)$ $E^{\ominus} = +0.35\ V$
$Ag^+(aq) + e^- \rightleftharpoons Ag(s)$ $E^{\ominus} = +0.81\ V$

EXERCISE 8.2

a *Method 1:* $2\overset{+5}{N}O_3^- \rightarrow \overset{+4}{N_2}O_4$ 1 e per N atom, 2 e's in all
Method 2: Total charge on left = 2(−1) + 4(+1) = +2
Total charge on right = 0
$\therefore$ 2 e's must be added.
$\mathbf{2NO_3^-(aq) + 4H^+(aq) + 2e^- \rightarrow N_2O_4(g) + 2H_2O(l)}$

b *Method 1:* $2H_2\overset{+4}{S}O_3 \rightarrow \overset{+2}{S_2}O_3^{2-}$ 2 e's per S atom, 4 e's in all
Method 2: Total charge on left = 2(+1) = +2
Total charge on right = −2
$\therefore$ 4 e's must be added.
$\mathbf{2H_2SO_3(aq) + 2H^+(aq) + 4e^- \rightarrow S_2O_3^{2-}(aq) + 3H_2O(l)}$

c *Method 1:* $\overset{+1}{IO^-} \rightarrow \overset{-1}{I^-}$ 2 e's added
Method 2: Total charge on left = −1
Total charge on right = −1 + 2(−1) = −3

∴ 2 e's must be added.

$$\mathbf{IO^-(aq) + H_2O(l) + 2e^- \rightarrow I^-(aq) + 2OH^-(aq)}$$

EXERCISE 8.3

a $Pt(s),H_2(g) \mid H^+(aq) \;\vdots\vdots\; Ag^+(aq) \mid Ag(s)$
Substituting into the expression:

$$\Delta E^{\circ} = E_R^{\circ} - E_L^{\circ}$$
$$\Delta E^{\circ} = +0.80\ V - 0.00\ V = \mathbf{+0.80\ V}$$

The positive e.m.f. means the cell reaction would take place from left to right as shown in the cell diagram:

$$Pt(s),H_2(g) \mid H^+(aq) \;\vdots\vdots\; Ag^+(aq) \mid Ag(s)$$
$$H_2(g) \rightarrow 2H^+(aq) + 2e^- \qquad Ag^+(aq) + e^- \rightarrow Ag(s)$$

Adding the first half-equation to twice the second (in order to transfer the same number of electrons) gives the overall reaction:

$$\mathbf{H_2(g) + 2Ag^+(aq) \rightarrow 2H^+(aq) + Ag(s)}$$

b $Ni(s) \mid Ni^{2+}(aq) \;\vdots\vdots\; Zn^{2+}(aq) \mid Zn(s)$
Substituting into the expression:

$$\Delta E^{\circ} = E_R^{\circ} - E_L^{\circ}$$
$$\Delta E^{\circ} = -0.76\ V - (-0.25\ V) = \mathbf{-0.51\ V}$$

The negative sign on the e.m.f. means the cell reaction would take place from right to left as shown in the cell diagram:

$$Ni(s) \mid Ni^{2+}(aq) \;\vdots\vdots\; Zn^{2+}(aq) \mid Zn(s)$$
$$Ni^{2+}(aq) + 2e^- \rightarrow Ni(s) \qquad Zn(s) \rightarrow Zn^{2+}(aq) + 2e^-$$

Adding these half-equations gives the overall reaction:

$$\mathbf{Zn(s) + Ni^{2+}(aq) \rightarrow Zn^{2+}(aq) + Ni(s)}$$

c $Ni(s) \mid Ni^{2+}(aq) \;\vdots\vdots\; [NO_3^-(aq) + 3H^+(aq)],[HNO_2(aq) + H_2O(l)] \mid Pt(s)$
Substituting into the expression:

$$\Delta E^{\circ} = E_R^{\circ} - E_L^{\circ}$$
$$\Delta E^{\circ} = +0.94\ V - (-0.25\ V) = \mathbf{+1.19\ V}$$

The positive e.m.f. means the cell reaction would take place from left to right as shown in the cell diagram:

$$Ni(s) \mid Ni^{2+}(aq) \;\vdots\vdots\; [NO_3^-(aq) + 3H^+(aq)],[HNO_2(aq) + H_2O(l)] \mid Pt(s)$$
$$Ni(s) \rightarrow Ni^{2+}(aq) + 2e^-$$
$$NO_3^-(aq) + 3H^+(aq) + xe^- \rightarrow HNO_2(aq) + H_2O(l)$$

To work out *x* in the right-hand half-cell:

$$\overset{+5}{NO_3^-} \rightarrow \overset{+3}{HNO_2} \qquad \text{2 e's must be added, i.e. } x = 2$$

Adding the two half-equations:

$$NO_3^-(aq) + 3H^+(aq) + 2e^- \rightarrow HNO_2(aq) + H_2O(l)$$
$$Ni(s) \rightarrow Ni^{2+}(aq) + 2e^-$$

gives the overall reaction:

$$\mathbf{Ni(s) + NO_3^-(aq) + 3H^+(aq) \rightarrow Ni^{2+}(aq) + HNO_2(aq) + H_2O(l)}$$

d $Pt(s) \mid 2H_2SO_3(aq),[4H^+(aq) + S_2O_6^{2-}(aq)] \;\vdots\vdots\; Cr^{3+}(aq) \mid Cr(s)$
Substituting into the expression:

$$\Delta E^{\circ} = E_R^{\circ} - E_L^{\circ}$$
$$\Delta E^{\circ} = -0.74\ V - (+0.57\ V) = \mathbf{-1.31\ V}$$

The negative e.m.f. means the cell reaction would take place from right to left as shown in the cell diagram:

$$Pt(s) \mid 2H_2SO_3(aq),[4H^+(aq) + S_2O_6^{2-}(aq)] \;\vdots\vdots\; Cr^{3+}(aq) \mid Cr(s)$$
$$S_2O_6^{2-}(aq) + 4H^+(aq) + xe^- \rightarrow 2H_2SO_3(aq) \quad Cr(s) \rightarrow Cr^{3+}(aq) + 3e^-$$

To work out *x* in the left-hand half-cell:

$$\overset{+5}{S_2O_6^{2-}} \rightarrow \overset{+4}{2H_2SO_3} \qquad \text{1 e per S atom, i.e. } x = 2$$

Adding the two half-equations:

$$3 \times [S_2O_6^{2-}(aq) + 4H^+(aq) + 2e^- \rightarrow 2H_2SO_3(aq)]$$
$$2 \times [Cr(s) \rightarrow Cr^{3+}(aq) + 3e^-]$$

gives the overall equation:

$$\mathbf{3S_2O_6^{2-}(aq) + 12H^+(aq) + 2Cr(s) \rightarrow 6H_2SO_3(aq) + 2Cr^{3+}(aq)}$$

EXERCISE 8.4

a The two half-equations are:

$$Br_2 + 2e^- \rightleftharpoons 2Br^-(aq)$$
$$2I^-(aq) \rightleftharpoons I_2(aq) + 2e^-$$

which combine to make the cell:

$$Pt(s) \mid I^-(aq),I_2(aq) \;\vdots\vdots\; Br_2(aq),Br^-(aq) \mid Pt(s)$$

Substituting into the expression:

$$\Delta E^{\circ} = E_R^{\circ} - E_L^{\circ}$$
$$\Delta E^{\circ} = +1.07\ V - (+0.54\ V) = +0.53\ V$$

Since ΔE° is positive, the reaction would be expected to 'go'.

b The two half-equations are:

$$Br_2(aq) + 2e^- \rightleftharpoons 2Br^-(aq)$$
$$2Cl^-(aq) \rightleftharpoons Cl_2(aq) + 2e^-$$

which combine to make the cell:

$$Pt(s) \mid Cl^-(aq),Cl_2(aq) \;\vdots\vdots\; Br_2(aq),Br^-(aq) \mid Pt(s)$$

Substituting into the expression:

$$\Delta E^{\circ} = E_R^{\circ} - E_L^{\circ}$$
$$\Delta E^{\circ} = +1.07\ V - (+1.36\ V) = -0.29\ V$$

Since ΔE° is negative, the reaction would not be expected to 'go'.

c The two half-equations are:

$$Zn(s) \rightleftharpoons Zn^{2+}(aq) + 2e^-$$
$$2Fe^{3+}(aq) + 2e^- \rightleftharpoons 2Fe^{2+}(aq)$$

which combine to make the cell:

$$Zn(s) \mid Zn^{2+}(aq) \;\vdots\vdots\; Fe^{3+}(aq),Fe^{2+}(aq) \mid Pt(s)$$

Substituting into the expression:

$$\Delta E^\ominus = E_R^\ominus - E_L^\ominus$$
$$\Delta E^\ominus = +0.77\ \text{V} - (-0.76\ \text{V}) = +1.53\ \text{V}$$

Since $\Delta E^\ominus$ is positive, the reaction would be expected to 'go'.

d The two half-equations are:

$$2MnO_4^-(aq) + 16H^+(aq) + 10e^- \rightleftharpoons 8H_2O(l) + 2Mn^{2+}(aq)$$
$$10Br^-(aq) \rightleftharpoons 5Br_2(aq) + 10e^-$$

which combine to make the cell:

$Pt(s) \mid 2Br^-(aq),Br_2(aq)$ ¦¦ $[MnO_4^-(aq) + 8H^+(aq)],[4H_2O(l) + Mn^{2+}(aq)] \mid Pt(s)$

(Note that the stoichiometric coefficients are reduced to the smallest whole numbers which give the correct ratio.)

Substituting into the expression:

$$\Delta E^\ominus = E_R^\ominus - E_L^\ominus$$
$$\Delta E^\ominus = +1.51\ \text{V} - (+1.07\ \text{V}) = +0.44\ \text{V}$$

Since $\Delta E^\ominus$ is positive, the reaction would be expected to 'go'.

e The two half-equations are:

$$2MnO_4^-(aq) + 16H^+(aq) + 10e^- \rightleftharpoons 8H_2O(l) + 2Mn^{2+}(aq)$$
$$5Cu(s) \rightleftharpoons 5Cu^{2+}(aq) + 10e^-$$

which combine to make the cell:

$Cu(s) \mid Cu^{2+}(aq)$ ¦¦ $[MnO_4^-(aq) + 8H^+(aq)],[4H_2O(l) + Mn^{2+}(aq)] \mid Pt(s)$

Substituting into the expression:

$$\Delta E^\ominus = E_R^\ominus - E_L^\ominus$$
$$\Delta E^\ominus = +1.51\ \text{V} - (+0.34\ \text{V}) = +1.17\ \text{V}$$

Since $\Delta E^\ominus$ is positive, the reaction would be expected to 'go'.

f The two half-equations are:

$$3S_2O_8^{2-}(aq) + 6e^- \rightleftharpoons 6SO_4^{2-}(aq)$$
$$2Cr^{3+}(aq) + 7H_2O(l) \rightleftharpoons Cr_2O_7^{2-}(aq) + 14H^+(aq) + 6e^-$$

which combine to make the cell:

$Pt(s) \mid [2Cr^{3+}(aq) + 7H_2O(l)],[Cr_2O_7^{2-}(aq) + 14H^+(aq)]$ ¦¦ $S_2O_8^{2-}(aq),2SO_4^{2-}(aq) \mid Pt(s)$

Substituting into the expression:

$$\Delta E^\ominus = E_R^\ominus - E_L^\ominus$$
$$\Delta E^\ominus = +2.01\ \text{V} - (+1.33\ \text{V}) = +0.68\ \text{V}$$

Since $\Delta E^\ominus$ is positive, the reaction would be expected to 'go'.

EXERCISE 8.5

a $Mn^{2+}(aq) + 2e^- \rightleftharpoons Mn(s)$ $E^\ominus = -1.19$ V
$Pb^{2+}(aq) + 2e^- \rightleftharpoons Pb(s)$ $E^\ominus = -0.13$ V

$Mn(s) + Pb^{2+}(aq) \rightarrow Mn^{2+}(aq) + Pb(s)$

b $S(s) + 2e^- \rightleftharpoons S^{2-}(aq)$ $E^\ominus = -0.48$ V
$I_2(aq) + 2e^- \rightleftharpoons 2I^-(aq)$ $E^\ominus = +0.54$ V

$S^{2-}(aq) + I_2(aq) \rightarrow S(s) + 2I^-(aq)$

(The equations were not tabulated in the correct order in the question.)

c $2H^+(aq) + 2e^- \rightleftharpoons H_2(g)$ $E^\ominus = 0.00$ V
$Ag^+(aq) + e^- \rightleftharpoons Ag(s)$ $E^\ominus = +0.80$ V

$H_2(g) + 2Ag^+(aq) \rightarrow 2H^+(aq) + 2Ag(s)$

EXERCISE 8.6

a The half-equations are listed in order of increasingly positive values of $E^\ominus$. Therefore, for the Fe^{3+},Fe^{2+} reaction to proceed from right to left, it must be combined with either of the half-reactions listed below it, and these will proceed from left to right. The substances which should be able to oxidise $Fe^{2+}(aq)$ to $Fe^{3+}(aq)$ are **acidified hydrogen peroxide solution** and **cobalt(III) ions**.

b $Fe^{3+}(aq) + e^- \rightleftharpoons Fe^{2+}(aq)$
$H_2O_2(l) + 2H^+(aq) + 2e^- \rightleftharpoons 2H_2O(l)$

$2Fe^{2+}(aq) + H_2O_2(l) + 2H^+(aq) \rightarrow 2Fe^{3+}(aq) + 2H_2O(l)$

$Fe^{3+}(aq) + e^- \rightleftharpoons Fe^{2+}(aq)$
$Co^{3+}(aq) + e^- \rightleftharpoons Co^{2+}(aq)$

$Fe^{2+}(aq) + Co^{3+}(aq) \rightarrow Fe^{3+}(aq) + Co^{2+}(aq)$

EXERCISE 8.7

a The equilibrium at the copper electrode is:

$$Cu^{2+}(aq) + 2e^- \rightleftharpoons Cu(s)$$

Reducing $[Cu^{2+}(aq)]$ would result in the equilibrium shifting to the left, producing more electrons. This would make the electrode less positive (more negative) so the **e.m.f. would be reduced.**

b By a similar argument to the above, the zinc electrode would become more negative, so the **e.m.f. would be increased.**

EXERCISE 8.8

a Diluting the copper ion solution results in a **decrease of e.m.f.**

b Diluting the zinc ion solution results in an **increase of e.m.f.**

c **Yes.** This is what was predicted in Exercise 8.7.

EXERCISE 8.9

a $Zn(s) \mid Zn^{2+}(aq) \;\vdots\vdots\; Cd^{2+}(aq) \mid Cd(s)$

$\Delta E^\ominus = E^\ominus(Cd^{2+} \mid Cd) - E^\ominus(Zn^{2+} \mid Zn)$

$= -0.40\ V - (-0.76\ V) = \mathbf{+0.36\ V}$

Since $\Delta E^\ominus$ is positive, the reaction proceeds from **left to right** as indicated in the cell diagram, i.e.:

$\mathbf{Zn(s) + Cd^{2+}(aq) \rightarrow Zn^{2+}(aq) + Cd(s)\ \Delta E^\ominus = +0.36\ V}$

$\Delta G^\ominus = -zF\Delta E^\ominus = -2 \times 9.65 \times 10^4\ C\ mol^{-1} \times 0.36\ V$

$= -6.9 \times 10^4\ C\ V\ mol^{-1} = \mathbf{-69\ kJ\ mol^{-1}}$

b $Cu(s) \mid Cu^{2+}(aq) \;\vdots\vdots\; Cd^{2+}(aq) \mid Cd(s)$

$\Delta E^\ominus = E^\ominus(Cd^{2+} \mid Cd) - E^\ominus(Cu^{2+} \mid Cu)$

$= -0.40\ V - 0.34\ V = \mathbf{-0.74\ V}$

Since $\Delta E^\ominus$ is negative, the reaction proceeds from **right to left** as indicated in the cell diagram above. When re-written in this way, $\Delta E^\ominus$ is positive.

$\mathbf{Cd(s) + Cu^{2+}(aq) \rightarrow Cd^{2+}(aq) + Cu(s)\ \Delta E^\ominus = +0.74\ V}$

$\Delta G^\ominus = -zF\Delta E^\ominus = -2 \times 9.65 \times 10^4\ C\ mol^{-1} \times 0.74\ V$

$= -1.43 \times 10^5\ C\ V\ mol^{-1} = \mathbf{-143\ kJ\ mol^{-1}}$

c $Ag(s) \mid Ag^+(aq) \;\vdots\vdots\; Cd^{2+}(aq) \mid Cd(s)$

$\Delta E^\ominus = E^\ominus(Cd^{2+} \mid Cd) - E^\ominus(Ag^+ \mid Ag)$

$= -0.40\ V - 0.80\ V = \mathbf{-1.20\ V}$

Since $\Delta E^\ominus$ is negative, the reaction proceeds from **right to left** as indicated in the cell diagram above. When re-written in this way, $\Delta E^\ominus$ is positive:

$\mathbf{Cd(s) + 2Ag^+(aq) \rightarrow Cd^{2+}(aq) + 2Ag(s) \qquad \Delta E^\ominus = +1.20\ V}$

$\Delta G^\ominus = -zF\Delta E^\ominus = -2 \times 9.65 \times 10^4\ C\ mol^{-1} \times 1.20\ V$

$= -2.32 \times 10^5\ C\ V\ mol^{-1} = \mathbf{-232\ kJ\ mol^{-1}}$

EXERCISE 8.10

a $Cu(s) \mid Cu^{2+}(aq) \;\vdots\vdots\; Br_2(aq),Br^-(aq) \mid Pt(s)$

$\Delta E^\ominus = E^\ominus(Br_2,Br^-) - E^\ominus(Cu^{2+} \mid Cu)$

$= +1.09\ V - 0.34\ V = +0.75\ V$

Since $\Delta E^\ominus$ is positive, the reaction proceeds from left to right as indicated in the cell diagram:

$\mathbf{Cu(s) + Br_2(aq) \rightleftharpoons Cu^{2+}(aq) + 2Br^-(aq) \qquad \Delta E^\ominus = +0.75\ V}$

b $zF\Delta E^\ominus = RT \ln K_c$

$$\therefore \ln K_c = \frac{zF\Delta E^\ominus}{RT} = \frac{2 \times 9.65 \times 10^4\ C\ mol^{-1} \times 0.75\ V}{8.31\ J\ K^{-1}\ mol^{-1} \times 298\ K} = 58.5\ C\ V\ J^{-1} = 58.5$$

$\therefore K_c = e^{58.5} = \mathbf{2.5 \times 10^{25}\ mol^2\ dm^{-6}}$

This very large value of K_c indicates that the reaction goes virtually to completion. (Note that the unit is not given by the calculation but by reference to the equation and the expression for K_c.)

EXERCISE 8.11

a $\mathbf{Fe^{2+}(aq) \rightleftharpoons Fe^{3+}(aq) + e^-}$

$\mathbf{Cr_2O_7^{2-}(aq) + 14H^+(aq) + 6e^- \rightleftharpoons 2Cr^{3+}(aq) + 7H_2O(l)}$

b $\mathbf{Pt(s) \mid Fe^{2+}(aq),Fe^{3+}(aq) \;\vdots\vdots\; [Cr_2O_7^{2-}(aq) + H^+(aq)],[Cr^{3+}(aq) + H_2O(l)] \mid Pt(s)}$

c $\Delta E^\ominus = E_R^\ominus - E_L^\ominus = 1.33\ V - 0.77\ V = \mathbf{0.56\ V}$

d $zF\Delta E^\ominus = RT \ln K_c$

$$\therefore \ln K_c = \frac{zF\Delta E^\ominus}{RT} = \frac{6 \times 9.65 \times 10^4\ C\ mol^{-1} \times 0.56\ V}{8.31\ J\ K^{-1}\ mol^{-1} \times 298\ K} = 130.9$$

$\therefore K_c = e^{130.9} = \mathbf{7.3 \times 10^{56}\ dm^{39}\ mol^{-13}}$

The unit is not given by the calculation, but by reference to the expression for K_c:

$$K_c = \frac{[Fe^{3+}(aq)]^6[Cr^{3+}(aq)]^2}{[Fe^{2+}(aq)]^6[Cr_2O_7^{2-}(aq)][H^+(aq)]^{14}}$$

EXERCISE 8.12

a $Pt(s) \mid Fe^{2+}(aq),Fe^{3+}(aq) \;\vdots\vdots\; Ag^+(aq) \mid Ag(s)$

$\Delta E^\ominus = E_R^\ominus - E_L^\ominus = 0.80\ V - 0.77\ V = +0.03\ V$

$zF\Delta E^\ominus = RT \ln K_c$

$$\therefore \ln K_c = \frac{zF\Delta E^\ominus}{RT} = \frac{1 \times 9.65 \times 10^4\ C\ mol^{-1} \times 0.03\ V}{8.31\ J\ K^{-1}\ mol^{-1} \times 298\ K} = 1.17$$

$\therefore K_c = e^{1.17} = \mathbf{3.2\ dm^3\ mol^{-1}}$

b $Pt(s) \mid [Cr^{3+}(aq),H_2O(l)],[Cr_2O_7^{2-}(aq),H^+(aq)] \;\vdots\vdots\; Cl_2(aq),Cl^-(aq) \mid Pt(s)$

$\Delta E^\ominus = E_R^\ominus - E_L^\ominus = 1.36\ V - 1.33\ V = +0.03\ V$

$zF\Delta E^\ominus = RT \ln K_c$

$$\therefore \ln K_c = \frac{zF\Delta E^\ominus}{RT} = \frac{6 \times 9.65 \times 10^4\ C\ mol^{-1} \times 0.03\ V}{8.31\ J\ K^{-1}\ mol^{-1} \times 298\ K} = 7.0$$

$\therefore K_c = e^{7.0} = \mathbf{1.1 \times 10^3\ mol^{16}\ dm^{-48}}$

End-of-chapter questions

8.1 **a** A **b** C

8.2 **a** D **b** E **c** D

8.3 **a** B **b** A

8.4 **a** $Cr_2O_7^{2-}(aq) + 14H^+(aq) + 6Fe^{2+}(aq) \rightleftharpoons 2Cr^{3+}(aq) + 6Fe^{3+}(aq) + 7H_2O(l)$

$\Delta E^\ominus = +0.56\ V$

Since $\Delta E^\ominus > 0$ the spontaneous reaction is from left to right. $Cr_2O_7^{2-}$ is the oxidant, Fe^{2+} is the reductant. H^+ is also necessary for the reaction to proceed.

b i) $\Delta E^\ominus = +2.03$ V
ii) Zn forms the battery casing and in the cell reaction Zn corrodes to form $[Zn(NH_3)_4]^{2+}$.
iii) Non-standard concentrations.
iv) $\Delta G^\ominus = -290$ kJ mol^{-1}

8.5 **a** H_2O_2 should oxidise $(C_4H_4O_6)^{2-}$
b $(C_4H_4O_6)^{2-} + 3H_2O_2 \rightarrow 2HCO_2^- + 2CO_2 + 4H_2O$
c Disproportionation occurs when the same element, whether free or combined, is simultaneously oxidised and reduced.
d Oxygen is both oxidised and reduced.

8.6 A, B and D

8.7 **a** $Cr^{3+}(aq) + e^- \rightleftharpoons Cr^{2+}(aq)$
$Ce^{4+}(aq) + e^- \rightleftharpoons Ce^{3+}(aq)$
$Cr_2O_7^{2-}(aq) + 14H^+(aq) + 6e^- \rightleftharpoons 2Cr^{3+}(aq) + 7H_2O(l)$
b i) $\Delta E^\ominus = +1.70\ \text{V} - (-0.41\ \text{V}) = +2.11$ V
ii) $\Delta E^\ominus = +1.70\ \text{V} - (+1.33\ \text{V}) = +0.37$ V
c i) $Cr^{2+}(aq) + Ce^{4+}(aq) \rightarrow Cr^{3+}(aq) + Ce^{3+}(aq)$
ii) $2Cr^{3+}(aq) + 7H_2O(l) + 6Ce^{4+}(aq) \rightarrow Cr_2O_7^{2-}(aq) + 14H^+(aq) + 6Ce^{3+}(aq)$

CHAPTER 9

EXERCISE 9.1

a In each case, $p = p^0X$

X_{hexane}	0.00	0.200	0.400	0.600	0.800	1.00
p_{hexane}/mmHg	0.00	74.0	149	223	298	372
$X_{heptane}$	0.00	0.200	0.400	0.600	0.800	1.00
$p_{heptane}$/mmHg	0.00	25.4	50.8	76.2	102	127

b and **c**

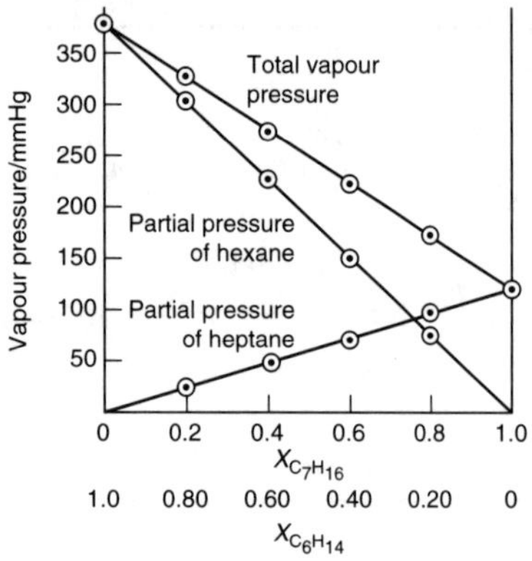

d 250 mmHg
e $X_{C_7H_{16}} = 0.7$ $X_{C_6H_{14}} = 0.3$

EXERCISE 9.2

a $p_T = p_A^0X_A + p_B^0X_B = (75\ \text{mmHg} \times \frac{1}{2}) + (130\ \text{mmHg} \times \frac{1}{2})$
= **103 mmHg**

b $p_T = p_A^0X_A + p_B^0X_B = (75\ \text{mmHg} \times \frac{3}{4}) + (130\ \text{mmHg} \times \frac{1}{4})$
= **89 mmHg**

c $p_T = p_A^0X_A + p_B^0X_B = (75\ \text{mmHg} \times \frac{1}{5}) + (130\ \text{mmHg} \times \frac{4}{5})$
= **119 mmHg**

EXERCISE 9.3

a In an equimolar liquid mixture $X_{hexane} = X_{heptane} = 0.5$
The total vapour pressure is given by the general formula:
$p = X_Ap^0_A + X_Bp^0_B$
$= (0.5 \times 56\,000\ \text{N m}^{-2}) + (0.5 \times 24\,000\ \text{N m}^{-2})$
= **40 000 N m^{-2}**

b The partial pressure of heptane in the vapour $= 0.5 \times 56\,000$ N m^{-2}
= **28 000 N m^{-2}**

In the vapour, $p_{heptane} = X_{heptane}\,p_{total}$

$$\therefore X_{heptane} = \frac{p_{heptane}}{p_{total}} = \frac{28\,000\ \text{N m}^{-2}}{40\,000\ \text{N m}^{-2}} = \mathbf{0.7}$$

EXERCISE 9.4

a p^0_A = **3.0 × 10^3 Pa** p^0_B = **10.0 × 10^3 Pa**

b i) $p_A = p^0_AX_A = 3.0 \times 10^3\ \text{Pa} \times 0.50$ = **1.5 × 10^3 Pa**
$p_B = p^0_BX_B = 10.0 \times 10^3\ \text{Pa} \times 0.50$ = **5.0 × 10^3 Pa**

ii) $$X'_A = \frac{p_A}{p_A + p_B} = \frac{1.5 \times 10^3\ \text{Pa}}{(1.5 + 5.0) \times 10^3\ \text{Pa}} = \frac{1.5}{6.5} = \mathbf{0.23}$$

iii) $$X'_B = \frac{p_B}{p_A + p_B} = \frac{5.0 \times 10^3\ \text{Pa}}{(1.5 + 5.0) \times 10^3\ \text{Pa}} = \frac{5.0}{6.5} = \mathbf{0.77}$$

(Alternatively, $X'_B = 1 - X'_A = 1.00 - 0.23 = 0.77$.)

c The mole fraction of B, the more volatile component, is **greater in the vapour phase** than in the liquid phase.

d X'_A = **0.73, 0.55, 0.31, 0.17, 0.07**
X'_B = **0.27, 0.45, 0.69, 0.83, 0.93**

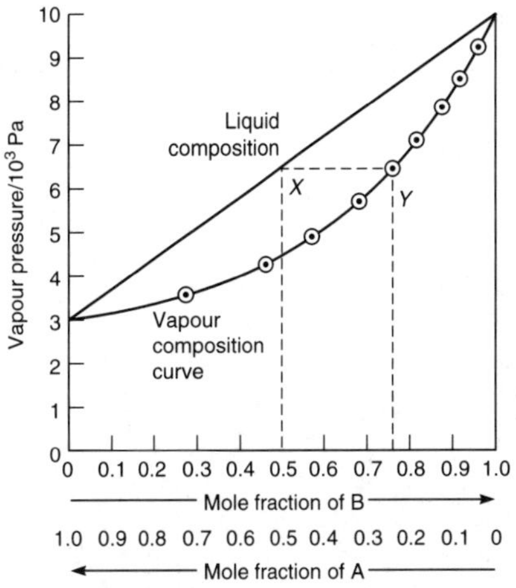

EXERCISE 9.5

a The total vapour pressure exerted by the mixture is **equal to the sum** of the separate saturation vapour pressures.

b The boiling point of the mixture is **98 °C** since the total vapour pressure exerted at this temperature equals atmospheric pressure.

c The boiling point of the mixture is **lower** than either of the boiling points of the pure liquids. Extrapolation of the vapour pressure curves for water and nitrobenzene shows that they will cross the dotted line of atmospheric pressure at temperatures higher than 98 °C.

d The mixture will steam distil at **98 °C**.

EXERCISE 9.6

$$\frac{\text{mass of bromobenzene}}{\text{mass of water}} = \frac{\text{r.m.m. of bromobenzene}}{\text{r.m.m. of water}} \times \frac{p_{\text{bromobenzene}}}{p_{\text{water}}}$$

$$= \frac{157 \times (101 - 85.8)\text{ kPa}}{18 \times 85.8\text{ kPa}} = \frac{1.55}{1}$$

$$\therefore \text{ \% by mass of bromobenzene} = \frac{1.55}{1 + 1.55} \times 100 = \mathbf{60.8\%}$$

EXERCISE 9.7

a The phenylamine–water mixture will steam distil at about **98 °C**.

b

$$\frac{\text{mass of phenylamine}}{\text{mass of water}} = \frac{\text{r.m.m. of phenylamine}}{\text{r.m.m. of water}} \times \frac{p_{\text{phenylamine}}}{p_{\text{water}}}$$

$$= \frac{93}{18} \times \frac{6}{94} = \frac{0.33}{1}$$

$$\therefore \text{ \% by mass of phenylamine} = \frac{0.33}{1.33} \times 100 = \mathbf{24.8\%}$$

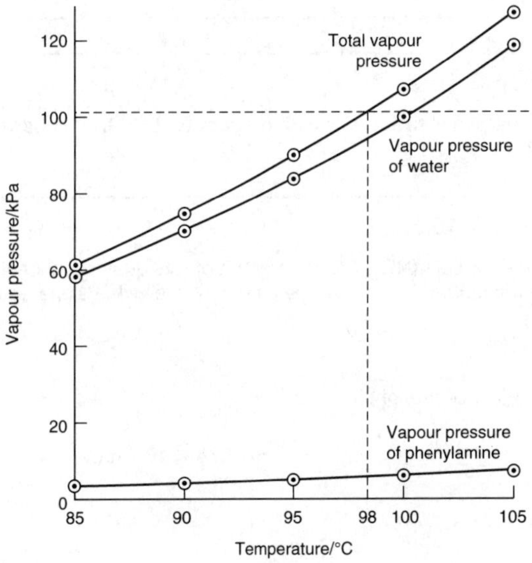

End-of-chapter questions

9.1 16 mmHg

9.2 **a** 0.389

b p(hexane) = 7850 Pa
p(heptane) = 3677 Pa

c 11527 Pa

d 0.681

e The difference in mole fraction of hexane is 0.292.

CHAPTER 10

EXERCISE 10.1

a

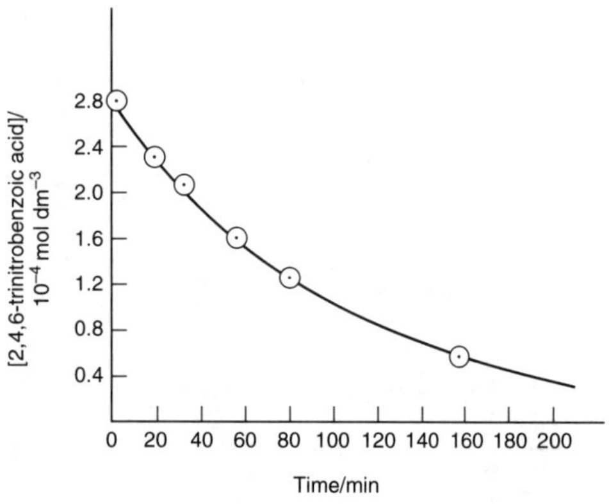

b

Time /min	Concentration /mol dm^{-3}	Slope /mol dm^{-3} min^{-1}	Rate /mol dm^{-3} min^{-1}
10	2.50×10^{-4}	$-\dfrac{2.74 \times 10^{-4}}{112.5}$	2.44×10^{-6}
50	1.66×10^{-4}	$-\dfrac{2.49 \times 10^{-4}}{151.5}$	1.64×10^{-6}
100	1.04×10^{-4}	$-\dfrac{2.04 \times 10^{-4}}{199.5}$	1.02×10^{-6}
150	0.63×10^{-4}	$-\dfrac{1.58 \times 10^{-4}}{249}$	0.635×10^{-6}

c

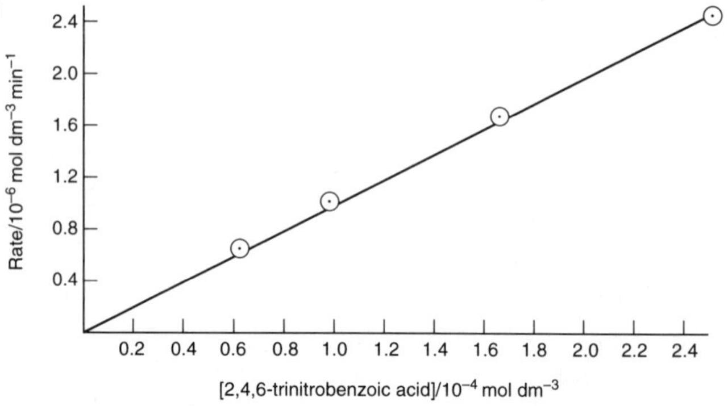

i) **Yes**, the graph does pass through the origin. There can be no reaction (zero rate) if there is no reactant present (zero concentration).

ii) **The rate of reaction is proportional to the concentration of 2,4,6-trinitrobenzoic acid.**
Rate $\propto$ **$[C_6H_2(NO_2)_3CO_2H]$** or **rate = $k[C_6H_2(NO_2)_3CO_2H]$**

EXERCISE 10.2

a The reaction is **first order with respect to 2,4,6-trinitrobenzoic acid**.

b The reaction is also **first order overall**, since there is only one concentration term in the rate equation.

EXERCISE 10.3

Since rate = $k[C_6H_2(NO_2)_3CO_2H]$, a graph of rate against concentration of 2,4,6-trinitrobenzoic acid will be a straight line with a slope equal to k.

$$\text{slope} = \frac{\Delta(\text{rate})}{\Delta(\text{concentration})}$$

$$= \frac{2.44 \times 10^{-6}\ \text{mol dm}^{-3}\ \text{min}^{-1}}{2.50 \times 10^{-4}\ \text{mol dm}^{-3}} = \mathbf{9.76 \times 10^{-3}\ min^{-1}}$$

EXERCISE 10.4

a

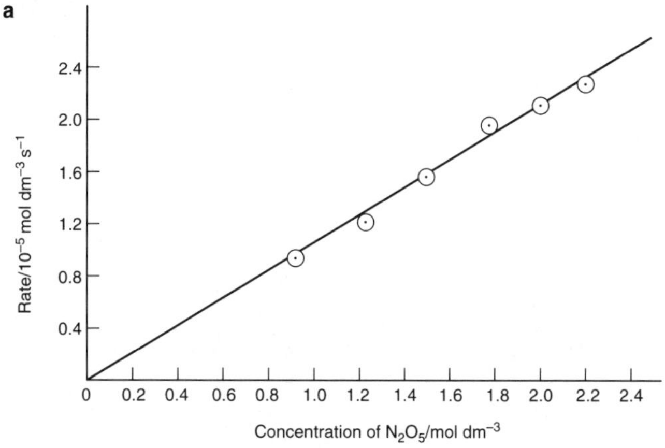

Since the graph of rate against $[N_2O_5]$ is a straight line, rate is directly proportional to concentration of N_2O_5.

$\therefore$ **rate = $k[N_2O_5]$**

b The reaction is **first order with respect to N_2O_5**.

c The value of the rate constant k is given by the slope of the graph, which is:

$$\text{slope} = \frac{\Delta(\text{rate})}{\Delta(\text{concentration})}$$

$$= \frac{2.5 \times 10^{-5}\ \text{mol dm}^{-3}\ \text{s}^{-1}}{2.5\ \text{mol dm}^{-3}} = \mathbf{1.0 \times 10^{-5}\ s^{-1}}$$

EXERCISE 10.5

a **First order with respect to $HgCl_2$. Second order with respect to $C_2O_4^{\ 2-}$.**

b The reaction is **third order overall**.

EXERCISE 10.6

a The order is **1 with respect to C_2H_4** and **1.5 with respect to I_2**.

b The **overall order of reaction is 2.5**.

EXERCISE 10.7

Since the reaction is third order overall, **D could not be correct**:

rate = $k[P][Q]^3$

as this shows a reaction which is fourth order overall.
Note that as $[P]^0 = 1$, B could be written as: rate = $k[Q]^3$.

EXERCISE 10.8

rate = $k[S_2O_8^{\ 2-}(aq)][I^-(aq)]$

$$\therefore\ k = \frac{\text{rate}}{[S_2O_8^{\ 2-}(aq)][I^-(aq)]}$$

Let rate = x mol dm^{-3} s^{-1}, $[S_2O_8^{\ 2-}(aq)] = y$ mol dm^{-3} and $[I^-(aq)] = z$ mol dm^{-3}

$$\therefore\ k = \frac{x\ \text{mol dm}^{-3}\ \text{s}^{-1}}{(y\ \text{mol dm}^{-3})(z\ \text{mol dm}^{-3})}$$

$$= \left[\frac{x}{yz}\right]\frac{\text{mol dm}^{-3}\ \text{s}^{-1}}{\text{mol}^2\ \text{dm}^{-6}}$$

$$= \left[\frac{x}{yz}\right]\text{dm}^3\ \text{mol}^{-1}\ \text{s}^{-1}$$

i.e. the unit is **$dm^3\ mol^{-1}\ s^{-1}$**.

EXERCISE 10.9

a The half-life of a reactant in a chemical reaction (or in radioactive decay) is the **time taken for the concentration of a substance (or the amount, if solid) to fall to half its initial value**.

b i) The half-life of X is **10 minutes**.
ii) **First order.**

c

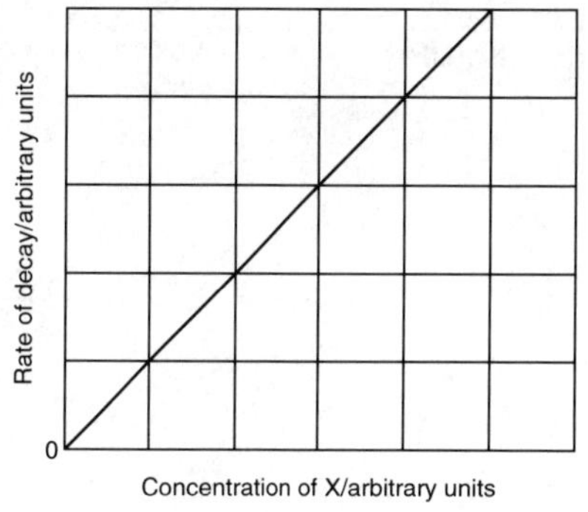

EXERCISE 10.10

a

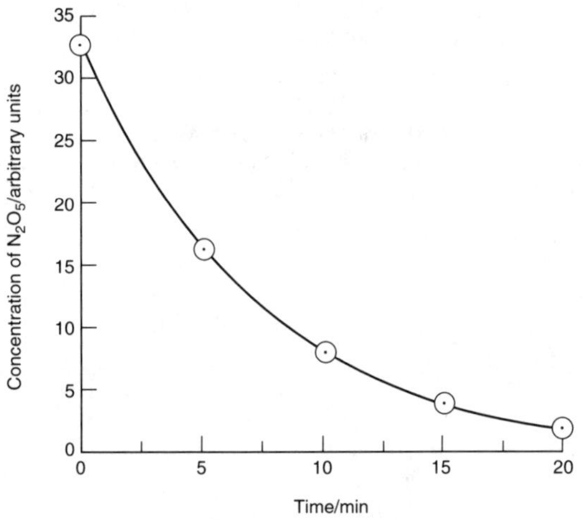

b Rate = $k[N_2O_5]$

$$\therefore\ k = \frac{\text{rate}}{[N_2O_5]}$$

$$= \frac{2.10 \times 10^{-5}\ \text{mol dm}^{-3}\ \text{s}^{-1}}{2.00\ \text{mol dm}^{-3}} = \mathbf{1.05 \times 10^{-5}\ s^{-1}}$$

EXERCISE 10.11

Since the reaction is first order, the half-life is constant.
Half the benzenediazonium chloride decomposes in 40 mins giving 40 cm^3 of gas.
A further half decomposes in a further 40 mins giving 20 cm^3 of gas.
A further half decomposes in a further 40 mins giving 10 cm^3 of gas.
Adding these figures shows that 70 cm^3 of gas is evolved in **120 mins**.

EXERCISE 10.12

a i) Consider experiments 1, 4 and 5, in which initial [B] is constant.
Doubling [A] doubles the rate, and tripling [A] triples the rate.
∴ with respect to A, reaction is **first order**.

ii) Consider experiments 1, 2 and 3, in which initial [A] is constant.
Doubling [B] multiplies rate by 4 (2^2).
Tripling [B] multiplies rate by 9 (3^2).
∴ with respect to B, reaction is **second order**.

b **Rate = $k[A][B]^2$**

c Rearranging the expression in **b** gives $k = \dfrac{\text{rate}}{[A][B]^2}$

$$\therefore\ k = \frac{2.0 \times 10^{-3}\ \text{mol dm}^{-3}\ \text{min}^{-1}}{(0.100\ \text{mol dm}^{-3})^3} = \mathbf{2.00\ mol^{-2}\ dm^{6}\ min^{-1}}$$

EXERCISE 10.13

a Consider experiments 1, 2 and 3, in which initial [NO] is constant.
Doubling $[H_2]$ doubles the rate, and tripling $[H_2]$ triples the rate.
∴ with respect to $[H_2]$, the reaction is **first order**, i.e. $m = 1$.

b Consider experiments 4, 5 and 6, in which initial $[H_2]$ is constant.
Doubling [NO] multiplies rate by 4 (2^2).
Tripling [NO] multiplies rate by 9 (3^2).
∴ with respect to [NO], the reaction is **second order**, i.e. $n = 2$.

EXERCISE 10.14

Hydrolysis of bromomethane

In experiments A, B and C, $[CH_3Br]$ is held constant; the effect of changing $[OH^-]$ is reflected in the rate. Thus, doubling $[OH^-]$ doubles the rate and tripling $[OH^-]$ triples the rate, so the reaction is first order with respect to OH^-.

In experiments D, E and F, $[OH^-]$ is held constant and $[CH_3Br]$ varies. Doubling $[CH_3Br]$ doubles the rate and tripling $[CH_3Br]$ triples the rate, so the reaction is first order with respect to CH_3Br:

∴ **rate = $k[CH_3Br][OH^-]$**

Hydrolysis of 2-bromo-2-methylpropane

In experiments A, B and C, $[(CH_3)_3CBr]$ is held constant and $[OH^-]$ varied. This has no effect on the rate. Therefore, the reaction is zero order with respect to OH^-; i.e. OH^- does not appear in the rate equation.

In experiments D, E and F, $[OH^-]$ is constant and $[(CH_3)_3CBr]$ varies. Doubling $[(CH_3)_3CBr]$ doubles the rate and tripling $[(CH_3)_3CBr]$ triples the rate; therefore, the reaction is first order with respect to $(CH_3)_3CBr$:

∴ **rate = $k[(CH_3)_3CBr]$**

EXERCISE 10.15

The graph is a straight line, which shows that the rate of reaction (as given by the slope) does not change with the concentration of ethylbenzene. This means that the reaction is **zero order with respect to $C_6H_5C_2H_5$**. This experiment alone gives no information about the order with respect to HNO_3 because $[HNO_3]$ was very much higher and therefore remained virtually constant.

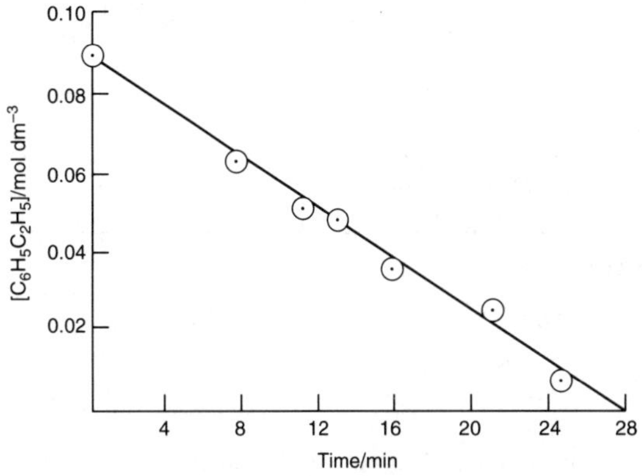

EXERCISE 10.16

a The hydrogencarbonate reacts with the acid catalyst and effectively stops the reaction to enable the titration to be done. It also prevents a reaction between acid and thiosulphate:

$$2H^+(aq) + S_2O_3^{2-}(aq) \rightarrow S(s) + SO_2(aq) + H_2O(l)$$

b Since the iodine concentration is proportional to the volume of thiosulphate solution used, the rate of change of iodine concentration is proportional to the slope of the graph. **This is constant over the period illustrated.**

c $\text{Rate} = \dfrac{\Delta V}{\Delta t} = \dfrac{(20 - 17)\ \text{cm}^3}{30\ \text{min}} = \mathbf{0.10\ cm^3\ min^{-1}}$

d The concentrations of propanone and acid are much greater than the initial concentration of iodine and can therefore be assumed to remain effectively constant. As the concentration of iodine falls, the rate of reaction remains constant; the reaction is therefore **zero order with respect to iodine**.

e Doubling the concentration should double the rate, i.e. **0.20 cm^3 min^{-1}**.

f & g There are two possible answers here. Either show the volume remaining constant over a period of 30 minutes, because the reaction, in the absence of a catalyst, proceeds too slowly to be measured; or show the volume decreasing slowly at first and then more rapidly over a longer period of time because there are a few hydrogen ions in any aqueous solution to catalyse the reaction and more are produced as the reaction proceeds.

The second alternative is theoretically sound but, in practice, the first corresponds to observation.

EXERCISE 10.17

a In each case the rate is obtained by substituting in the equation:

$$\text{rate} = k[H_2(g)][I_2(g)]$$

A Rate = 8.58×10^{-5} mol^{-1} dm^3 s^{-1} $\times$ 0.010 mol dm^{-3} $\times$ 0.050 mol dm^{-3}
= **4.29×10^{-8} mol dm^{-3} s^{-1}**

B Rate = 8.58×10^{-5} mol^{-1} dm^3 s^{-1} $\times$ 0.020 mol dm^{-3} $\times$ 0.050 mol dm^{-3}
= **8.58×10^{-8} mol dm^{-3} s^{-1}**

C Rate = 8.58×10^{-5} mol^{-1} dm^3 s^{-1} $\times$ 0.020 mol dm^{-3} $\times$ 0.010 mol dm^{-3}
= **1.72×10^{-7} mol dm^{-3} s^{-1}**

b Considering experiments A and B, **doubling the initial concentration of hydrogen** at constant initial concentration of iodine **doubles the rate of reaction.**

Considering experiments B and C, **doubling the initial concentration of iodine** at constant initial concentration of hydrogen **doubles the rate of reaction**.

EXERCISE 10.18

a

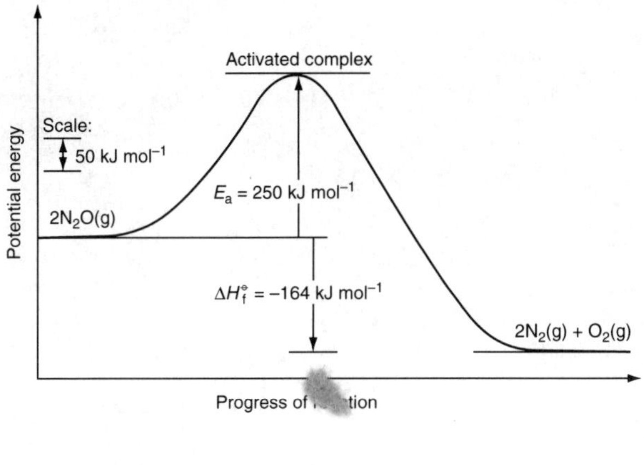

b The activation energy* of the back reaction = (250 + 164) kJ mol^{-1} = **+414 kJ mol^{-1}**.

*It would be better to call this quantity an 'energy barrier' since it is not quite the same as the activation energy used in the Arrhenius equation. However, the distinction is rarely made at A-level.

EXERCISE 10.19

a Fraction of molecules with $E > 55.0\ \text{kJ mol}^{-1} = e^{-E/RT}$

$$= e^{-(55\,000\ \text{J mol}^{-1})/(8.31\ \text{J K}^{-1}\ \text{mol}^{-1})(308\ \text{K})}$$

$$= e^{-21.5} = 4.60 \times 10^{-10}$$

Number $= L \times e^{-E/RT}$

$$= 6.02 \times 10^{23}\ \text{mol}^{-1} \times 4.60 \times 10^{-10}$$

$$= \mathbf{2.77 \times 10^{14}\ mol^{-1}}$$

b $$\frac{\text{number at 308 K}}{\text{number at 298 K}} = \frac{2.77 \times 10^{14}}{1.37 \times 10^{14}} = \mathbf{2.02}$$

∴ twice as many molecules exceed the activation energy at 308 K as do at 298 K.

EXERCISE 10.20

a The variables in the equation:

$$\ln k = \ln A - \frac{E_a}{RT}$$

are ***k*** and ***T***.

b ***y* is analogous to ln *k*, *m* to $-E_a/R$, *x* to 1/*T* and *c* to ln *A*.**

c The slope, $-E_a/R$, is always negative and $1/T$ is always positive.
ln *k* may be positive or negative according to the magnitude of *k* (it is usually negative, corresponding to small values of *k*). The plot may take either of the two forms shown, with (2) the more probable.

d The intercept on the vertical axis is equal to ln *A*. If the value of ln *A* is obtained by reading the graph then *A* can be calculated. The slope of the straight line is equal to $-E_a/R$. If the slope of the line is calculated from the graph then E_a can also be calculated, since *R* is a known constant.

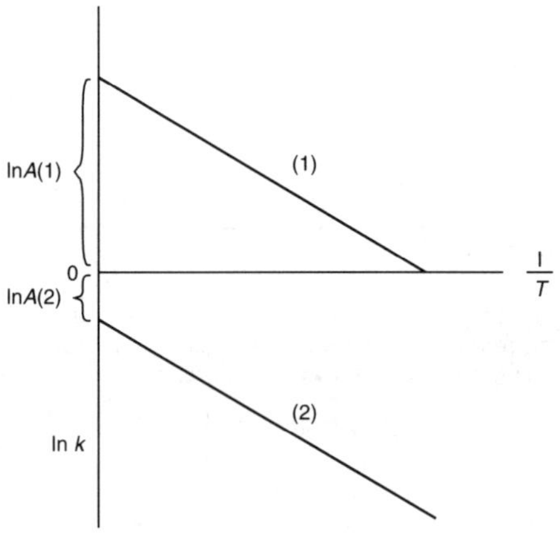

EXERCISE 10.21

a

Temperature (*T*) /K	1/*T* K^{-1}	Rate constant (*k*) /dm^3 mol^{-1} s^{-1}	ln *k*
633	1.58×10^{-3}	1.78×10^{-5}	−10.94
666	1.50×10^{-3}	1.07×10^{-4}	−9.14
697	1.43×10^{-3}	5.01×10^{-4}	−7.60
715	1.40×10^{-3}	1.05×10^{-3}	−6.86
781	1.28×10^{-3}	1.51×10^{-2}	−4.19

Since $\ln k = \frac{-E}{R} \times \left(\frac{1}{T}\right) + \ln A$, the slope of the graph $= -\frac{E}{R}$

but the slope $= \dfrac{-11.0 + 2.45}{(1.58 - 1.20) \times 10^{-3}\ \text{K}^{-1}} = -2.25 \times 10^4\ \text{K}$

$$\therefore\ E = -R \times \text{slope} = -8.31\ \text{J K}^{-1}\ \text{mol}^{-1} \times (-2.25 \times 10^4\ \text{K})$$

$$= 1.87 \times 10^5\ \text{J mol}^{-1} = \mathbf{187\ kJ\ mol^{-1}}$$

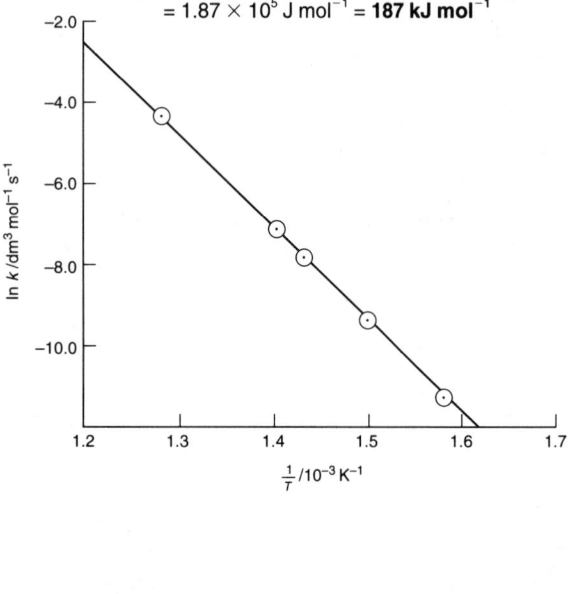

b Writing k_1 and k_2 for the rate constants at 300 K and 310 K respectively:

$$\ln\left(\frac{k_2}{k_1}\right) = \ln k_2 - \ln k_1 = -\frac{E}{R}\left(\frac{1}{310\ \text{K}} - \frac{1}{300\ \text{K}}\right)$$

$$= \frac{-187 \times 1000\ \text{J mol}^{-1}}{8.31\ \text{J K}^{-1}\ \text{mol}^{-1}}\left(\frac{300\ \text{K} - 310\ \text{K}}{310\ \text{K} \times 300\ \text{K}}\right) = 2.42$$

$$\therefore\ \frac{k_2}{k_1} = e^{2.42} = \mathbf{11.2}$$

c $$\frac{\text{kinetic energy at 310 K}}{\text{kinetic energy at 300 K}} = \frac{310\ \text{K} \times \text{constant}}{300\ \text{K} \times \text{constant}} = \mathbf{1.03}$$

d A very small increase in kinetic energy of a fixed mass of gas (such as HI) results in a great increase in the rate of reaction. This arises partly from the fact that collisions are more frequent at higher temperatures, but mostly from the fact that a greater proportion of the collisions have sufficient energy to cause a reaction to occur. The minimum energy for reaction to occur is called the activation energy.

EXERCISE 10.22

Temperature/°C	30	36	39	45	51
Temperature, T/K	303	309	312	318	324
Time t/s	204	138	115	75	55
$\ln\frac{1}{t}$	−5.32	−4.93	−4.74	−4.32	−4.01
$\frac{1}{T}$/10^{-3} K^{-1} (or 10^3 K/T)	3.30	3.24	3.21	3.14	3.09

$$\text{Slope of graph} = \frac{\Delta y}{\Delta x} = \frac{-5.26-(-3.86)}{(3.290-3.070)\times 10^{-3}\,\text{K}^{-1}} = -6.36\times 10^{3}\,\text{K}$$

$$\text{But slope} = \frac{-E_a}{R}$$

$$\therefore E_a = -\text{slope}\times R$$
$$= 6.36\times 10^{3}\,\text{K}\times 8.31\,\text{J K}^{-1}\,\text{mol}^{-1}$$
$$= \mathbf{52.9\ kJ\ mol^{-1}}$$

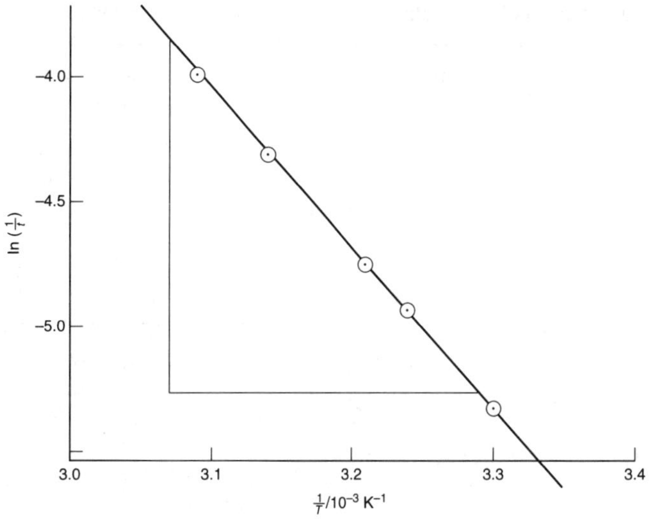

EXERCISE 10.23

Temperature/°C	15	19.5	26	35	42
Temperature, T/K	288	292.5	299	308	315
Time, t/s	10.0	7.0	5.0	3.5	2.5
ln(1/t)	−2.30	−1.95	−1.61	−1.25	−0.92
$\frac{1}{T}$/10^{-3} K^{-1}	3.472	3.419	3.344	3.247	3.175

$$\text{Slope} = \frac{-1.96-(-1.04)}{(3.420-3.200\times 10^{-3}\,\text{K}^{-1}} = \frac{-0.92}{0.220\times 10^{-3}\,\text{K}^{-1}} = -4180\,\text{K}$$

$$\text{But the slope} = -\frac{E_a}{R}$$

$$\therefore E_a = -\text{slope}\times R$$
$$= 4180\,\text{K}\times 8.31\,\text{J K mol}^{-1} = \mathbf{35\ kJ\ mol^{-1}}$$

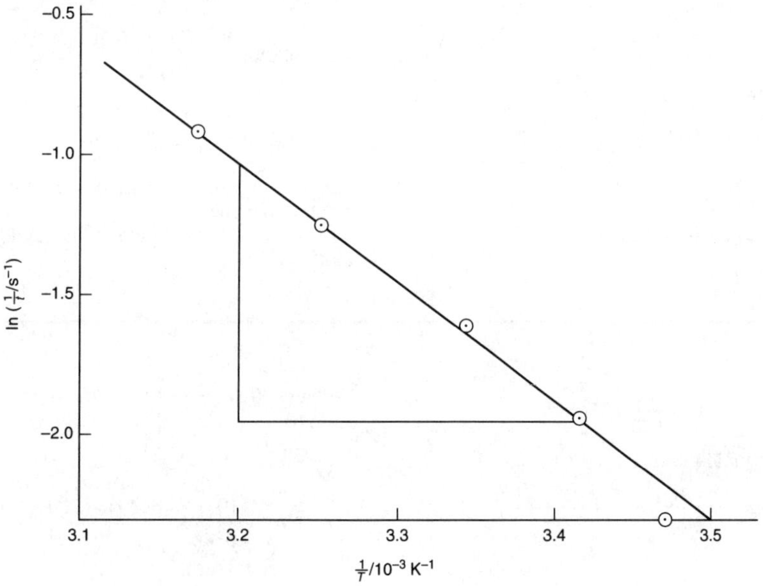

EXERCISE 10.24

Rate constants (k) are obtained by plotting loss in mass against time.

Experiment 1 at 0.5 °C			Experiment 2 at 13.5 °C			Experiment 3 at 25 °C		
Time /min	**Mass /mg**	**Loss /mg**	**Time /min**	**Mass /mg**	**Loss /mg**	**Time /min**	**Mass /mg**	**Loss /mg**
0	130	0	0	130	0	0	130	0
20	120	10	10	115	15	4	116	14
60	98	32	15	106	24	8	103	27
80	86	44	30	86	44	12	87	43

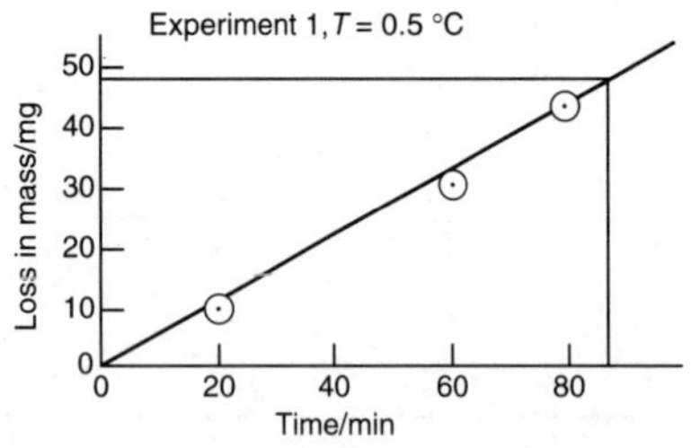

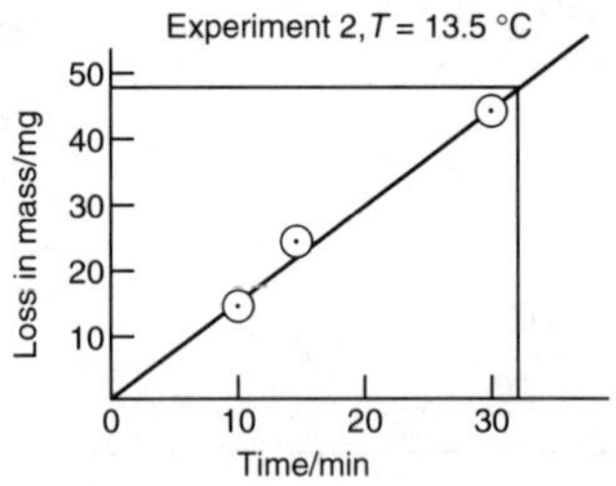

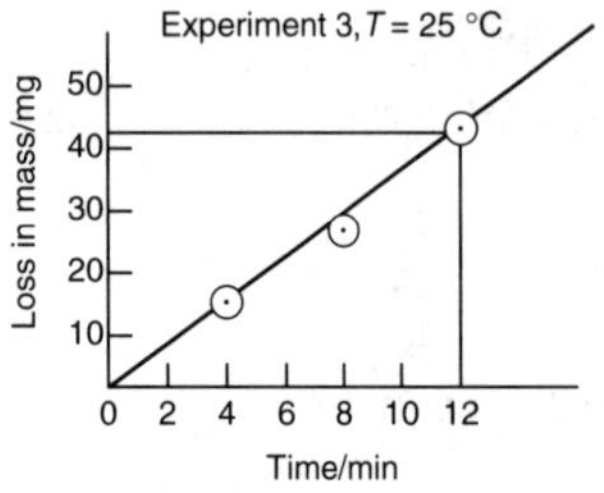

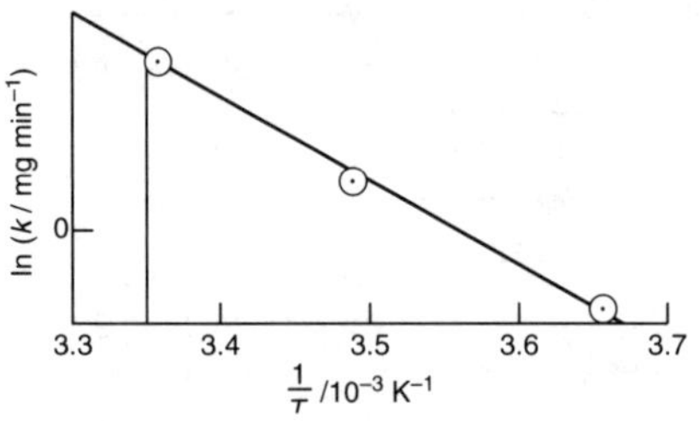

In each experiment, the straight line graph shows that the rate of loss of mass is constant. The rate constant, k, is the slope of the line.

1. $k = \dfrac{48.0\text{ mg}}{88.0\text{ min}} = \mathbf{0.55\ mg\ min^{-1}}$
2. $k = \dfrac{48.0\text{ mg}}{32.5\text{ min}} = \mathbf{1.48\ mg\ min^{-1}}$
3. $k = \dfrac{43.0\text{ mg}}{12.0\text{ min}} = \mathbf{3.58\ mg\ min^{-1}}$

Activation energy is obtained by plotting $\ln k$ against $\dfrac{1}{T}$.

$k = B \times e^{-E/RT}$ (B is a constant)

$$\therefore \ln k = \ln B - \frac{E}{RT}$$

$$\text{Slope of graph} = \frac{(-0.69 - 1.34)}{(3.67 - 3.35) \times 10^{-3}\text{ K}^{-1}} = -6.34 \times 10^3\text{ K} = \frac{-E}{R}$$

$$\therefore E = 8.31\text{ J K}^{-1}\text{ mol}^{-1} \times 6.34 \times 10^3\text{ K} = \mathbf{52.7\ kJ\ mol^{-1}}$$

If the cobalt were not rotated, the rates of reaction (and hence the rate constants) would be slightly smaller, due to the time taken for $S_2O_8^{2-}$ ions to diffuse towards the metal surface. Activation energy would be unaffected.

EXERCISE 10.25

$$\ln k_1 = \ln A - \frac{E_a}{RT_1} \quad \text{and} \quad \ln k_2 = \ln A - \frac{E_a}{RT_2}$$

Subtracting the first equation from the second:

$$\ln\left(\frac{k_2}{k_1}\right) = -\frac{E_a}{R}\left(\frac{1}{T_2} - \frac{1}{T_1}\right) \quad \text{or} \quad \ln\left(\frac{k_2}{k_1}\right) = \frac{E_a}{R}\left(\frac{T_2 - T_1}{T_1T_2}\right)$$

Substituting values into this equation:

$$\ln\left(\frac{116}{8.58}\right) = \frac{E_a}{8.31\text{ J K}^{-1}\text{ mol}^{-1}}\left(\frac{700\text{ K} - 647\text{ K}}{647\text{ K} \times 700\text{ K}}\right)$$

$$\therefore E_a = \ln\left(\frac{116}{8.58}\right) \times \frac{647\text{ K} \times 700\text{ K}}{53\text{ K}} \times 8.31\text{ J K}^{-1}\text{ mol}^{-1}$$

$$= \mathbf{185\ kJ\ mol^{-1}}$$

EXERCISE 10.26

a E_a for the forward reaction (uncatalysed) is **184 kJ mol^{-1}**.
E_a for the back reaction (uncatalysed) is **236 kJ mol^{-1}**.

b E_a for the forward reaction (catalysed) is **59 kJ mol^{-1}**.
E_a for the back reaction (catalysed) is **111 kJ mol^{-1}**.

c The enthalpy change is **±52 kJ mol^{-1}**.

EXERCISE 10.27

a Initial rate = slope at $t = 0 \simeq$ **6.3×10^{-7} mol dm^{-3} s^{-1}**
Rate at $t = 50$ s $\simeq$ **4.5×10^{-6} mol dm^{-3} s^{-1}**
Thus, the rate of reaction increases seven-fold over the first 50 seconds. This increase in the rate indicates that one of the products of the reaction is acting as a catalyst. The catalyst is likely to be the transition metal ions, Mn^{2+}(aq).

b Two effects are operating during this reaction: the production of catalyst and the decrease in concentration of reactants. At the start of the reaction the first effect dominates; catalyst is being produced and speeds up the rate of the reaction as long as there is an adequate supply of reactant. Towards the end of the reaction, however, the second effect becomes more important. The concentration of reactants has fallen to a low level and so the rate of reaction decreases, even though there is an adequate supply of catalyst.

End-of-chapter questions

10.1 84 days
10.2 16 800 years
10.3 **a** Mainly to keep [acid] effectively constant but also to disperse heat.
b i) $[HCl(aq)]^2 \propto 1/t$ or $t \propto 1/[HCl(aq)]^2$
ii) Second order.
10.4 **a** rate $\propto [OH^-(aq)]^2$
b rate $\propto [PH_2O_2^-(aq)]$
c rate = $k[OH^-(aq)]^2[PH_2O_2^-(aq)]$
d mol^{-2} dm^6 min^{-1} or dm^6 min^{-1} mol^{-2}
10.5 **a** rate = $k[CH_3Br][OH^-]$ $k = 8.13 \times 10^{-3}$ mol^{-1} dm^3 s^{-1}
b $E_a = 88.7$ kJ mol^{-1}
10.6 **b** First order with respect to Br^-(aq) and BrO_3^-(aq).
Probably second order with respect to H^+(aq). However, the slope of the log graph is 1.8 which could mean that there are two mechanisms operating at once giving a fractional order.
d rate = $k[Br^-(aq)][BrO_3^-(aq)][H^+(aq)]^2$
10.7 32
10.8 **a** Curve A: 1.82 unit of concentration min^{-1}
Curve B: 1.25 unit of concentration min^{-1}
Curve C: 0.625 unit of concentration min^{-1}
b i) Iodide ions
ii) First order
c i) 0.75 min
ii) 1.5 min; First order
10.9 **b** i) $x = 1$
ii) -5.0×10^{-4} s^{-1}
c 2772 s
10.10 119 kJ mol^{-1}
10.11 **a** First order (constant half-life considering $[C_4H_9Cl]$ only, so change in $[OH^-]$ has no effect).
b rate = $k[C_4H_9Cl]$
10.12 **a** $2I^-(aq) + S_2O_8^{2-}(aq) \rightleftharpoons 2SO_4^{2-}(aq) + I_2(aq)$
b Second order
c 2.04×10^{-3} dm^3 mol^{-1} s^{-1}

APPENDIX

EXERCISE A1

a 3
b 4
c 2
d 3
e 3
f 5
g 4
h 3 or 4
i 3, 4, 5 or 6

EXERCISE A2

a 208 g

b 0.649 dm^3

EXERCISE A3

a 3.4 m

b 76 cm^3

EXERCISE A4

a 3.0×10^{-3} mol

b 67.8 g

EXERCISE A5

a 0.48 mol

b 2 g cm^{-3}

Relative Atomic Masses

(To three significant figures)

An asterisk * indicates that the element has no stable isotope and no fixed isotopic composition – M_r then refers to the isotope with longest half-life.

Element		Z	M_r	Element		Z	M_r
Actinium	Ac	89	*227	Mercury	Hg	80	201
Aluminium	Al	13	27.0	Molybdenum	Mo	42	95.9
Americium	Am	95	*243	Neodymium	Nd	60	144
Antimony	Sb	51	122	Neon	Ne	10	20.2
Argon	Ar	18	39.9	Neptunium	Np	93	*239
Astatine	At	85	*210	Nickel	Ni	28	58.7
Arsenic	As	33	74.9	Niobium	Nb	41	92.9
Barium	Ba	56	137	Nitrogen	N	7	14.0
Berkelium	Bk	97	*247	Nobelium	No	102	*255
Beryllium	Be	4	9.01	Osmium	Os	76	190
Bismuth	Bi	83	209	Oxygen	O	8	16.0
Boron	B	5	10.8	Palladium	Pd	46	106
Bromine	Br	35	79.9	Phosphorus	P	15	31.0
Cadmium	Cd	48	112	Platinum	Pt	78	195
Caesium	Cs	55	133	Plutonium	Pu	94	*239
Calcium	Ca	20	40.1	Polonium	Po	84	*210
Californium	Cf	98	*252	Potassium	K	19	39.1
Carbon	C	6	12.0	Praseodymium	Pr	59	141
Cerium	Ce	58	140	Promethium	Pm	61	*145
Chlorine	Cl	17	35.5	Protactinium	Pa	91	*231
Chromium	Cr	24	52.0	Radium	Ra	88	*226
Cobalt	Co	27	58.9	Radon	Rn	86	*222
Copper	Cu	29	63.5	Rhenium	Re	75	186
Curium	Cm	96	*247	Rhodium	Rh	45	103
Dysprosium	Dy	66	162	Rubidium	Rb	37	85.5
Einsteinium	Es	99	*254	Ruthenium	Ru	44	101
Erbium	Er	68	167	Rutherfordium	Rf	104	*260
Europium	Eu	63	152	Samarium	Sm	62	150
Fermium	Fm	100	*253	Scandium	Sc	21	45.0
Fluorine	F	9	19.0	Selenium	Se	34	79.0
Francium	Fr	87	*223	Silicon	Si	14	28.1
Gadolinium	Gd	64	157	Silver	Ag	47	108
Gallium	Ga	31	69.7	Sodium	Na	11	23.0
Germanium	Ge	32	72.6	Strontium	Sr	38	87.6
Gold	Au	79	197	Sulphur	S	16	32.1
Hafnium	Hf	72	178	Tantalum	Ta	73	181
Helium	He	2	4.00	Technetium	Tc	43	*98.9
Holmium	Ho	67	165	Tellurium	Te	52	128
Hydrogen	H	1	1.01	Terbium	Tb	65	159
Indium	In	49	115	Thallium	Tl	81	204
Iodine	I	53	127	Thorium	Th	90	232
Iridium	Ir	77	192	Thulium	Tm	69	169
Iron	Fe	26	55.8	Tin	Sn	50	119
Krypton	Kr	36	83.8	Titanium	Ti	22	47.9
Lanthanum	La	57	139	Tungsten	W	74	184
Lawrencium	Lr	103	*257	Uranium	U	92	238
Lead	Pb	82	207	Vanadium	V	23	50.9
Lithium	Li	3	6.94	Xenon	Xe	54	131
Lutetium	Lu	71	175	Ytterbium	Yb	70	173
Magnesium	Mg	12	24.3	Yttrium	Y	39	88.9
Mendelevium	Md	101	*257	Zinc	Zn	30	65.4
				Zirconium	Zr	40	91.2